NEUROSCIENCES IN MUSIC PEDAGOGY

NEUROSCIENCES IN MUSIC PEDAGOGY

WILFRIED GRUHN
FRANCES H. RAUSCHER
EDITORS

Nova Biomedical Books
New York

LIBRARY OF CONGRESS CATALOGING-IN-PUBLICATION DATA

Neurosciences in music pedagogy / [edited by] Frances Rauscher and Wilfried Gruhn.
p. ; cm.
Includes bibliographical references and index.
ISBN-13: 978-1-60021-834-7 (hardcover : alk. paper)
ISBN-10: 1-60021-834-2 (hardcover : alk. paper)
1. Music--Physiological aspects. 2. Musical ability in children. 3. Music--Psychological aspects. 4. Cognitive neuroscience. 5. Neurobiology.
[DNLM: 1. Child Psychology. 2. Music--psychology. 3. Brain--growth & development. 4. Child Development. 5. Learning--physiology. WS 350 N4945 2007] I. Rauscher, Frances. II. Gruhn, Wilfried, 1939-
ML3820.N48 2007
781'.11--dc22 2007022704

Published by Nova Science Publishers, Inc. ✢ New York

CONTENTS

PREFACE

When four of the authors in this book presented a symposium on "Neuroscience in Music Pedagogy" at the 8th International Conference on Music Perception and Cognition (ICMPC) in Evanston, USA, we were faced with the unexpected interest of music educators who were looking for concrete and substantial information on neuromusical research relevant to music learning and teaching. This was the starting point for our consideration of how to transmit topical knowledge and recent findings in neurosciences to the needs of music educators. When Nova Science Publications offered the opportunity to publish a book on this topic, we gratefully and enthusiastically started to plan a comprehensive view of neuromusical research and its potential applications to music pedagogy.

However, we have to take into consideration that (1) knowledge as such is not transferable; we cannot force children to learn or push synapses to grow; we can only provide a stimulating environment and environmental conditions that enhance and support learning, and (2) knowledge acquisition is governed by factors that are not fully under conscious control and can hardly be influenced externally (Roth, 2006). Nevertheless, children learn and are extremely curious and eager to learn. Their cortex is the organ where new experiences and knowledge are processed by interconnected neurons (mental representations) which become activated when a similar sensorial input is perceived. These structural and functional changes compose the experience-driven plasticity of the brain.

Brain researchers have investigated when and how dendritic spines sprout, how neurons communicate with each other, how neural networks develop, and where brain activations are located; however, there is in fact nothing really new that brain researchers can tell educators about teaching that they did not know before. Brain research has nevertheless attracted growing interest because it can explain, support and ground pedagogical common sense on empirical data and can prevent educators from developing counterproductive teaching attitudes.

That is why so many people, for different reasons, are highly interested in neuromusical research. Since musicians have become a favoured model of brain plasticity in neurosciences, pedagogical expectations arose that education could benefit from music, and that neurosciences could underpin this assumption with solid and robust research data. Therefore, a new branch of education theory appeared which attempts to relate teaching strategies and learning modalities to the hard facts of brain research. This new learning theory, which is named "Neuropedagogy" or "Neurodidactics" referring to brain-based learning in the anglophone world, came under debate in Europe recently (Caspary, 2006; Herrmann, 2006; Preiss, 1998; Spitzer, 2006). The philosophy behind this approach is to stimulate a change of teachers' professional behaviour which, then, might initiate a paradigm change of schooling. Therefore, teachers should acquire at least some basic knowledge about the mental processes involved in learning and the mental state of their pupils so that they can adapt their teaching to the most efficient way of how children learn more appropriately, i.e. to adjust the curriculum to the mental conditions of learning instead of adjusting the children to the political conditions of the curriculum.

The link between research in general and brain research in particular and the methods and means of teaching shall be discussed in a broader context at the end of the book (chapter 10). We should be careful not to exaggerate the expectation that neuronal findings can easily and immediately be applied to classroom teaching. But nevertheless, a better knowledge of the processing structures of the brain in different situations and for different ages is important, and we hope that we can provide up-to-date information about the effects of learning on the brain and the structural and functional changes associated with music listening and music making. Of course, this cannot solve educational problems in general, but it might support teachers in their educational efforts of applying the best possible ways of teaching according to the cognitive needs of their students.

To link the different aspects of teaching and learning with some areas of brain research, we have asked experts from various disciplines to join together in this interdisciplinary handbook to discuss relevant research questions from their different points of view. In the following chapters, therefore, psychologists and educators collaborate with neuroscientists in many different ways, whether working on separate sections of a chapter or presenting a common text based on mutual discussions of the actual problems. By this procedure, we want to illuminate various aspects of brain research relevant to and potentially connected with music education.

First, a general overview on the extant neuromusical research and its broad variety of topics is presented which intends to provide a rough orientation regarding the huge number of single studies (chapter 1). The next four chapters are specifically devoted to learning. First, the peculiarities of learning in childhood will be described, in which all events and learning environments contribute to the development of immature, preliminary and unspecific neuronal networks (chapter 2). This chapter emphasizes the importance of the quality of parenting and schooling for experience-driven brain plasticity. Therefore, this chapter supplies the foundation for establishing a more general framework of learning within which a psychobiological model of early musical behaviour can be developed. Innate communicative musicality is understood as a natural behaviour essential to being human (chapter 3). This is followed by a research report on the empirical evidence of neural correlates of music learning and playing musical instruments (chapter 4). Musicians provide an excellent model to investigate brain plasticity and the effects of learning with regard to acquiring complex motor and auditory skills as well as multimodal skills through continuous practice from early childhood on. The last chapter in this package focuses on motor learning and instrumental training (chapter 5). This chapter provides relevant information from research on motor development, motor control, sensorimotor interaction, and motor coordination involved in practice and performance. It also describes the neuroanatomy and neurophysiology of motor systems.

The following three chapters deal with more specific aspects of music learning and the psychological and cognitive effects related to music, such as the neuroscientific aspects of emotion, intelligence and giftedness. A report of research studies and an introduction into the research on music and emotion reflect the increasing interest in identifying the neural mechanisms underlying emotional responses to music (chapter 6). Here it is stated that musical emotions recruit networks of emotion processing that are involved in visual and auditory perception. The hypothesized benefits of music on non-musical domains are critically revisited by reviewing the research on the effect of music instruction on spatial abilities, arithmetic, and other cognitive skills, including visual-motor integration and verbal/reading performance (chapter 7). Neurobiological aspects of giftedness and musical talent are viewed from two different perspectives: First, from experimental findings on musicians and musically talented individuals, and second, with regard to the musical skills of individuals with disabilities. Research on musical savants stresses that several component skills contribute to the emergence of exceptional musical abilities (chapter 8).

Finally, the increasing number of professional and amateur musicians with specific diseases and disabilities who are treated within the field of music medicine attracts attention towards special music learning needs of individuals with many kinds of disabilities and impairments, such as people with various physical and intellectual disabilities, with developmental disorders like dyslexia, Attention Deficit Syndrome and Savant Syndrome and with chromosomal aberrations like Down Syndrome and Williams Syndrome (chapter 9).

The concluding chapter (chapter 10) comes back to formal music learning and debates new approaches to music instruction based on the results from the neurobiology of learning. It draws attention to findings of studies that investigate the impact of music learning on the structure of mental processing, namely the development and differentiation of mental musical representations.

A broad collection of interdisciplinary contributions to the neurosciences in music pedagogy inevitably causes some overlap between chapters and duplications because the same phenomena are likely to appear in different contexts and research questions such as the phenomena of mirror neurons (chapters 4 and 10), amusia (chapters 6 and 9) or musical savants (chapters 8 and 9) among others. We are convinced that this is not only acceptable, but it is rather appreciated because it illustrates the complexity, interactivity, and interdependence of the presented perspectives, and it offers the opportunity to approach the same aspect from different starting points. Accordingly, wherever an overlap appears, a cross-reference is indicated for those who wish to step deeper into that issue. However, we have tried to avoid an overload of too many cross-references in the text. We rather want to present a stringent and consistent overview of the actual research situation and results from neurosciences under the given perspective of music learning, music making, and music listening. For this, each chapter can be read separately and independently and presents its information in a way that is understandable in itself. Readers are invited to read the chapters selectively and choose those parts first that they are especially interested in. Furthermore, those who are not specialised in neuroscience and brain research find brief explanations of terms and techniques frequently mentioned in the chapters at the end of the book in a glossary.

The information on neuroscience and learning assembled in this handbook aims to provide our readers with seminal knowledge up to date, which may contribute to a better understanding of the teaching and learning process, and which can hopefully facilitate the distribution of results from neuromusical research relevant and beneficial to music pedagogy.

Freiburg and Oshkosh, Spring 2007

REFERENCES

Caspary, R. (Ed.). (2006). *Lernen und Gehirn. Der Weg zu einer neuen Pädagogik.* Freiburg: Herder.

Herrmann, U. (Ed.). (2006). *Neurodidaktik. Grundlagen und Vorschläge für gehirngerechtes Lehren und Lernen*. Weinheim: Beltz.

Preiss, G. (Ed.). (1998). *Neurodidaktik. Theoretische und praktische Beiträge*. Herbolzheim: Centaurus.

Roth, G. (2006). Möglichkeiten und Grenzen von Wissensvermittlung und Wissenserwerb. In R. Caspary (Ed.), *Lernen und Gehirn* (pp. 54 - 69). Freiburg: Herder.

Spitzer, M. (2006). Medizin für die Schule. Plädoyer für eine evidenzbasierte Pädagogik. In R. Caspary (Ed.), *Lernen und Gehirn* (pp. 23 - 35). Freiburg: Herder.

ACKNOWLEDGEMENTS

We are grateful to Nova Science publishers for its consistent support of this enterprise. We also thank our friends and colleagues at the University of Freiburg and the University of Wisconsin Oshkosh who provided help and encouragement at critical moments of this project. We appreciate the long hours that Dace Almane and Amber Corry spent in the library, researching some of the "technical" terms that appear in the glossary. Special thanks go to Sean Hinton, whose thoughtful input made this volume much better than it would otherwise have been. Most of all, we are grateful to the contributors themselves, who wrote and rewrote their chapters, put up with our many delays and sometimes awkward requests, and finally brought a fine book into existence. Thanks to one and all.

Frances H. Rauscher
Wilfried Gruhn

In: Neurosciences in Music Pedagogy
Editors: W. Gruhn and F. Rauscher, pp. 1-25
ISBN: 978-1-60021-834-7

Chapter I

NEUROMUSICAL RESEARCH: AN OVERVIEW OF THE LITERATURE

Richard D. Edwards and Donald A. Hodges

ABSTRACT

With growing interest in the neuroscience of music among both neuroscientists and music educators, the task of reviewing the extant neuromusical research has become more exciting, if not a bit more complicated due to the diversity of topics and the increasing number of available studies. We provide an overview of neuromusical research by discussing its historical foundations (especially with the advancements of imaging technologies), support from ancillary areas (anthropology, ethnomusicology, ethology, and music psychology), support from fetal and infant responses to music, and support from studies of special musicians (prodigies, savants, Williams Syndrome musicians, and Alzheimer's patients). The main section presents findings and implications from recent neuroimaging studies by dividing the research into five categories: (1) Perception and Cognition, (2) Affective Responses, (3) Musical Performance, (4) Learning, and (5) Genetic Factors. Several broad conclusions regarding the state of human musicality are presented in the concluding section. Among these conclusions, perhaps the most valuable evidence that neuromusical research currently holds for educators is that musicality is a birthright of all people and that music processing is inherent to some degree in all humans.

People have long wondered about the mind. Before Descartes proposed "I think, therefore, I am" or even before Plato and Aristotle ever contemplated the psyche, the oldest brain map on record was being drawn upon papyrus in Egypt almost 5,000 years ago[1]. Today, as neuroscientists draw increasingly detailed maps of the brain, they are pursuing an age-old

[1] The Edwin Smith Surgical papyrus describes several case studies of neurological disorders and offers the oldest recorded use of the word "brain" (Minagar, Ragheb, & Kelley, 2003)

curiosity. Recently, a fascinating agenda gaining the attention of the neuroscience community has been the way the brain engages in musical processes. Perhaps the rising interest in neuromusical research is due to music's status as a human constant. Evidence of music has been confirmed in every civilization throughout history (Chailley, 1964) and based on research in anthropology (Merriam, 1964), enthnomusicology (Blacking, 1973), and psychology (Gardner, 1983), there is strong evidence to suggest that not only is music a cultural invariant, but that every person is born with the potential for some form of meaningful musical experiences.

While the connection between musical behaviors and brain processes has been apparent, advancements in imaging technologies are leading to more sophisticated investigations of music processing than ever before. Though valuable evidence can be gleaned from ancillary areas such as anthropology or from observational or behavioral studies, many exciting discoveries have come from measuring brain activations directly. Currently, neuroscience has at its disposal a broad array of imaging tools and ongoing refinements are making these tools even more powerful.

The purpose of this chapter is to survey the neuromusical research literature and to report on the general findings. No attempt to cite every study is made, as they are now far too numerous. Rather, broad, general conclusions are drawn and supported by a few representative studies. The chapter begins with three brief sections—an historical perspective, support from ancillary areas, and support from special musician populations—and the main section presents findings from brain imaging studies and related neuromusical literature.

A Brief Historical Perspective

German neurologists conducted some of the pioneering work beginning in the latter half of the 19th century (Henson, 1977), including August Knoblauch, who coined the term amusia (i.e., the loss of musical abilities) and began to make clinical diagnoses (Johnson & Graziano, 2003). However, experiments involving direct measurements of brain activations to musical stimuli did not occur until much later. Three types of studies relevant to an understanding of music and the brain began to appear near mid-20th century. Two clinicians prepared a battery of tests to determine the extent of amusia following brain damage (Botez & Wertheim, 1959; Wertheim & Botez, 1961). Studies using dichotic listening tasks began appearing in the 1960s in an attempt to determine contributions of the left and right hemispheres toward musical behaviors (e.g., Kimura, 1964)[2]. Finally, experiments on musical behaviors using the electroencephalogram (EEG) began to be published in the 1970s (Wagner & Hannon, 1975). One of the earliest, if not the first, comprehensive reviews of music and brain research was published in 1977 (Critchley & Henson, 1977).

The advent of more powerful imaging technologies, such as PET, MRI, and fMRI have led to a steep rise in neuromusical research as can be seen in Figure 1. While redundancy

[2] Sergent (1993) subsequently found the limitations of this technique so considerable as to call into question any relevant findings.

among databases may account for some of the large numbers, it is clear that more and more neuroscientists are including music as a topic of interest.

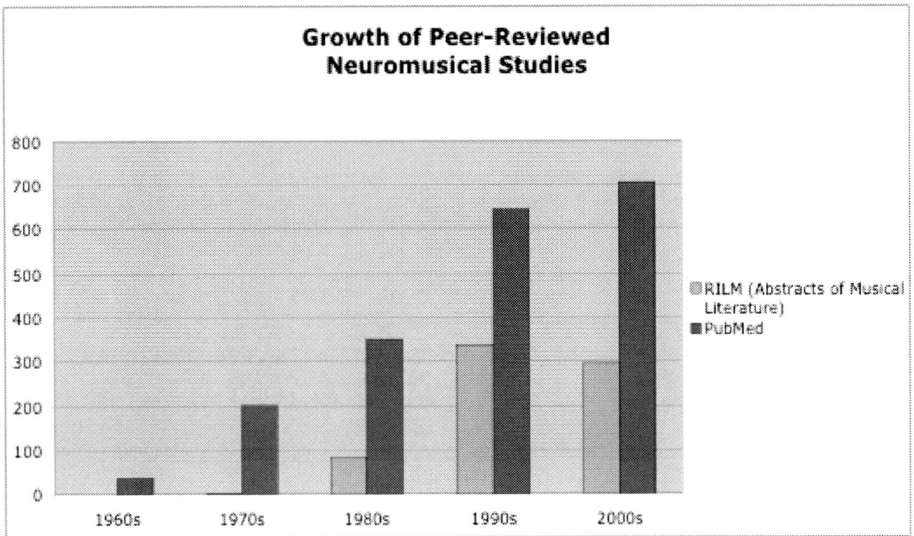

Figure 1. A graphic representation of the rising number of neuromusical studies (*y* axis) found in a keyword query for "music and brain" in two research databases of scientific journals during the past 45 years (*x* axis). Figures do not account for possible overlap (i.e., redundancy between databases).

Increasing attention to neuromusical research has led to three recent conferences and the attendant publications of proceedings (Avanzini et al., 2003; 2005; Zatorre & Peretz, 2001). As can be seen in Table 1, 170 articles have appeared in these three publications alone.

Table 1. Articles published from recent conference proceedings

Title of Book	NYAS Annals Vol.	Year	# of articles
The Biological Foundations of Music	930	2001	48
Neurosciences of Music	999	2003	69
Neurosciences of Music II	1060	2006	53
		Total	170

Another view of the scope of this literature can be seen in Table 2. Numbers may be taken as rough approximations of the frequency with which different technologies have been used to study music and the brain. Factors such as cost, access to equipment, and length of time the technology has been available play into the overall numbers.

Table 2. Keyword searches in selected online databases. A keyword search chart of potential neuromusical studies.

Neuromusical keyword search tallies (as of 12/4/05)	*PubMed*	*PsycINFO*	*RILM**	*CAIRSS**	*ERIC*	*Totals*
"Music" and "Brain"	951	691	830	1456	176	4104
"Music" and "EEG" (Electroencephalography)	293	138	81	115	5	632
"Music" and "MRI" (Magnetic Resonance Imaging)	190	14	13	26	1	244
"Music" and "fMRI" (functional Magnetic Resonance Imaging)	184	22	11	0	0	217
"Music" and "PET" (Positron Emission Tomography)	37	24	77	8	36	182
"Music" and "ERP" (Event-Related Potentials)	30	36	21	48	1	136
"Music" and "MEG" (Magnetoencephalography)	35	18	63	3	3	122
"Music" and "TMS" (Transcranial Magnetic Stimulation)	9	3	11	0	0	23
"Music" and "NIRS" (Near-Infra-red Spectroscopy)	2	0	0	0	0	2
"Music" and "DTI" (Diffusion Tensor Imaging)	2	0	0	0	0	2
Totals						5664

RILM* and CAIRSS* are limited to music related research and thus, keyword searches within these databases were conducted using only neuroscience terms without using the keyword "music"

SUPPORT FROM ANCILLARY AREAS

Anthropology and Ethnomusicology

Anthropologists and ethnomusicologists have provided abundant evidence that all groups of human beings, always and everywhere, are musical (Chailley, 1964; Hood, 1971; Merriam, 1964; Nettl, 1983) and the ubiquity of music in all the world's cultures caused Blacking (1973) to claim that music is a species-specific trait. Lomax (1968) examined 233 cultures worldwide with a specialized coding and analysis technique called cantometrics. He determined that music, especially singing, is a universal form of human behavior. Although work on the genetic basis of musicality is just now beginning (Baharloo, Service, & Risch, 2000; Gregersen et al., 2000), anthropologists and ethnomusicologists provide strong support that there is a biological basis for musicality.

Ethology

The earth is filled with the sounds of animals and many of these sounds indicate sophisticated processes at work. In the rainforests of Borneo, male tree-hole frogs adjust the frequency of their calls over a wide range to match the resonating point of the logs in which they build their nests (Lardner & Lakim, 2002). Because it rains frequently, the logs fill up with varying levels of water thus changing the resonating point. Female tree-hole frogs routinely select males that do the best job of emitting a resonant sound.

Male humpback whales create extended vocalizations (i.e., compose songs) that are shared and recognized by members of a given pod (Gray et al., 2001). Over the course of a breeding season this song is varied so that by the next season it is completely changed (Payne, 2000). Whale vocalizations utilize many features that bear similarities to human music, such as improvisation, imitation, rhythm patterns, phrases, pitch intervals, formal structures, and even rhyming schemes.

Nearly half of the 9,000 species of birds are songbirds who, like whales, invest their songs with many of the same characteristics as human music (Gray et al., 2001; Whaling, 2000). Although males are the primary singers, antiphonal singing or duetting involves both males and females (Slater, 2000). In duetting, a male and female bird alternate phrases in an exchange so tightly interwoven it can sound as if only one bird is singing. Apes also engage in duetting, although singing, in general, is practiced perhaps by as little as 11% of primate species (Geissmann, 2000).

Granted that animals make sounds; what does this have to do with human musicality? A number of scholars have written intriguing accounts of an evolutionary basis of musicality (Wallin, Merker, & Brown, 2000). As with the literature from anthropology and ethnomusicology, studies of animal soundmaking provide strong circumstantial support for neural mechanisms in the human brain dedicated to music processing and musical behaviors.

Music Psychology

The field of music psychology has provided considerable information that provides circumstantial evidence about the brain's role in musical behavior. Journals, such as *Psychology of Music* and *Music Perception*, conferences and their attendant proceedings, such as the *Society for Music Perception and Cognition*, and books such as *Psychological Foundations of Musical Behavior* (Radocy & Boyle, 2003), *The Psychology of Music* (Deutsch, 1999), and *Handbook of Music Psychology* (Hodges, 1996) have led to an extensive knowledge base.

Much is known about the cognitive processes involved in specific musical operations and this information provides a strong foundation for neuroscientific investigations. For example, music conductors were faster and more accurate in pitch discrimination, temporal order judgments, and in spatially locating targets by sound than untrained controls (Hodges, Hairston, & Burdette, 2005). These same subjects also demonstrated a benefit from the combination of auditory and visual information that was not observed in control subjects when localizing visual targets. Subsequently, brain regions known as convergence zones for the integration of sensory input were identified as potential areas underlying the conductors' superior multisensory temporal order judgments.

Support from Indirect Approaches

Fetal and Infant Responses to Music

Studying fetal and infant responses is useful in that the role of learning is minimized in comparison to older subjects. Considerable evidence indicates that during the last trimester, a fetus responds to musical sounds (Lecanuet, 1996). Likewise, infants selectively respond to music at very early ages (Fassbender, 1996; Panneton, 1985), express preferences for consonance over dissonance (Trainor, Tsang, & Cheung, 2002) and possess many musical processing skills (e.g., detection of changes in melody, in terms of pitches, rhythms, tempo, and contour) (Trehub, 2001; 2003; 2004). In turn, infant preverbal speech and singing includes musical qualities such as timbre modulation, melodic contour, and timing (Fridman, 1973; Papousek, 1996; Trevarthen & Malloch, 2002). Research on fetal and infant responses to music provides strong confirmation of inherent neural networks that subserve musical processing.

Studies of Special Musicians

Special musicians include musical prodigies, savant syndrome musicians, Williams Syndrome musicians, and Alzheimer's musicians. In each case, it would be difficult, if not impossible, to account for musical behaviors exhibited without the presence of relevant brain structures.

There are numerous musicological studies attesting to the brilliance of the young Mozart, along with more recent scientific explorations of his genius (Banks & Turner, 1991; Gedo, 1986). Révész (1925/1970) conducted a psychological study of a 20th century musical prodigy, Erwin Nyiregyházi. Gardner (1983) contends that music emerges earlier than any other "gift" and this is certainly exemplified in precocious violin students who may be as young as two years old (Suzuki, 1983).

Savant syndrome and Williams Syndrome musicians represent cognitively-impaired individuals who, despite severe limitations in other domains, display astonishing musical skills (Levitin & Bellugi, 1998; Miller, 1989). Musicians with Alzheimer's disease may also fall into this category (Crystal, Grober, & Masur, 1989; Sacks, 1999). Providing further information about the role of the brain is the fact that some Alzheimer's patients can sing when they can no longer speak coherently (Johnson & Ulatowska, 1995).

General Findings from Neuromusical Research

Modern neuroscience has access to a wide variety of technologies and protocols to study the brain. These include studies of brain damage (i.e., connecting lesion sites to deficits in performance) as well as imaging tools such as positron emission tomography (PET), magnetic resonance imaging (MRI), functional MRI (fMRI), electroencephalography (EEG), event related potentials (ERP), transcranial magnetic stimulation (TMS), magnetoencephalography (MEG), and diffusion tensor imaging (DTI). Each of these approaches has strengths and weaknesses and it is always important to pool findings from different approaches for a more complete picture. Rather than organize the literature according to protocol, findings are arranged under the following rubrics: (1) Perception and Cognition, (2) Affective Responses, (3) Musical Performance, (4) Learning, and (5) Genetic Factors.

Perception and Cognition

Perception is based on the sensory information that is gathered by the brain regarding one's external and internal environment. On the other hand, the highest order of nervous function draws upon memory, emotion, and cognition for complex thought processes (Shepherd, 1994). To date, the majority of neuromusical research has focused on music perception since most of the musical stimuli measured thus far with human subjects are no more advanced than a musical phrase (Peretz & Zatorre, 2003). Before observing something as complicated as music at the cognitive level, many researchers have acknowledged the initial investigative value of using a bottom-up approach whereby each element of music processing is studied separately (Zatorre & McGill, 2005). The general idea to this reductionist approach is that by studying the specific neural substrates of music processing's discrete components (e.g., pitch, rhythm, or timbre), the foundation will be set for future studies to explore the gestalt of these discrete components in holistic musical experiences.

But identifying music's discrete components presents an intriguing dilemma because some of these components may be strictly delegated to the processing of music (e.g., melodic pitch relationships or metric rhythm patterns), while other components such as long-term memory are activated by multiple cognitive systems. Thus, in the current review of neuromusical perception studies, it is important to recognize that conclusions are frequently based on brain activations in response to a discrete musical element (e.g., pitch or rhythm) and not a holistic musical experience. For example, several neuroimaging studies support the common observation that right hemispheric regions are engaged in the perception of pitch (Kohlmetz et al., 2003; Kuriki et al., 2005; Peretz & Zatorre, 2005; Schneider et al., 2005; Shahin et al., 2003; Warrier & Zatorre, 2004) and that left hemispheric regions are engaged in the perception of rhythm (Bengtsson & Ullen, 2006; Di Pietro et al., 2004; Schneider et al., 2005; Vuust et al., 2005). However, this evidence only offers an informative insight into *part* of the human musical experience.

The entirety of music processing is much more complicated than an examination of the brain's hemispheric or localized parts—an idea that has been strengthened in the last few years by neuroimaging studies revealing widespread bilateral brain activity during discrete music processing tasks (Bunzeck et al., 2005; Kristeva et al., 2003; Kuck et al., 2003; Lo & Fook-Chong, 2004; Lo, Fook-Chong, Lau, & Tan, 2003; Popescu, Otsuka, & Ioannides, 2004; Satoh et al., 2003) and even some cases of more holistic musical experiences (e.g., piano performance of Bach) (Fox, 2001; Parsons, 2001; Parsons et al., 2005). Yet more data is needed, especially for generative holistic musical experiences such as composition and improvisation.

The available neuroimaging research of both discrete and holistic musical processes is still too limited to make many strong conclusions about the nature of music cognition and whether it is (a) specifically localized to distinct neural networks (i.e., modularity), (b) made up of shared neural networks that are associated with other brain processes (i.e., connectionism), or (c) perhaps a little bit of both. Recent studies addressing the level of music cognition have explored the possibility of connections between music and language processing (Koelsch et al., 2003; Koelsch et al., 2004; Levitin & Menon, 2003; Saffran, 2003; Schon, Magne, & Besson, 2004), or the role of brain regions mediating pleasure, autonomic and cognitive processes which contribute to the enjoyment and ubiquity of human musical experiences (Khalfa et al., 2005; Menon & Levitin, 2005).

Another important observation is that brain activation sites can be altered in response to changes in three variables: musical stimuli (e.g., "real" music as opposed to MIDI-generated chord sequences), tasks (e.g., holistic listening versus discrete features detection), and subjects (e.g., trained versus untrained). For example, musical training appears to increase the areas of brain activation during music processing (Cui et al., 2005; Koelsch et al., 2005; Schneider et al., 2005; Seung et al., 2005) as well as increase the efficiency of brain activity during musical tasks (Haslinger et al., 2004; Meister et al., 2005). Furthermore, while most people regardless of their musical experience are able to identify deviations from expected melodic, harmonic, and rhythmic outcomes (i.e., mismatched negativity), musically trained individuals appear to have an enhanced ability to detect these musical deviations (Besson, Faita, & Requin, 1994; Fujioka et al., 2004).

Changing any one of these experimental variables (stimuli, subjects, or tasks) may affect the location of brain activation sites; therefore, to say that musical element X (e.g., pitch) is associated with region Y (right auditory cortex) in every situation may not be true. Also, it should be noted that often there are activation sites that have been "subtracted out" due to the analysis process. In many protocols, activations in a baseline condition (e.g., passive listening to a musical passage) are subtracted from the activations in a task condition (e.g., listening for pitch changes in a similar musical passage). The resultant findings of brain activations during the task condition therefore do not show all the activations, only those "beyond" or in addition to the control condition. Thus, it is not so much that certain brain regions are inactive in the task condition, rather it is that they are active in a variety of conditions.

Pitch Perception

While certain aspects of pitch perception have been found to occur in the brainstem (Gulick, Gescheider, & Frisina, 1989), multiple neuroimaging studies have consistently observed that the right secondary auditory cortex is the area responsible for various types of pitch perception such as pitch discrimination (Schneider et al., 2005; Seung et al., 2005), amplitude modulation (i.e., changes in loudness) (Hart, Palmer, & Hall, 2003), and spectral shape awareness (i.e., timbre perception) (Hall et al., 2002; Kohlmetz et al., 2003; Schneider et al., 2005; Thivard et al., 2000; Warrier & Zatorre, 2004). Finally, studies into the perception of simultaneous different pitches (i.e., harmony) observed bilateral activation during music processing (Koelsch et al., 2002; Koelsch & Mulder, 2002; Maess et al., 2001; Satoh et al., 2003).

Rhythm Perception

In contrast to pitch perception, neuroimaging studies of temporal processes (i.e., rhythm perception) often involve the activation of regions in the left hemisphere (Bengtsson et al., 2005; Di Pietro et al., 2004; Schneider et al., 2005; Vuust et al., 2005). This was suspected even before neuroimaging was available based on observations that it is easier for more people to tap a complex, syncopated rhythm with the right hand than with the left, even when left-handed subjects are observed as well (Ibbotson & Morton, 1981). More recent findings have confirmed the left hemispheric dominance of temporal grouping through neuroimaging studies of rhythmic tapping exercises (Sakai et al., 1999) and brain lesion studies (Di Pietro et al., 2004; Penhune, Zatorre, & Feindel, 1999) as well as identifying the involvement of the cerebellum in rhythmic awareness (Janata & Grafton, 2003; Penhune, Zatorre, & Evans, 1998).

While the perception of rhythm has frequently implicated neural regions in the left hemisphere, metric grouping processes (i.e., beat perception) have been observed in the right hemisphere (Li et al., 2000; Penhune et al., 1999) or even bilaterally (Kuck et al., 2003) further strengthening the argument for holistic music processes occurring throughout the brain.

Affective Responses

For the average music lover, emotional responses to music are often cited when people attempt to describe why they value music. Philosophical investigations of emotion have produced several famous treatises on this complex relationship (Langer, 1967; Meyer, 1956; Reimer, 1989) and a recent book has brought the topic to the fore in music psychology (Juslin & Sloboda, 2001). Yet among neuroscientists, emotion and music has not received much attention, perhaps because of the difficulties involved. However, interest in this topic has increased of late.

Understandably, studying affective responses to music is problematic given the subjective nature of emotional experiences. Music is known to have a wide range of physiological effects on the human body including changes in heart rate, respiration, blood pressure, skin conductivity, skin temperature, muscle tension, and biochemical responses (Bartlett, 1996). Responses to music that change the body's chemistry have been of great interest in the medical field for the therapeutic benefits that musical experiences bring to patients. A small but intriguing body of research suggests that musical experiences combined with imagery strengthen the immune system by promoting the release of stress-reducing biochemicals such as interleukin-1 (Barlett, Kaufman, & Smeltekop, 1993) or by controlling the release of stress-related biochemicals such as cortisol (Tanioka et al., 1987) and immunoglobulin A (Tsao et al., 1991). The use of music is proving very successful for alleviating pain in patients, speeding up recovery time, and reducing drug dosages up to 50 percent (Spintge, 1992).

Music's effect on the release of neurotransmitters in the brain is gaining interest as well. Serotonin is a neurotransmitter commonly associated with feelings of satisfaction from expected outcomes, and dopamine is associated with feelings of pleasure based on novelty or newness. In a study of neurochemical responses to pleasant and unpleasant music, serotonin levels were significantly higher when subjects were exposed to music they found pleasing (Evers & Suhr, 2000). Another study with subjects exposed to pleasing music found that dopamine levels increased while connectivity between areas of the brain responsible for mediating reward, autonomic, and cognitive processes was observed (Menon & Levitin, 2005). Even rats with hypertension were able to reduce their blood pressure and increase their dopamine levels when they were exposed to Mozart (Sutoo & Akiyama, 2004).

In what was perhaps the earliest imaging study on this topic, subjects underwent PET scans while listening to a musical passage that varied in levels of dissonance (Blood et al., 1999). Paralimbic and neocortical regions covaried with the degree of perceived pleasantness/unpleasantness. Subsequently, Blood and Zatorre (2001) determined that pleasing music resulting in "chills" activated areas of the brain believed to be involved in the regulation of reward and motivation (e.g., the basal forebrain, brainstem, and the orbitofrontal cortex). In a third study, musically untrained subjects listened to unfamiliar music that they reported as having enjoyed (Brown, Martinez, & Parsons, 2004). Bilateral activations were distributed widely throughout limbic and paralimbic regions. These were stronger in the left hemisphere, which is consistent with hypotheses about positive emotions being more strongly registered on the left. Overall, brain regions activated were those concerned with emotion, reward, or motivation.

The musical activation of areas involved in mediating biological responses for rewarding stimuli (e.g., food or sex) is sparking new interest in emotion research because music appears to connect the rational parts of the modern brain with the survival-based systems of the primordial brainstem (Zatorre & McGill, 2005). Perhaps the importance that music has achieved throughout humanity is based on the way it appeals to both our feelings and our intellect.

Musical Performance

The act of making music is so intensely physical that neurologist Frank Wilson (1986) referred to musicians as small-muscle athletes. The sensorimotor cortex is responsible for interpreting incoming sensory information and controlling the muscles throughout the body. In conjunction, the basal ganglia control large groups of muscles in cooperative functions, and the cerebellum regulates intricate muscle movements and stores habituated motor patterns. The brain is highly adaptable and with repetitive training, the brain's homunculus (i.e., sensorimotor map) is reorganized accordingly (Kaas, 1991). For example, long-term musical training has been found to increase the area of the motor cortex responsible for controlling the fingers of violinists (Elbert et al., 1995) and pianists (Meister et al., 2005).

Very few imaging studies have been conducted while musicians were in the act of performing. Resultant activation sites for music making are extensive and diffuse. These include:

- Pianists (Parsons et al., 2005): primary motor cortex, corresponding somatosensory areas, inferior parietal cortex, supplementary motor area, motor cingulate, bilateral superior and middle temporal cortex, right thalamus, anterior and posterior cerebellum, superior and middle temporal cortex, planum polare, thalamus, basal ganglia, posterior cerebellum, dorsolateral premotor cortex, right insula, right supplementary motor area, lingual gyrus, and posterior cingulate. Also noted were strong deactivations in posterior cingulate, parahippocampus, precuneus, prefrontal, middle temporal, and posterior cerebellar cortices. Mental rehearsal of a piano exercise activates the same motor cortex areas as performing the actual piano exercise (Pascual-Leone et al., 1995).

- Violinists (Kristeva et al., 2003; Langheim et al., 2002; Nirkko et al., 2000): bilateral primary and secondary sensorimotor areas, supplementary motor and premotor areas, bilateral superior parietal lobule, right inferior frontal gyrus, bilateral mid-frontal gyri, and bilateral lateral cerebellum, bilateral frontal opercular, primary auditory cortex Many of these same areas, though with some differences, were activated during imagined performances.
- Singers (Brown et al., 2004): primary and secondary auditory cortices, primary motor cortex, frontal operculum, supplementary motor area, insula, posterior cerebellum, and basal ganglia.

Changing variables such as musical instrument or tasks performed results in changes in activation sites. Thus with so few studies in the literature, these findings perhaps hint at rather than definitively identify brain regions involved in musical performance. Perhaps a more important concept to be realized is that musical performance, indeed all musical behaviors, are subserved by widely-distributed, but locally-specialized neural networks.

Listening to music also generates motor responses such as toe-tapping and head nodding. Thaut and colleagues have harnessed this natural response in helping Parkinsonian and stroke patients regain motor control (McIntosh et al., 1997; McIntosh, Thaut, & Rice, 1996; Thaut et al., 1996). In fact, Rhythmic Auditory Stimulation (RAS), which involves rhythmic entrainment, has been listed as one of five research-supported treatments for motor rehabilitation (Hummelsheim, 1999; Mauritz, 2002).

Using fMRI, investigators determined that the cerebellum is involved when conscious processing is required for rhythmic tasks (e.g., tapping to a rhythm) but is not involved in subconscious processing, as when rhythmic cues elicit motor entrainment (Molinari et al., 2003; Thaut, 2003). Using MEG and PET, Thaut (2003) confirmed these findings by showing that rhythmic processing engages widely distributed cortical and subcortical neural networks. These investigations are laying a foundation for an understanding of why musical rhythms are effective in entraining motor behaviors.

Learning

The role of the brain in music learning is an exceedingly complex one and much more awaits discovery before definitive statements can be made. However, progress is being made at a rapid pace. Studies relating to music learning can be organized, somewhat arbitrarily, into those that deal with pruning and plasticity, and memory.

Neural Pruning and Plasticity

From birth on, genetic instructions and life experiences work together to sculpt the brain to its eventual adult configuration. In the first years of life, there is a massive overproduction of synapses by as much as 50% (Berk, 2004; Stiles, 2000). Throughout childhood, these synaptic connections are either strengthened with repeated use or deleted due to lack of use in a process called neural pruning (Chugani, Phelps, & Mazziota, 1993; Gopnik, Meltznoff, & Kuhl, 2001). Different regions of the brain shed synapses at different rates and at different times (Stiles, 2000; Thompson et al., 2000; Webb, Monk, & Nelson, 2001).

The anatomy and physiology of the brain are affected by a person's experiences throughout life (Stiles, 2000). Shaping neural pathways through life experiences is known as plasticity and these changes can occur either by positive influences (e.g., learning and training) or by negative influences (e.g., injury and illness) (Nelson & Bloom, 1997). For example, when one area of the brain is damaged, nearby neural pathways are sometimes able to reorganize and assume the responsibilities of the damaged areas (Taupin, 2006; N. Ward, 2005). Conversely, as behaviors are learned over time, the morphology of the brain changes.

Musicians' brains are models of neuroplasticity (Muente, Altenmueller, & Jaencke, 2002; Pantev et al., 2001; Ross, Olson, & Gore, 2003; Schlaug, 2001) as changes have been observed in such brain structures as the auditory cortex (Schlaug et al., 1995a), the corpus callosum (Lee, Chen, & Schlaug, 2003; Schlaug et al., 1995b), cerebellum (Hutchinson et al., 2003), gray matter (Gaser & Schlaug, 2003; Sluming et al., 2002), and the motor cortex (Elbert et al., 1995; Schlaug, 2001). In general, these changes are more pronounced when subjects started studying music seriously before the age of seven.

Four-year old children engaged in daily classical music listening activities for six months were found to have significant increases in brain activity as compared to controls (Malyrenko et al., 2003). Also, four-year olds receiving Suzuki training had greater auditory cortex responses to tonal stimuli than untrained children (Trainor, Shahin, & Roberts, 2003). Children aged four to six who received musical training exhibited EEG patterns during music listening activities, suggesting increased cognitive activity and greater relaxation than untrained children (Flohr, Persellin, & Miller, 1996).

Studying music beginning at an early age causes increases in the left auditory association cortex (Schlaug et al., 1995; Zatorre et al., 1998); alternatively it is possible that left–right ratios are a result of neural pruning in the right auditory association cortex among musicians (Keenan et al., 2001). Strengthening of neural connections is seen in the fact that auditory cortex in both hemispheres responding to piano tones is 25% larger among experienced musicians, again, the effect being greater for those who begin musical studies at an early age (Pantev et al., 1998). Instrumentalists (e.g., violinists and trumpeters) are more responsive to the tones of their own instrument (Pantev et al., 2001). Numerous studies have shown differences in electrical brain responses between trained and untrained musicians (Altenmueller et al., 2000; Faita & Besson, 1994; Lopez et al., 2003; Nager et al., 2003; Tervaniemi & Huotilainen, 2003). In general, musicians show faster and stronger electrical responses than controls, reflecting a greater ability to process musical information and to complete musical tasks successfully.

Based on their review of neuromusical research, Peretz and Zatorre write: "Musicians appear to recruit more neural tissue or to use it more efficiently than do nonmusicians" (2005, p.105). While it is becoming clear that musical processes affect neural development in various ways, a clearer understanding of the influence these experiences have on neural development and brain organization will require more longitudinal studies of musically-trained subjects for longer periods of time such as recording brain changes across the span of time a child receives music lessons. Such a study is now underway (Schlaug et al., 2005).

While genetic influences are likely at play as well (see subsequent discussion), this body of research demonstrates the effect of music learning experiences on the brain. The notion of neural plasticity in response to musical experiences was confirmed in an experiment by Bangert and Altenmueller (2003) in which they observed increases in motor cortex activations in as little as 20 minutes in beginners who received piano instruction. This is in contrast to professional pianists who showed less activation in primary and secondary motor cortex than controls, suggesting greater efficiency (Jaencke, Shah, & Peters, 2000). It appears that once the task is learned, and perhaps habituated (e.g., scales), fewer neural resources are required.

Memory

The experiences of life are stored in the brain in two ways: short term (i.e., working memory) and long-term memory. Investigations into the forms of musical working memory indicate that pitch recognition and tonal memory engage the right temporal cortex (Zatorre & Samson, 1991) and areas of the frontal cortex (Gaab et al., 2003). These findings suggest that musical working memory may be a specialized subsystem of general working memory (Marin & Perry, 1999).

In terms of long-term musical memory processes (e.g., recognizing familiar melodies), activation of the frontal cortex and left inferior temporal lobe is a key difference from musical working memory (Platel et al., 2003; Platel et al., 1997). Using musical imagery has been a useful method of measuring long-term musical memory by taking brain scans of subjects as they imagine (but do not hear) a familiar tune. Apparently, imagery accesses the perceptual systems that are involved in music cognition as demonstrated by activations of the secondary auditory cortices during imagined melodic rehearsals (Halpern & Zatorre, 1999; Zatorre et al., 1996), tonal sequences (Penhune et al., 1998; Yoo, Lee, & Choi, 2001) or even just isolated pitches (Halpern et al., 2004; Pepper, 2005). Furthermore, activation of the auditory cortex in the absence of acoustical stimuli suggests that musical memory is a subjective experience (Peretz & Zatorre, 2005).

Since it is one's memories that serve as the basis for subjectivity, much interest has been dedicated to how musical memories are formed, stored, and retrieved, and how this process may be implicated not only in the recognition of familiar musical stimuli, but in a greater sense, how musical memories may connect to affective responses, non-musical memories, and even serve to trigger personally reflective states of consciousness. It is interesting to note that the rostral medial prefrontal cortex (RMPFC), a region of the brain recently implicated in the working memory tracking of melodic tonality during multiple modulations (Janata et al., 2002), is also the region that has been associated with personal reflections of self-knowledge (Kelley et al., 2002), the cognitive evaluation and control of emotions (Ochsner & Gross, 2005), and the ability to maintain attentional monitoring of external stimuli in conjunction with non-stimulus internal goals (e.g., abstract thoughts) (Gilbert, Frith, & Burgess, 2005).

There is also evidence that the RMPFC is one of the last regions of the brain to deteriorate in Alzheimer's patients (Thompson et al., 2003), an observation made even more fascinating given the way that Alzheimer's patients are sometimes able to engage in musical activities long after other cognitive functions have been lost (Crystal et al., 1989; Cuddy & Duffin, 2005; Johnson & Ulatowska, 1995). Janata (2005) suggests that the significance of the RMPFC may be a neural point at which some form of music processing interacts with autobiographical memories.

Genetic Factors

While music processing is common among all human beings, the extent of a person's musical capacity is not simply based on a *tabula rasa* in which everyone learns to be musical from the same blank slate. Although every brain has the same basic anatomy, the complex interaction of nature and nurture (i.e., genetic expressions and environmental experiences) combine to produce the unique neural organization of each human brain (Oerter, 2003). Determining the exact degree of influence from these varying factors is what remains unknown.

To illustrate the challenge of measuring the difference between nature and nurture, consider that while all humans have the capacity to be musical to some degree, it is widely accepted that Mozart probably had a greater genetic potential for music processing than the average person. However, even Mozart's natural disposition for music would not have flourished if he had not been given the opportunity to nurture his brain through musical exposure, training, and practice.

The development of Absolute Pitch (AP) is one example of music processing that seems to be determined by both genetic and environmental influences. Research showing that AP is a hereditary trait (Baharloo et al., 2000; Drayna et al., 2001; Gregersen et al., 2000) is balanced by evidence that explores the effect of musical training on the development of AP (Takeuchi & Hulse, 1993; W. Ward, 1999; Zatorre, 2003). Additionally, research into extreme degrees of congenital amusia (Peretz et al., 2002) and musical savants or prodigies may enhance the understanding of genetic factors for musical development.

Conclusion

Neuromusical research—studying musical experiences with direct brain imaging techniques (e.g., PET, fMRI, etc.)—is a rapidly growing field. More and more studies are being published in a wider variety of venues and gaps in the knowledge base are slowly being closed. New techniques, such as diffusion tensor imaging (DTI), are being added to the arsenal available to researchers. Once relegated to the fringe of neuroscience, studying music has gained credibility and perhaps even a little cachet.

Such progress notwithstanding, there are still many topics in need of exploration. Disparities in findings have yet to be reconciled. Technologies still place severe limitations on the kinds of musical experiences that can be studied effectively. A complete description that connects micro (genetic instructions) to macro levels (observable behaviors) in a smooth, comprehensive accounting of musical processes is not yet feasible.

Considering the status to occupy a mid-position between research still in its infancy and a mature, full-blown depiction, how can the current state of knowledge be summarized? The following represent general statements that are supported by data:

- Some degree of musicality is a birthright of all human beings.
- Musical expressiveness and responsiveness appear at birth (even before birth) and, given appropriate learning opportunities and reinforcements, develop at a natural pace throughout childhood and into adulthood.
- Musicality is highly resilient and persists to some degree even in individuals who are afflicted with physical, cognitive, or emotional impairment.
- Genetic instructions and life experiences work together to shape the "musical brain."
- The brain changes and adapts in response to music learning experiences such that the brains of adult musicians show marked differences when compared to controls.
- Music is subserved by widely-distributed neural networks, with locally-specific nodal points contributing to the overall experience.
- Although some locally-specific neural substrates have been identified, activation sites may change in response to changes in subject, stimuli, or task variables.
- Certain aspects of musical experience are supported by disassociated (i.e., distinctive, non-shared) neural networks.
- The vast majority of neuromusical studies have been conducted with subjects (both trained and untrained) familiar with Western music. Considerably more research is needed with those engaged in non-Western musical experiences. Short of multi-cultural investigations, universal explanations of the musical brain are not possible.

The degree to which this information influences music pedagogy or the teaching/learning of nonmusical subjects (e.g., mathematics) is the subject of the remainder of this book, along with many other relevant topics. This is an exciting time in neuroscience with new findings coming online at a rapid pace. Significant contributions to our understanding of the music teaching/learning process are already being made. While music practitioners may not have all the answers they would like at the moment, they are encouraged to stay current with the literature. The future of music education will undoubtedly be impacted in significant ways by the swift progress of neuromusical research.

REFERENCES

Altenmueller, E. O., Gruhn, W., Liebert, G., & Parlitz, D. (2000). The impact of music education on brain networks: evidence from EEG-studies. *International Journal of Music Education, 35*, 47-53.

Avanzini, G., Faienza, C., Minciacchi, D., Lopez, L., & Majno, M. (2003). *The Neurosciences and Music* (Vol. 999). New York: The New York Academy of Sciences.

Avanzini, G., Lopez, L., Koelsch, S., & Majno, M. (2005). *The Neurosciences and Music II: From Perception to Performance* (Vol. 1060): The New York Academy of Sciences.

Baharloo, S., Service, S., & Risch, N. (2000). Familial aggreagation of absolute pitch. *American Journal of Human Genetics., 67*, 755-758.

Bangert, M., & Altenmueller, E. (2003). Mapping perception to action in piano practice: a longitudinal DC-EEG study. *BMC Neuroscience, 4*(26), 14.

Banks, C., & Turner, J. (1991). *Mozart: prodigy of nature*. New York: Pierpont Morgan Library.

Barlett, D., Kaufman, D., & Smeltekop, R. (1993). The effects of music listening and perceived sensory experiences on the immune system as measured by interleukin-1 and cortisol. *Journal of Music Therapy, 30*(4), 194-209.

Bartlett, D. (1996). Physiological responses to music and sound stimuli. In D. Hodges (Ed.), *Handbook of music psychology* (2nd ed.). San Antonio: IMR Press.

Bengtsson, S. L., Nagy, Z., Skare, S., Forsman, L., Forssberg, H., & Ullen, F. (2005). Extensive piano practicing has regionally specific effects on white matter development. *Nature Neuroscience, 8*(9), 1148-1150.

Bengtsson, S. L., & Ullen, F. (2006). Dissociation between melodic and rhythmic processing during piano performance from musical scores. *Neuroimage, 30*(1), 272-284.

Berk, L. (2004). *Development through the lifespan* (3 ed.). New York: Allyn & Bacon.

Besson, M., Faita, F., & Requin, J. (1994). Brain waves associated with musical incongruities differ for musicians and non-musicians. *Neuroscience Letters, 168*(1-2), 101-105.

Blacking, J. (1973). *How Musical is Man?* Seattle: University of Washington Press.

Blood, A., & Zatorre, R. (2001). Intensely pleasurable responses to music correlate with activity in brain regions implicated in reward and emotion. *Proceedings of the National Academy of Sciences, 98*(II), 818-823.

Blood, A. J., Zatorre, R., Bermudez, P., & Evans, A. C. (1999). Emotional responses to pleasant and unpleasant music correlate with activity in paralimbic brain regions. *Nature Neuroscience, 2*(4), 382-387.

Botez, M., & Wertheim, N. (1959). Expressive aphasia and amusia following right frontal lesion in a right-handed man. *Journal of Experimental Pscyhology, 65*, 103-105.

Brown, S., Martinez, M., Hodges, D., Fox, P., & Parsons, L. (2004). The song system of the human brain. *Cognitive Brain Research, 20*, 363-375.

Brown, S., Martinez, M. J., & Parsons, L. M. (2004). Passive music listening spontaneously engages limbic and paralimbic systems. *NeuroReport, 15*(13), 2033-2037.

Bunzeck, N., Wuestenberg, T., Lutz, K., Heinze, H. J., & Jaencke, L. (2005). Scanning silence: mental imagery of complex sounds. *Neuroimage, 26*(4), 1119-1127.

Chailley, J. (1964). *40,000 years of music*. New York: Farrar, Straus, and Giroux.

Chugani, H., Phelps, M., & Mazziota, J. (1993). Positron emission tomography study of human brain functional development. In M. Johnson (Ed.), *Brain development and cognition,* (pp. 125-143). Cambridge, MA: Blackwell Publishers.

Critchley, M., & Henson, R. (1977). *Music and the Brain*. Springfield, IL: C.C. Thomas.

Crystal, H., Grober, E., & Masur, D. (1989). Preservation of musical memory in Alzheimer's disease. *Journal of Neurology, Neruosurgery, and Pschiatry, 52*, 1415-1416.

Cuddy, L. L., & Duffin, J. (2005). Music, memory, and Alzheimer's disease: Is music recognition spared in dementia, and how can it be assessed? *Medical Hypotheses, 64*(2), 229-235.

Cui, H. W., Zhang, S. Z., Di, H. B., Liu, H., Zhu, Y. H., Zhang, Q. W., et al. (2005). [Functional MRI of human brain in musicians and non-musicians]. *Zhejiang Da Xue Xue Bao Yi Xue Ban, 34*(4), 326-330.

Deutsch, D. (1999). *The Psychology of Music* (2nd ed.). New York: Academic Press.

Di Pietro, M., Laganaro, M., Leemann, B., & Schnider, A. (2004). Receptive amusia: temporal auditory processing deficit in a professional musician following a left temporo-parietal lesion. *Neuropsychologia, 42*(7), 868-877.

Drayna, D., Manichaikul, A., de Lange, M., Snieder, H., & Spector, T. (2001). Genetic correlates of musical pitch recognition in humans. *Science, 291*, 1969-1971.

Elbert, T., Pantev, C., Wienbruch, C., Rockstroh, B., & Taub, E. (1995). Increased cortical representation of the fingers of the left hand in string players. *Science, 270*, 305-307.

Evers, S., & Suhr, B. (2000). Changes of the neurotransmitter serotonin but not of hormones during short time music perception. *European Archives of Psychiatryand Clinical Neuroscience, 250*(3), 144-147.

Faita, F., & Besson, M. (1994). *Electrophysiological index of musical expectancy: Is there a repetition effect on the event-related potentials associated with musical incongruities?* Paper presented at the Proceedings of the 3rd international conference for music perception and cognition, Liege, Belgium.

Fassbender, C. (1996). Infants' auditory sensitivity towards acoustic parameters of speech and music. In I. Deliège & J. A. Sloboda (Eds.), *Musical beginnings: origins and development of musical competence* (pp. 56-87). New York: Oxford University Press.

Flohr, J., Persellin, D., & Miller, D. (1996). *Children's electrophysical responses to music.* Paper presented at the 22nd International Society for Music Education World Conference, Amsterdam, The Netherlands.

Fox, N., W. Crum, R. Scahill, J. Stevesn, J. Janssen, and M. Rossor. (2001). Imaging of onset and progression of Alzheimer's disease with voxel-compression mapping of serial magnetic resonance images. *The Lancet, 358*(9277), 201-205.

Fridman, R. (1973). The first cry of the newborn: Basis for the child's future musical development. *Journal of Research in Music Education, 21*(3), 264-269.

Fujioka, T., Trainor, L. J., Ross, B., Kakigi, R., & Pantev, C. (2004). Musical training enhances automatic encoding of melodic contour and interval structure. *Journal of Cognitive Neuroscience, 16*(6), 1010-1021.

Gaab, N., Gaser, C., Zaehle, T., Jaencke, L., & Schlaug, G. (2003). Functional anatomy of pitch memory: an fMRI study with sparse temporal sampling. *Neuroimage, 19*, 1417-1426.

Gardner, H. (1983). *Frames of mind: the theory of multiple intelligences.* New York: BasicBooks.

Gaser, C., & Schlaug, G. (2003). Brain structures differ between musicians and non-musicians. *Journal of Neuroscience, 23*(27), 9240-9245.

Gedo, J. (1986). Portrait of the artist as adolescent prodigy: Mozart and the Magic Flute. *Medical Problems of Performing Artists, 1*(3), 107-111.

Geissmann, T. (2000). Gibbon songs and human music from an evolutionary perspective. In N. Wallin, B. Merker & S. Brown (Eds.), *The origins of music* (pp. 103-123). Cambridge: The MIT Press.

Gilbert, S. J., Frith, C. D., & Burgess, P. W. (2005). Involvement of rostral prefrontal cortex in selection between stimulus-oriented and stimulus-independent thought. *European Journal of Neuroscience, 21*(5), 1423-1431.

Gopnik, A., Meltznoff, A., & Kuhl, P. (2001). *The scientist in the crib*. New York: Perennial.

Gray, P., Krause, B., Atema, J., Payne, R., Krumhansl, C., & Baptista, L. (2001). The music of nature and the nature of music. *Science, 29*, 52-54.

Gregersen, P., Kowalsky, E., Kohn, N., & Marvin, E. (2000). Early childhood music education and predisposition to absolute pitch: teasing apart genes and environment. *American Journal of Human Genetics., 98*, 280-282.

Gulick, W., Gescheider, G., & Frisina, R. (1989). *Hearing: physiological acoustics, neural coding, and psychoacoustics*. New York: Oxford University Press.

Hall, D., Johnsrude, I., M., H., A., P., Akeroyd, M., & A., S. (2002). Spectral and temporal processing in human auditory cortex. *Cerebral Cortex, 12*, 140-149.

Halpern, A., & Zatorre, R. (1999). When that tune runs through your head: a PET investigation of auditory imagery for familiar melodies. *Cerebral Cortex, 9*, 697-704.

Halpern, A., Zatorre, R., Bouffard, M., & Johnson, J. (2004). Behavioral and neural correlates of perceived and imagined musical timbre. *Neuropsychologia, 42*(9), 1281-1292.

Hart, H., Palmer, A., & Hall, D. (2003). Amplitude and frequency-modulated stimuli activate common regions of human auditory cortex. *Cerebral Cortex, 13*, 773-781.

Haslinger, B., Erhard, P., Altenmueller, E., Hennenlotter, A., Schwaiger, M., Grafin von Einsiedel, H., et al. (2004). Reduced recruitment of motor association areas during bimanual coordination in concert pianists. *Human Brain Mapping, 22*(3), 206-215.

Henson, R. (1977). Neurological aspects of musical experience. In M. Critchley & R. Henson (Eds.), *Music and the brain: studies in the neurology of music* (pp. 3-21). Springfield, IL: C.C. Thomas.

Hodges, D. (Ed.). (1996). *Handbook of music psychology* (2 ed.). San Antonio: IMR Press.

Hodges, D., Hairston, W. D., & Burdette, J. H. (2005). Aspects of multisensory perception: the integration of visual and auditory information in musical experiences. *Annals of the. New York Academy of Scences., 1060*, 175-185.

Hood, M. (1971). *The ethnomusicologist*. New York: McGraw-Hill.

Hummelsheim, H. (1999). Rationales for improving motor function. *Current Opinion in Neurology, 12*, 697-701.

Hutchinson, S., Lee, L. H., Gaab, N., & Schlaug, G. (2003). Cerebellar volume of musicians. *Cerebral Cortex, 13*(9), 943-949.

Ibbotson, N., & Morton, J. (1981). Rhythm and dominance. *Cognition, 9*, 125-138.

Janata, P., Birk, J. L., Van Horn, J. D., Leman, M., Tillmann, B., & Bharucha, J. J. (2002). The cortical topography of tonal structures underlying Western music. *Science, 298*(5601), 2167-2170.

Janata, P., & Grafton, S. T. (2003). Swinging in the brain: shared neural substrates for behaviors related to sequencing and music. *Nature Neuroscience, 6*(7), 682-687.

Jaencke, L., Shah, N. J., & Peters, M. (2000). Cortical activations in primary and secondary motor areas for complex bimanual movements in professional pianists. *Brain Res Cogn Brain Res, 10*(1-2), 177-183.

Johnson, J., & Graziano, A. (2003). August Knoblauch and amusia: a nineteenth-century cognitive model of music. *Brain and Cognition, 51*(1), 102-114.

Johnson, J., & Ulatowska, H. (1995). The nature of the tune and text in the production of songs. In R. Pratt & R. Spintge (Eds.), *Music Medicine II* (pp. 153-168). St. Louis: MMB Music.

Juslin, P., & Sloboda, J. (Eds.). (2001). *Music and emotion: theory and research.* New York: Oxford University Press.

Kaas, J. H. (1991). Plasticity of sensory and motor maps in adult mammals. *Annual Review of Neuroscience, 14*, 137-167.

Keenan, J. P., Thangaraj, V., Halpern, A. R., & Schlaug, G. (2001). Absolute pitch and planum temporale. *Neuroimage, 14*(6), 1402-1408.

Kelley, W. M., Macrae, C. N., Wyland, C. L., Caglar, S., Inati, S., & Heatherton, T. F. (2002). Finding the self? An event-related fMRI study. *Journal of Cognitive Neuroscience, 14*(5), 785-794.

Khalfa, S., Schon, D., Anton, J. L., & Liegeois-Chauvel, C. (2005). Brain regions involved in the recognition of happiness and sadness in music. *NeuroReport, 16*(18), 1981-1984.

Kimura, D. (1964). Left-right differences in the perception of melodies. *Quarterly Journal of Experimental Psychology, 16*, 355-358.

Koelsch, S., Fritz, T., Schulze, K., Alsop, D., & Schlaug, G. (2005). Adults and children processing music: an fMRI study. *Neuroimage, 25*(4), 1068-1076.

Koelsch, S., Grossmann, T., Gunter, T. C., Hahne, A., Schroger, E., & Friederici, A. D. (2003). Children processing music: electric brain responses reveal musical competence and gender differences. *Journal of Cognitive Neuroscience, 15*(5), 683-693.

Koelsch, S., Gunter, T. C., v Cramon, D. Y., Zysset, S., Lohmann, G., & Friederici, A. D. (2002). Bach speaks: a cortical "language-network" serves the processing of music. *Neuroimage, 17*(2), 956-966.

Koelsch, S., Kasper, E., Sammler, D., Schulze, K., Gunter, T., & Friederici, A. D. (2004). Music, language and meaning: brain signatures of semantic processing. *Nature Neuroscience, 7*(3), 302-307.

Koelsch, S., & Mulder, J. (2002). Electric brain responses to inappropriate harmonies during listening to expressive music. *Clinical Neurophysiology, 113*(6), 862-869.

Kohlmetz, C., Muller, S. V., Nager, W., Muente, T. F., & Altenmueller, E. (2003). Selective loss of timbre perception for keyboard and percussion instruments following a right temporal lesion. *Neurocase, 9*(1), 86-93.

Kristeva, R., Chakarova, V., Schulte-Mönting, J., & Spreer, J. (2003). Activation of cortical areas in music execution and imagining: a high-resolution EEG study. *Neuroimage, 20*(3), 1872-1883.

Kuck, H., Grossbach, M., Bangert, M., & Altenmueller, E. (2003). Brain processing of meter and rhythm in music: Electrophysiological evidence of a common network. *Annals of the New York Academy of Sciences, 999*, 244-253.

Kuriki, S., Isahai, N., & Ohtsuka, A. (2005). Spatiotemporal characteristics of the neural activities processing consonant/dissonant tones in melody. *Experimental Brain Research, 162*(1), 46-55.

Langer, S. (1967). *Mind: An Essay on Human Feeling* (Vol. 1). Baltimore: The Johns Hopkins Press.

Langheim, F., Callicott, J., Mattay, V., Duyn, J., & Weinberger, D. (2002). Cortical systems associated with covert music rehearsal. *Neuroimage, 16*(4), 901-908.

Lardner, B., & Lakim, M. (2002). Tree-hole frogs exploit resonance effects. *Nature, 420*, 475.

Lecanuet, J. (1996). Prenatal auditory experience. In I. Deliège & J. A. Sloboda (Eds.), *Musical beginnings: origins and development of musical competence* (pp. 3-34). New York: Oxford University Press.

Lee, D. J., Chen, Y., & Schlaug, G. (2003). Corpus callosum: musician and gender effects. *NeuroReport, 14*(2), 205-209.

Levitin, D., & Bellugi, U. (1998). Musical abilities in individuals with Williams syndrome. *Music Perception, 15*(4), 357-389.

Levitin, D., & Menon, V. (2003). Musical structure is processed in "language" areas of the brain: a possible role for Brodmann Area 47 in temporal coherence. *Neuroimage, 20*(4), 2142-2152.

Li, E., Weng, X., Han, Y., Wu, S., Zhuang, J., Chen, C., et al. (2000). Asymmetry of brain functional activation: fMRI study under language and music stimulation. *Chin Med J (Engl), 113*(2), 154-158.

Lo, Y. L., & Fook-Chong, S. (2004). Ipsilateral and contralateral motor inhibitory control in musical and vocalization tasks. *Experimental Brain Research, 159*(2), 258-262.

Lo, Y. L., Fook-Chong, S., Lau, D. P., & Tan, E. K. (2003). Cortical excitability changes associated with musical tasks: a transcranial magnetic stimulation study in humans. *Neuroscience Letters, 352*(2), 85-88.

Lomax, A. (1968). *Folk Song Style and Culture*. New Brunswick, NJ: Transaction Books.

Lopez, L., Jurgens, R., Diekmann, V., Becker, W., Ried, S., Grozinger, B., et al. (2003). Musicians versus nonmusicians. A neurophysiological approach. *Annals of the New York Academy of Sciences, 999*, 124-130.

Maess, B., Koelsch, S., Gunter, T. C., & Friederici, A. D. (2001). Musical syntax is processed in Broca's area: an MEG study. *Nature Neuroscience, 4*(5), 540-545.

Malyrenko, T., Kuraev, G., Malyrenko, Y., Khvatova, M., Romanonva, N., & Gurina, V. (2003). The development of brain electrical activity in 4-year-old children by long-term sensory stimulation with music. *Human Physiology, 22*(1), 76-81.

Marin, O., & Perry, D. (1999). Aneurological aspects of music perception and performance. In D. Deutsch (Ed.), *The psychology of music*. San Diego: Academic.

Mauritz, K. (2002). Gait training in hemiplegia. *Eurpean Journal of Neurology, 9*, 23-29.

McIntosh, G., Brown, S., Rice, R., & Thaut, M. (1997). Rhythmic auditory-motor facilitation of gait patterns in patients with parkinsonís disease. *Journal of Neurology, Neurosurgery, and Psychiatry, 62*, 22-26.

McIntosh, G., Thaut, M., & Rice, R. (1996). Rhythmic auditory stimulation as an entrainment and therapy technique: Effects on gait of stroke and Parkinsonian's patients. In *MusicMedicine* (Vol. 2, pp. 145-152). St. Louis, MO: MMB Music.

Meister, I., Krings, T., Foltys, H., Boroojerdi, B., Muller, M., Topper, R., et al. (2005). Effects of long-term practice and task complexity in musicians and nonmusicians performing simple and complex motor tasks: implications for cortical motor organization. *Human Brain Mapping, 25*(3), 345-352.

Menon, V., & Levitin, D. (2005). The rewards of music listening: Response and physiological connectivity of the mesolimbic system. *Neuroimage, 28*(1), 175-184.

Merriam, A. (1964). *The anthropology of music*. Chicago: Northwestern University Press.

Meyer, L. (1956). *Emotion and meaning in music*. Chicago: University of Chicago Press.

Miller, L. (1989). *Musical savants: Exceptional skill and mental retardation*. Hillsdale, NJ: Laurence Erlbaum.

Minagar, A., Ragheb, J., & Kelley, R. E. (2003). The Edwin Smith surgical papyrus: description and analysis of the earliest case of aphasia. *Journal of Medical Biography, 11*(2), 114-117.

Molinari, M., Leggio, M., De Maritn, M., Cerasa, A., & Thaut, M. (2003). Neurobiology of rhythmic motor entrainment. *Annals of the New York Academy of Sciences, 999*, 313-321.

Muente, T., Altenmueller, E., & Jaencke, L. (2002). The musician's brain as a model of neuroplasticity. *Nature Neuroscience, 3*, 473-378.

Nager, W., Kohlmetz, C., Altenmueller, E., Rodriguez-Fornells, A., & Muente, T. F. (2003). The fate of sounds in conductors' brains: an ERP study. *Brain Research. Cognitive Brain Research, 17*(1), 83-93.

Nelson, C., & Bloom, F. (1997). Child development and neuroscience. *Child Development, 68*(5), 970-987.

Nettl, B. (1983). *The study of ethnomusicology*. Urbana: University of Illinois Press.

Nirkko, A., Baader, A., Loevblad, K.-O., Milani, P., & Wiesendanger, M. (2000). Cortical representation of music production in violin players: behavioral assessment and functional imaging of finger sequencing, bimanual coordination and music specific brain activation. *Neuroimage, 11*(5, Supplement 1), S106.

Ochsner, K. N., & Gross, J. J. (2005). The cognitive control of emotion. *Trends in Cognitive Sciences, 9*(5), 242-249.

Oerter, R. (2003). Biological and psychological correlates of exceptional performance in development. *Annals of the New York Academy of Sciences,, 999*, 451-460.

Panneton, R. (1985). Prenatal auditory experience with melodies: Effects on postnatal auditory preferences (Doctor of Philosophy, The University of North Carolina at Greensboro). *Dissertation Abstracts International*, 47/09-B, 3984. University Microfilms No. 8701333.

Pantev, C., Oostenveld, R., Engelien, A., Ross, B., Roberts, L. E., & Hoke, M. (1998). Increased auditory cortical representation in musicians. *Nature, 392*(6678), 811-814.

Pantev, C., Roberts, L., Schulz, M., Engelien, A., & Ross, B. (2001). Timbre-specific enhancement of auditory cortical representations in musicians. *NeuroReport, 1*, 169-174.

Papousek, M. (1996). Intuitive parenting: A hidden source of musical stimulation in infancy. In I. Deliège & J. A. Sloboda (Eds.), *Musical beginnings: origins and development of musical competence* (pp. 88-112). New York: Oxford University Press.

Parsons, L. (2001). Exploring the functional neuroanatomy of music performance, perception, and comprehension. *Annals of the . New York. Academy of Sciences, 930*, 211-231.

Parsons, L., Sergent, J., Hodges, D., & Fox, P. (2005). The brain basis of piano performance. *Neuropsychologia, 43*(2), 199-215.

Pascual-Leone, A., Nguyet, D., Cohen, L. G., Brasil-Neto, J. P., Cammarota, A., & Hallett, M. (1995). Modulation of muscle responses evoked by transcranial magnetic stimulation during the acquisition of new fine motor skills. *Journal of Neurophysiology, 74*(3), 1037-1045.

Payne, K. (2000). The progressively changing songs of humpback whales: A window on the creative process in a wild animal. In N. Wallin, B. Merker & S. Brown (Eds.), *The origins of music* (pp. 135-150). Cambridge: The MIT press.

Penhune, V., Zatorre, R., & Evans, A. (1998). Cerebellar contributions to motor timing: a PET sutdy of auditory and visual rhythm reproduction. *Journal of Cognitive Neuroscience, 10*, 752-765.

Penhune, V., Zatorre, R., & Feindel, W. (1999). The role of auditory cortex in retention of rhythmic patterns as studied in patients with temporal lobe removals including Heschl's gyrus. *Neuropsychologia, 37*(3), 315-331.

Pepper, T. (2005, February 21). Inside the head of an applicant. *Newsweek, CXLV*.

Peretz, I., Ayotte, J., Zatorre, R., Mehler, J., Ahad, P., Penhune, V. B., et al. (2002). Congenital amusia: a disorder of fine-grained pitch discrimination. *Neuron, 33*(2), 185-191.

Peretz, I., & Zatorre, R. (2003). *The cognitive neuroscience of music*. Oxford: Oxford University Press.

Peretz, I., & Zatorre, R. (2005). Brain organization for music processing. *Annual Review of Psychology, 56*, 89-114.

Platel, H., Baron J., Desgranges B, Bernarnd, F., & Eustache, F. (2003). Semantic and episodic memory for music are subserved by distinct neural networks. *Neuroimage, 20*, 244-256.

Platel, H., Price, C., Baron, J., Wise, R., Lambert, J., Frakowiak, R., et al. (1997). The structural components of music perception: A functional anatomical study. *Brain, 20*(2), 229-243.

Popescu, M., Otsuka, A., & Ioannides, A. A. (2004). Dynamics of brain activity in motor and frontal cortical areas during music listening: a magnetoencephalographic study. *Neuroimage, 21*(4), 1622-1638.

Radocy, R. E., & Boyle, J. D. (2003). *Psychological foundations of musical behavior*. Springfield, IL: Charles C. Thomas.

Reimer, B. (1989). *A philosophy of music education* (2 ed.). Englewood Cliffs, NJ: Prentice-Hall.

Révész, G. (1925/1970). *The psychology of a musical prodigy*. New York: Harcourt, Brace & Company, Inc. Reprinted Westport, Conn., Greenwoood Press [1970].

Ross, D. A., Olson, I. R., & Gore, J. C. (2003). Cortical plasticity in an early blind musician: an fMRl study. *Magnetic Resonance Imaging, 21*(7), 821-828.

Sacks, O. (1999). *Awakenings*. New York: Vintage Books.

Saffran, J. R. (2003). Musical learning and language development. *Annals of the New York. Academy of Sciences,, 999*, 397-401.

Sakai, K., Hikosaka, O., Miyauchi, S., Takino, R., Tamada, T., Iwata, N. K., et al. (1999). Neural representation of a rhythm depends on its interval ratio. *Journal of Neuroscience, 19*(22), 10074-10081.

Satoh, M., Takeda, K., Nagata, K., Hatazawa, J., & Kuzuhara, S. (2003). The anterior portion of the bilateral temporal lobes participates in music perception: a positron emission tomography study. *American Journal of Neuroradiology, 24*(9), 1843-1848.

Schlaug, G. (2001). The brain of musicians: A model for functional and structural changes. *Annals of the New York Academy of Sciences, 930*, 281-299.

Schlaug, G., Jaencke, L., Huang, Y., & Steinmetz, H. (1995a). In vivo evidence of structural brain asymmetry in musicians. *Science, 267*(5198), 699-701.

Schlaug, G., Jaencke, L., Huang, Y., Staiger, J. F., & Steinmetz, H. (1995b). Increased corpus callosum size in musicians. *Neuropsychologia, 33*, 1047-1055.

Schlaug, G., Norton, A., Overy, K., & Winner, E. (2005). Effects of music training on the child's brain and cognitive development. *Ann. N. Y. Acad. Sci., 1060*, 219-230.

Schneider, P., Sluming, V., Roberts, N., Scherg, M., Goebel, R., Specht, H. J., et al. (2005). Structural and functional asymmetry of lateral Heschl's gyrus reflects pitch perception preference. *Nature Neuroscience, 8*(9), 1241-1247.

Schon, D., Magne, C., & Besson, M. (2004). The music of speech: music training facilitates pitch processing in both music and language. *Psychophysiology, 41*(3), 341-349.

Sergent, J. (1993). Mapping the musician brain. *Human Brain Mapping, 1*(1), 20-38.

Seung, Y., Kyong, J. S., Woo, S. H., Lee, B. T., & Lee, K. M. (2005). Brain activation during music listening in individuals with or without prior music training. *Neurosci Res, 52*(4), 323-329.

Shahin, A., Bosnyak, D. J., Trainor, L. J., & Roberts, L. E. (2003). Enhancement of neuroplastic P2 and N1c auditory evoked potentials in musicians. *Journal of Neuroscience, 23*(13), 5545-5552.

Shepherd, G. (1994). *Neurobiology* (3 ed.). New York: Oxford University Press.

Slater, P. (2000). Birdsong repertoires: Their origins and use. In N. Wallin, B. Merker & S. Brown (Eds.), *The origins of music* (pp. 49-63). Cambridge: The MIT Press.

Sluming, V., Barrick, T., Howard, M., Cezayirli, E., Mayes, A., & Roberts, N. (2002). Voxel-based morphometry reveals increased gray matter density in Broca's area in male symphony orchestra musicians. *Neuroimage, 17*(3), 1613-1622.

Spintge, R. (1992). The neurophsyiology of emotion and its therapeutic applications in music therapy and music medicine. In C. Maranto (Ed.), *Applications of music in medicine*. Washington, D. C.: National Association for Music Therapy.

Stiles, J. (2000). Neural plasticity and cognitive development. *Developmental Neuropscyhology, 18*(2), 237-272.

Sutoo, D., & Akiyama, K. (2004). Music improves dopaminergic neurotransmission: demonstration based on the effect of music on blood pressure regulation. *Brain Research, 1016*(2), 255-262.

Suzuki, S. (1983). *Nurture by Love: The Classic Approach to Talent Education* (2 ed.). Miami: Summy-Birchard.

Takeuchi, A., & Hulse, S. (1993). Absolute pitch. *Psychological Bulletin, 113*, 345-361.

Tanioka, F., Takazawa, T., Kamata, S., Kudo, M., Matsuki, A., & Oyama, T. (1987). Hormonal effect of anxiolytic music in patients during surgical operations under epidural anesthesia. In R. Spintge & R. Droh (Eds.), *Music in medicine*. Berlin: Springer-Verlag.

Taupin, P. (2006). Adult neurogenesis and neuroplasticity. *Restorative Neurology and Neuroscience, 24*(1), 9-15.

Tervaniemi, M., & Huotilainen, M. (2003). The promises of change-related brain potentials in cognitive neuroscience of music. *Annals of the New York Academy of Sciences., 999*, 29-39.

Thaut, M. (2003). Neural basis of rhythmic timing networks in the human brain. *Annals of the New York Academy of Sciences, 999*, 364-373.

Thaut, M., McIntosh, G., Rice, R., Rathbun, J., & Brault, J. (1996). Rhythmic auditory stimulation in gait training for Parkinson's disease patients. *Movement Disorders, 11*(2), 193-200.

Thivard, L., Belin, P., Zibovicius, M., Poline, J., & Samson, Y. (2000). A cortical region sensitive to auditory spectral motion. *NeuroReport, 11*, 2969-2972.

Thompson, P., Giedd, J., Woods, R., MacDonald, D., Evans, A., & Toga, A. (2000). Growth patterns in the developing brain detected by using continuum mechanical tensor maps. *Nature, 404*(9), 190-193.

Thompson, P., Hayashi, K. M., de Zubicaray, G., Janke, A. L., Rose, S. E., & Semple, J. (2003). Dynamics of gray matter loss in Alzheimer's disease. *Journal of Neuroscience, 23*(3), 994-1005.

Trainor, L., Shahin, A., & Roberts, L. (2003). Effects of musical training on the auditory cortex in children. *Annals of the New York Academy of Sciences, 999*, 506-513.

Trainor, L., Tsang, C., & Cheung, V. (2002). Preference for consonance in two-month old infants. *Music Perception, 20*(2), 185-192.

Trehub, S. (2001). Musical predispositions in infancy. *Annals of the New York Academy of Sciences., 930*, 1-16.

Trehub, S. (2003). The developmental origins of musicality. *Nature Neuroscience, 6*(7), 669-673.

Trehub, S. (2004). Foundations: Music perception in infancy. In J. Flohr (Ed.), *The musical lives of young children* (pp. 24-29). Upper Saddle River, NJ: Prentice-Hall.

Trevarthen, C., & Malloch, S. (2002). Musicality and music before three: Human vitality and invention shared with pride. *Zero to Three, 23*(1), 10-18.

Tsao, C. C., Gordon, T. F., Maranto, C. D., Lerman, C., & Murasko, D. (1991). The effects of music and directed biological imagery on immune response S-IgA. In M. C. D. (Ed.), *Applications of music in medicine*. Washington, D. C.: National Association for Music Therapy.

Vuust, P., Pallesen, K. J., Bailey, C., van Zuijen, T. L., Gjedde, A., Roepstorff, A., et al. (2005). To musicians, the message is in the meter pre-attentive neuronal responses to incongruent rhythm are left-lateralized in musicians. *Neuroimage, 24*(2), 560-564.

Wagner, M., & Hannon, R. (1975). Effect of music and biofeedback on alpha brainwave rhythms and attentiveness. *Journal of Research in Music Education, 23*(1), 3-13.

Wallin, N. L., Merker, B., & Brown, S. (2000). *The origins of music*. Cambridge, MA: MIT Press.

Ward, N. (2005). Neural plasticity and recovery of function. *Progress in Brain Research, 150*, 527-535.

Ward, W. (1999). Absolute pitch. In D. Deutsch (Ed.), *The psychology of music* (2nd ed.). New York: Academic.

Warrier, C. M., & Zatorre, R. J. (2004). Right temporal cortex is critical for utilization of melodic contextual cues in a pitch constancy task. *Brain, 127*(Pt 7), 1616-1625.

Webb, S., Monk, C., & Nelson, C. (2001). Mechanisms of postnatal neurobiological development: Implications for human development. *Developmental Neuropsychology, 19*(2), 147-171.

Wertheim, N., & Botez, M. (1961). Receptive amusia: A clinical analysis. *Brain, 84*, 19-30.

Whaling, C. (2000). What's behind a song? The neural basis of song learning in birds. In N. Wallin, B. Merker & S. Brown (Eds.), *The origins of music* (pp. 65-76). Cambridge: The MIT Press.

Wilson, F. (1986). *Tone deaf and all thumbs*. New York: Viking.

Yoo, S., Lee, C., & Choi, B. (2001). Human brain mapping of auditory imagery; event-related functional MRI study. *NeuroReport, 12*, 3045-3049.

Zatorre, R. (2003). Music and the brain. *Annals of the. New York Academy of Sciences, 999*, 4-14.

Zatorre, R., & McGill, J. (2005). Music, the food of neuroscience? *Nature, 434*(7031), 312-315.

Zatorre, R., & Peretz, I. (2001). *The biological foundations of music* (Vol. 999). New York: The New York Academy of Sciences.

Zatorre, R., Perry, D., Beckett, C., Westbury, C., & Evans, A. (1998). Functional anatomy of musical processing in listeners with absolute pitch and relative pitch. *Proceedings of the National Academy of Sciences, 95*, 3172-3177.

Zatorre, R., Halpern, A., Perry, D., Meyer, E., & Evans, A. (1996). Hearing in the mind's ear: a pet investigation of musical imagery and perception. *Journal of Cognitive Neuroscience, 8*, 29-46.

Zatorre, R., & Samson, S. (1991). Role of the right temporal neocortex in retention of pitch in auditory short-term memory. *Brain, 114*(Pt. 6), 2403-2417.

In: Neurosciences in Music Pedagogy
Editors: W. Gruhn and F. Rauscher, pp. 27-51
ISBN: 978-1-60021-834-7

158

Chapter II

BORN TO LEARN: EARLY LEARNING OPTIMIZES BRAIN FUNCTION

Anna Katharina Braun and Joerg Bock

ABSTRACT

Early experience plays a powerful role in shaping adult brain circuitry and behavior. Learning in childhood differs from adult learning in that experience and learning events are used to optimize immature, preliminary and unspecific neuronal networks in the brain, in particular those that are part of the "reward" system, the limbic system. This learning- and experience-induced shaping of the brain could be compared to the "formatting" of a computer hard drive, in order to determine the "hardware´s" (or "wetware´s") capacity for future learning and memory formation. This developmental neuronal reorganization determines not only cognitive, but even more emotional competence through such experience-driven reorganization of neuronal networks.

HOW DO NEURONAL NETWORKS DEVELOP THEIR CAPACITY FOR INFORMATION PROCESSING?

At birth the brain has already most of its neurons in place. Only very few neurons will be formed postnatally (a process called "neurogenesis") and in some brain regions, such as the hippocampus, an area which is involved in spatial learning and memory formation, neurons are even created throughout adulthood. Interestingly, animal experiments revealed that neurogenesis is activated by environmental stimulation and learning. Thus, experience and learning may act as a "fountain of youth" in the brain and thereby might alter and perhaps increase its functional capacity. Many brain systems are more or less functional at birth, but still require fine-tuning during postnatal development. Brain centers which are essential for survival of the organism, such as the brainstem in which areas trigger breathing or heartbeat, are fully functional and do not require further functional perfection. On the other hand, sensory and motor systems (i.e. visual, auditory, somatosensory, speech areas) and in particular higher cognitive and emotional systems, such as the limbic and the prefrontal cortical systems, gradually optimize their functional capacity by experience- and learning-induced reorganization of neuronal networks. Different brain systems, however, do not mature in parallel and at the same speed. While the sensory systems develop relatively early and reach their full capacity in humans during the first years of life, the limbic system and the prefrontal cortex, i.e. regions that are essential for creating our emotional inner world and which we use for higher cognitive tasks and for the storage and retrieval of memory, develop late and, compared to other systems, quite slowly. The prefrontal cortex, i.e. the cortical region behind our forehead that is exceptionally large in primates (monkeys and humans), develops in humans until the age of 20 years or older.

This slow developmental time course of synaptic networks, in particular those that are relevant for achieving our emotional and cognitive competences, has advantages but also disadvantages. The biological advantage is that the brain can be optimally adapted to the environment in which the individual is raised in order to develop behavioral skills and strategies, which are essential for survival. A child who lives in the desert of Negev, for example, needs to learn different skills and therefore will develop different neuronal networks in his/her brain than a child who is born in New York City. The disadvantage of this pronounced experience-driven brain plasticity, however, is that synaptic networks also adapt to adverse or deprived environments, which may result in behavioral dysfunctions. Thus, the quality of parenting and education, wherever it takes place, within the family or at school, leaves its "footprints" in the child´s brain.

The Old Dispute: What is Innate and What is Learned?

From the neurobiological perspective this dispute is obsolete. Studies in animal models revealed that the establishment and maintenance of functional neuronal networks in the developing brain is achieved by a complex, well orchestrated interaction between genetic and environmental factors. In the juvenile brain the genetic and molecular "machinery" of nerve cells (and glial cells) is highly active and at "full speed". Thus, experience and learning can "use", i.e. activate or deactivate this machinery and thereby significantly affect the maturation of neuronal functions and the establishment and reorganization of their synaptic connectivities (Comery et al., 1995). The genetic predisposition provides the individual framework within which social and cognitive behaviors and competences can develop. The genetic predisposition can be positively influenced by adequate and stabilizing environmental factors, but it can also be influenced in an adverse direction in response to negative experience, inadequate stimulation or deprivation. Viewing the genes as the keyboard of a piano, the environment would be the pianist who plays on it, and depending on which and how many keys he hits, the result will either be a highly complex symphony (reflecting a stimulating environment) or a simple melody (reflecting a deprived environment) or chaos (reflecting inconsistent, random or meaningless information from the environment). The image of the environment being the piano player on the gene-“piano” also illustrates that even though the genetic predisposition (the quality and spectrum of the keyboard) is excellent, the capacity of the brain might not develop optimally if environmental stimulation is not adequate. Vice versa, it becomes also clear that even if the piano keyboard is of less quality and perhaps may have defective keys (which would reflect genetic damage like in Down´s or fragile X syndrome) one can still acquire remarkable capacities if the brain is “educated” in a stimulating environment.

Before birth the genetically determined molecular programs dominate, and due to the relative paucity of environmental influences *in utero* they are modulated only to a rather low extent. Such relatively "rigid" genetic and molecular programming provides a safety mechanism which ensures normal development of the brain even under suboptimal conditions (e.g. malnutrition, stress of the mother, etc). Starting at birth, the balance between endogenous and exogenous factors shifts towards environmental influences that gradually gain more influence on genetic programming (Figure 1). Now the new experiences from sensory, motor and emotional stimuli start to activate or deactivate the genetic and molecular programs and thereby “fine tune” and adapt functional neuronal networks to fit the individual's environment.

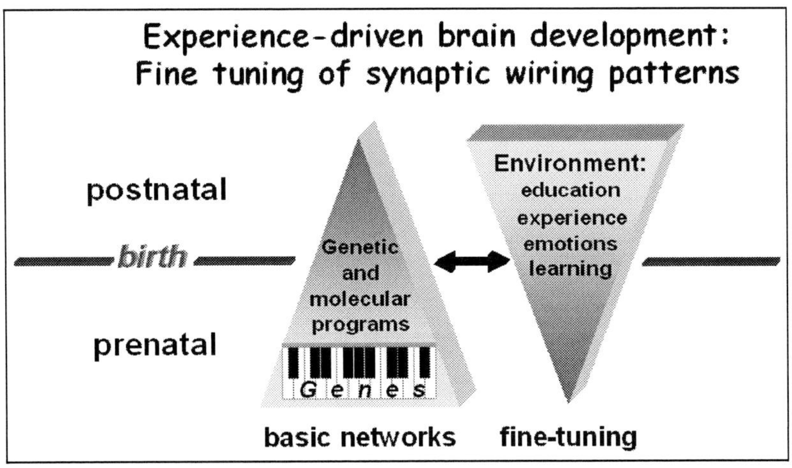

Figure 1. Interaction of genetic predisposition and environmental factors: the advantage of this mechanism is that the brain can be optimally adapted to its environment. The disadvantage of this mechanism is that the brain also adapts to adverse environments and deprivation. This might lead to “functional scars” which limit the brain´s functional capacity.

All signals from the environment that are detected by sensory systems are processed by synaptic networks which are available and “ready” at birth. On the cellular level, as outlined before, this signal transduction process involves electrical as well as chemical mechanisms, which not only register environmental information, but also activate highly complex intracellular events. The chemical signal, which is used at the synapse to transmit signals between neurons, also reaches the neuronal nucleus, which contains the genetic material that can be activated or deactivated in response to the signal transduction process.

One of the unsolved questions in neuroscience is when, and moreover, how do neuronal networks reorganize their synaptic connectivity in response to experience? There is evidence that in the human brain many functional circuits are optimized with the first four to six years of life. This implies that preschool education has a much higher impact in brain development than previously appreciated. On the other hand, it becomes now more and more evident that the human brain has a second major time window, - puberty/adolescence -, during which many brain circuits, in particular the prefrontal and limbic brain regions, undergo quite dramatic changes in wiring. And even later, during adulthood and aging, the human brain is particularly capable of undergoing considerable remodeling of synaptic connections in response to learning and memory formation, which in fact provides the substrate for its capacity for life-long learning. It appears, though, that the synaptic changes, which are known to occur during learning and memory formation in the adult brain, are much more subtle, and perhaps also of a different nature than those that occur in the immature, juvenile brain. The less severe synaptic changes of the mature brain might be due to the fact that on the cellular level the genetic-molecular "machinery" that drives neuronal changes is less active than in the immature brain in which the neurons are still developing and differentiating. In addition, the adult brain works more efficiently because it can retrieve information, which was previously stored (e.g. during childhood), and it can process old and new information using its highly optimized neuronal networks.

Electric Currents and Drugs: How do Nerve Cells in the Brain Conduct and Process Information?

Which factors play a role in brain development? What drives brain development? Which mechanisms are involved in the generation, migration and differentiation of nerve cells and their highly complex synaptic connections? These neurobiological questions are closely linked to the burning questions asked by parents and professional educators: What is innate and what is learned? Can brain functions retard due to inappropriate environmental stimulation, and how can education optimize the developing brain?

We will first look at the functional principles of the brain, and its signal-transducing units, the neurons. Already before birth neurons start to grow out their axons, i.e. cable-like structures, in order to make contact with other neurons. Through these intercellular contacts, the synapses, neurons transmit signals that carry visual, auditory, somatosensory, olfactory and other information from the environment by using electrical as well as chemical cues. Each neuron can receive as well as transmit signals in parallel, i.e. it can simultaneously act as "sender" and as "recipient" of information. The axon of the "sending" neuron forms synapses on the widely branched dendrites (see Figure 2) of thousands of other "receiving" neurons, whose dendrites act similar to antennae, which not only collect the incoming signals, but which also integrate and carry them to the cell body of the neuron, from where the signal, if it is strong enough, enters the axon and is in turn communicated to other neurons. Along the axon the signal is conveyed via miniature electric currents, which are very small, but which as sums of millions of nerve cells can be measured by electrodes fixed to the scalp (electroencephalogram or magnetoencephalogram). Upon reaching the tip of the axon, where one or several endings, called "boutons" (because of their round, button-like appearance) are formed, the electric signal is transformed into a chemical signal. Each bouton contains small vesicles containing a chemical substance, the neurotransmitter. Via complicated mechanisms, the electric current causes the release of transmitter from the vesicles into the narrow synaptic cleft that divides the sending axonal bouton from the dendrite of the receiving neuron. The neurotransmitter molecules diffuse through the narrow synaptic cleft towards the membrane of the receiving dendrite, where they bind to molecules, the receptors, which are specific for each type of neurotransmitter, quite similar to a key fitting into its matching key hole. Via this transmitter-receptor binding mechanism the receiving neuron is "informed" about the arrival of the signal. In addition, - and this is most relevant in the context of brain development, learning and memory formation -, this synaptic transduction triggers genetic and molecular events in the neuron, which changes its growth and reorganizes its synaptic contacts with other neurons. Thus, each learning process restructures the neuronal networks in the brain in a dramatic (juvenile) or more subtle (adult) way. In particular, within the limbic system (see below) learning and memory formation is achieved by either adding more synapses or, - and this appears to be specifically pronounced in the young brain-, by "weeding out" supernumerary, inefficient or redundant synaptic contacts.

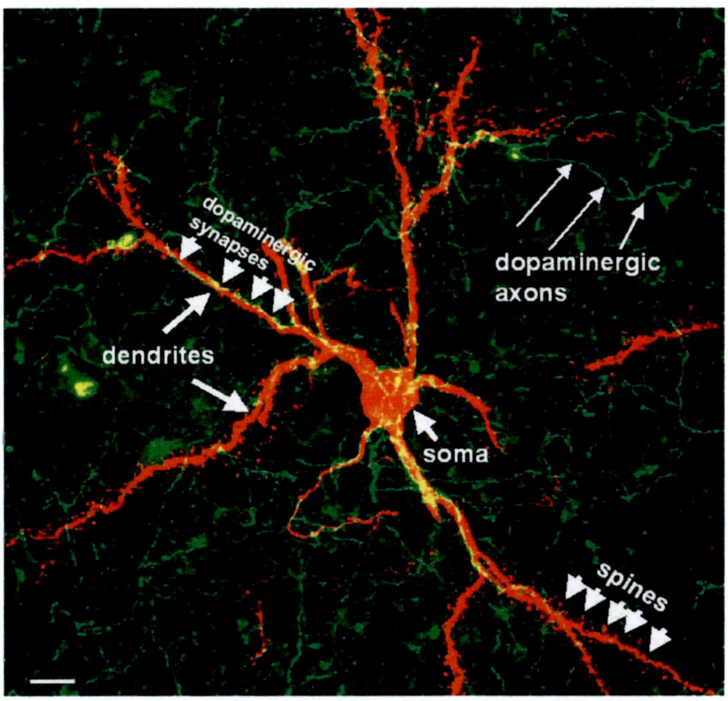

Figure 2. Neuron (red) labeled by immunohistochemistry, using an antibody against dopamine- and cyclic AMP-regulated phosphoprotein with a molecular weight of 32 *kDa* (DARPP-32, a protein which was discovered by Paul Greengard and co-workers). DARPP-32 is only expressed in neurons receiving dopaminergic input, which in this image are seen as green fluorescent fibers (arrows). The DARPP-32-expressing neuron displays numerous spine synapses (arrows) as well as en-passant dopaminergic synapses (arrows pointing on the double-labelled yellow structures) on its dendrites. (Photograph by Martin Metzger, now at University Sao Paolo, Brazil).

Proliferation and pruning of synaptic connections are achieved via a tough, Darwinian-type of selection principle of "use it or lose it". During the first years of life the synapses compete against each other for survival, and synapses survive this "synaptic war" only if they are frequently activated, whereas those synapses that are rarely used, are weakened and eventually eliminated from the network. This synaptic reorganization process in the brain, in particular the pruning of synapses, might be compared to a sculptor who models a statue by removing material from the raw stone. Also, this principle makes it clear that the capacity of the brain is not so much determined by the quantity of its synapses, but rather through the quality, i.e. specificity of its synaptic connections.

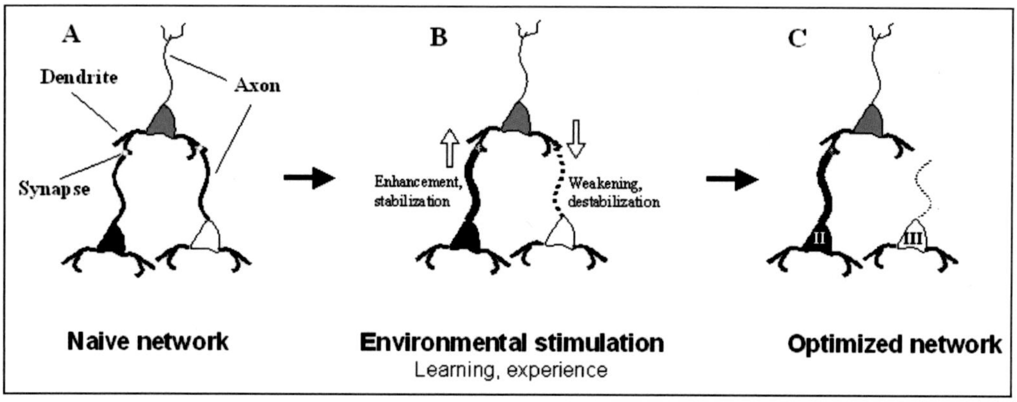

Figure 3. Illustration of synaptic selection in response to environmental stimulation, learning and memory formation.

ARE "CRITICAL" DEVELOPMENTAL PHASES REALLY CRITICAL?

Undoubtedly, there are postnatal time windows for the development of brain and behavior, the "sensitive" phases or "critical" periods, which are recognized by neuroscientists as well as by developmental psychologists. In the early studies by the groups of Arnold Scheibel, Mark Rosenzweig and William Greenough the influence of differing environmental conditions on the development of parts of the sensory and motor cortex was experimentally analyzed (Scheibel et al., 1975; 1976; Rosenzweig & Bennett, 1996; Turner & Greenough, 1985). These animal experiments revealed that early sensory or motor deprivation results in alterations of synaptic connectivities in their respective cortical regions. For example, it was discovered that neurons in the visual cortex of rats reared in an impoverished environment displayed significantly fewer synapses and reduced dendritic trees (smaller "antennae" which might reflect reduced capacity to receive signals) than neurons of rats reared in a more complex, highly stimulating environment. Similar effects were found in the somatosensory and motor cortex of monkeys reared in environments of different complexity during the first six months of life. Similar to the findings in the more simple mammals, neurons in the brain of monkeys reared in an impoverished environment displayed reduced synaptic densities compared to monkeys reared in more complex environments (Bryan & Riesen, 1989). Thus it can be concluded that the experience-dependent maturation of brain functions follow common, evolutionary old mechanisms that most likely also apply to the human brain.

For the functional maturation of sensory and motor systems the critical developmental time windows are relatively well defined and "rigid", i.e. once these time windows are closed it is almost impossible to correct failures which have occurred earlier. If the visual, auditory and somatosensory systems do not receive sensory stimulation during their critical developmental time windows, these brain circuits will never achieve their full functionality, in other words, these systems will miss the opportunity to "learn" how to see, hear and feel. However, in other brain systems the existence of sensitive or "critical" time windows does not at all mean that learning is no longer possible once these time windows are "closed" (if they ever close completely…)! In particular, the learning-pathways, i.e. the limbic system and the prefrontal cortical regions, do not have strictly defined time windows; these systems maintain a certain degree of plasticity throughout life, which is most likely the reason we can learn until very old age. Learning just takes longer once the optimal time windows are no longer wide open, and everyone knows from his/her own experience that, for example, learning a new language proceeds much faster in children than in adults.

The phases of brain development correlate quite well with phases of efficient learning capacity. The previously described pruning of synapses, similar to what has been described for the activity-dependent maturation of sensory cortical regions (Goodman & Shatz, 1993; Singer, 1995) and the innervation of skeletal muscle fibers (Changeux & Danchin, 1976) might be one mechanism or "trick" that is used by the juvenile brain to achieve high speed "turbo" learning. In contrast, the mechanism preferentially used for learning and memory formation in the adult brain is the formation of new synapses, and this process most likely takes more time and requires more energy consumption.

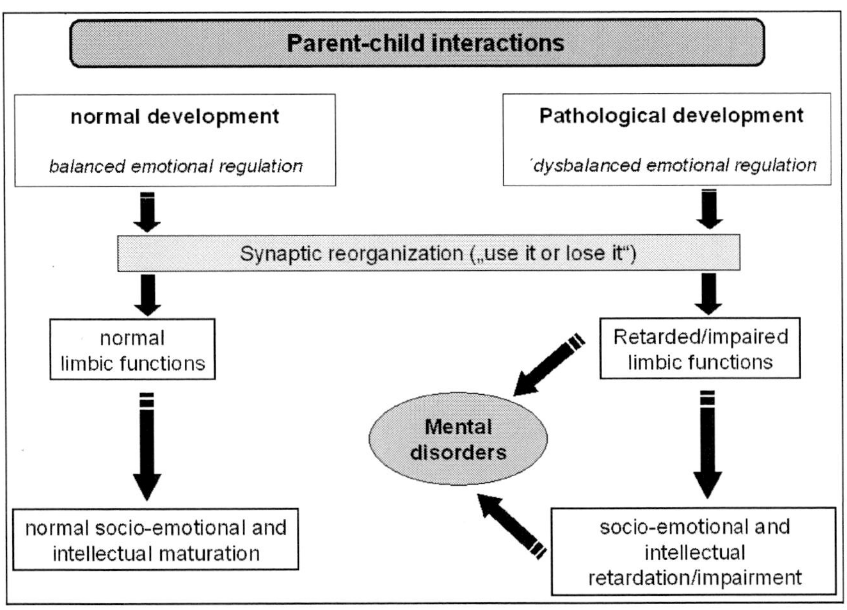

Figure 4. Impact of the emotional environment on the development of brain and behavior.

For human brain development, the critical time windows during which environmental influences are starting to interfere with the functional maturation of neural circuits occur before the children enter school, i.e. before they are exposed to professional education. For example, the functional maturation of the visual and auditory systems occurs during the first years of life, i.e. a time period that correlates with the perfecting of visual and acoustic precision. The auditory system develops earlier than the visual system, but also at a slower pace since, due to its continuous improvement in detecting and distinguishing language signals, it is critically involved in the process of speech learning and, related to this, later in development also for learning to read and write. Accordingly, an auditory cortex that receives no or only rudimentary exposure to and training to process sounds of human speech will be impaired in its precision for this task. This will inevitably have a strong impact on the acquisition of speech later in childhood. We can only imitate sounds that we can acoustically identify with high precision, and even more importantly, only the continuous feedback by critically listening to our own voice will perfect our speech skills (regarding the phonological loop see chapter 10). Furthermore, the child might later face problems in learning to write and read. if the acquisition of language is incomplete due to inappropriate stimulation. For example, problems might arise in cases where most auditory information was provided by watching TV or videos consisting of image sequences and speech episodes that are way too fast for the juvenile brain, instead of being provided by individual and interactive communication with caregivers. Since letters and syllables are visual codes for the basic acoustic features of language, the lack of the skill of precise seeing, hearing and speaking runs the risk that the child will develop dyslexia.

Impact of Early Emotional Experience on Behavioral Development

In his pioneering studies, Rene Spitz analyzed the behavior of newborn babies who had lost their parents and were raised in an orphanage. He observed a generally retarded psychological development in these children. Many of these orphans withdrew from their environment, i.e. they did not respond to caregivers trying to communicate with them, they often lay passively staring in their beds, and they were whiny and sad when separated from a caregiver (Emde et al., 1965; Spitz, 1945). Long-term studies by Skeels revealed that the critical factor for healthy psychological development in orphans is a stable emotional relationship with an attachment figure rather than intellectual stimulation (Skeels, 1966). Similar to Spitz, Skeels compared the development of orphaned babies who were raised in the regular regime of children´s homes with children from the same orphanage who had additional exposure to daily contact with "mothers" who were inmates of a nearby home for mentally challenged or difficult-to-educate girls and young women. Each of the women developed a deep emotional bond to "her" baby, and it turned out that these children later performed much better in school and developed normal social skills and emotional responses.

Pioneers in the establishment of animal models to systematically study the effects of early parental loss, inspired by the findings of Rene Spitz, were the Harlows in the 50's and 60's and later the group of Stephen Suomi (Harlow & Harlow, 1962; Suomi, 1991; 1997). In a series of experiments it was shown that young monkeys reared without contact with any other monkeys developed permanent behavioral disturbances that were strikingly similar to what Spitz had previously observed in human orphans. The deprived monkeys showed a strong reduction of vocalizations and play behavior. Moreover, the isolated monkeys showed increased anxiety behavior and decreased exploration of their environments. Females raised in isolation showed disturbed reproductive behavior later in life, and the few females that gave birth to an offspring often ignored or abused it.

If Harlow's monkey experiments might be considered cruel, what can we say about the human deprivation "experiment" in Romanian orphanages? Scientific studies confirmed what the untrained eye could see: These children were in the third to tenth percentile for physical growth, grossly delayed in motor and mental development. They rocked and grasped themselves like Harlow's monkeys, and grew up with odd social behaviors. As these children grew older, many of the orphans became homeless. With inappropriate social interactions, typically they act superficially friendly, but are unable to form permanent attachments. More recent studies with Romanian orphans who were adopted at the age of four support these findings (O'Connor & Rutter, 2000; Rutter et al., 2001). These studies revealed that the adoptees′ intellectual and emotional development was not related to the education and social state of the adoptive parents. Rather, it seemed that the establishment of an emotional attachment to the psychological parents improved the cognitive and socio-emotional development of the children. Recent clinical studies also revealed a correlation between the early loss or separation from one or both parents and the outbreak of affective illnesses (Canetti et al., 2000).

Impact of Early Emotional Experience on Brain Development

Although clinical as well as experimental animal studies revealed the importance of early experience and learning for mental development, the analyses were mainly restricted to the behavioral and endocrine level, and only recently, animal models were established to investigate the consequences of early experience and learning processes on brain development. It is tempting to speculate that emotional "templates", which are coded in the wiring of limbic networks, are continuously modified during childhood, and again during adolescence and determine emotional perception and regulation throughout life. Furthermore, neuroanatomical investigations of Huttenlocher (Huttenlocher, 1979; Huttenlocher & Dabholkar, 1997) and Rakic and colleagues (Rakic et al., 1994) in human and non-human primates have clearly shown that each cortical region has specific time windows of synaptic proliferation and of synaptic pruning, which most likely reflect synaptic reorganization in the course of experience-dependent competitive synaptic selection. The biological advantage for this synaptic competition mechanism is the optimal adaptation of functional brain systems to a given environment, and thereby increases the chance for survival and reproduction.

The disadvantage of such experience-driven synaptic reorganization, however, lies in the enhanced vulnerability of brain systems during developmental time periods of pronounced neuronal plasticity, during which adverse (or the complete lack of) environmental influences equally can "imprint" in the brain and result in retarded, dysfunctional and "defective" synaptic wiring patterns. Early traumatic experience, such as repeated exposure to brief separation from the mother or both parents, chronic social isolation or the loss of one or both parents have also been shown to induce dramatic region-specific synaptic changes in the developing brain, in particular in limbic regions, and these synaptic manifestations of traumatic memories may represent the neurological substrate for the development of pathological behaviors and mental disorders in humans.

Our findings in a new animal model, *Octodon degus* (degu, trumpet tailed rat), support this hypothesis. The degu, a semi-precocial lagomorph that is becoming increasingly popular as a laboratory animal, displays the same principal brain anatomy as common laboratory rodents. However, compared to other laboratory rats or mice this species displays closer similarities to human and non-human primate behavior and development, such as the presence of cortisol in the blood and the maturity of their sensory systems, which allows them to perceive and respond to familiar and novel stimuli from their environment immediately after birth. Similar to human babies (DeCaspar & Fifer, 1980), the newborn degu pups learn to recognize and to respond to their mothers' vocalizations within the first days of life, and this vocal communication appears to be an important component for the establishment and maintenance of the emotional attachment to the parents. Whereas common laboratory rodents appear not to exhibit a filial attachment of the kind shown by young primates or humans, in the biparental species *Octodon degus* the pups have been shown to develop a strong attachment to both parents. This feature makes them a particularly important animal model for understanding attachment in humans.

How do stressful or traumatic environmental influences experienced during early developmental phases affect the structural and neurochemical development of the brain, in particular the maturation of the limbic system? A relatively straight-forward experimental approach to address this question is to expose the newborn animals to maternal or parental separation, the experience of being repeatedly or chronically separated from their parents and siblings. For newborn pups the separation from the family represents an extremely traumatic experience that is associated with high levels of stress and anxiety.

The first question that arises is how does the brain of young degus respond to acute phases of parental separation? Using functional imaging in one-week-old degus during acute separation, we were able to identify the brain areas that are responsive to this type of acute stress. A 45-minute period of separation from the family and the home cage significantly decreased metabolic activity in most limbic cortical and subcortical regions, as well as in some sensory cortical areas (Braun & Bock, 2003).

It remains to be further investigated whether this acute downregulation of brain activity may be the starting point of chronic hypofunction of limbic regions, similar to prefrontal hypofunction detected in Romanian orphans using positron emission tomography (PET) (Chugani et al., 2001), as well as in patients suffering from attention deficit hyperactivity syndrome, schizophrenia or criminal aggression (Raine et al., 1997; Rubia et al., 1999; Brower & Price, 2001; Manoach, 2003).

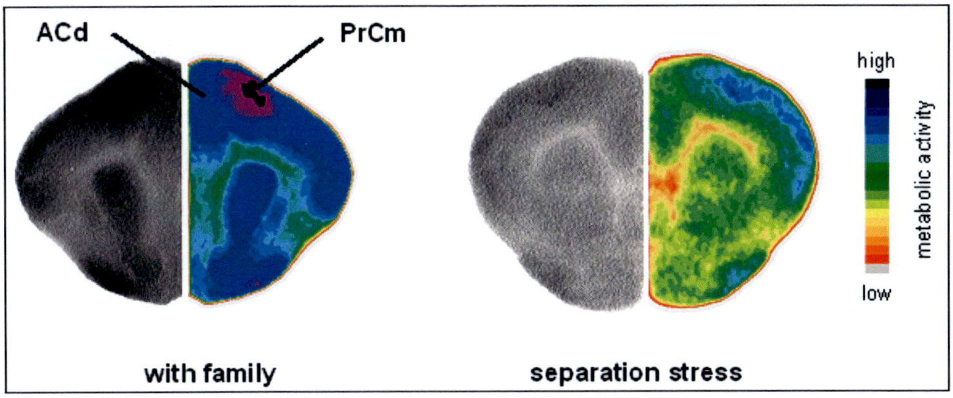

Figure 5. Metabolic activity (2-FDG uptake) in the forebrain of 8-day-old degu pups. Left image: forebrain activity in a degu pup while together with the family. Right image: forebrain activity in a degu pup during separation from the parents and siblings. Original autoradiograms (left half of the sections) and pseudocolor images (right half of the sections) from coronal sections at the level of the prefrontal cortex. ACd = anterior cingulate cortex, PrCm = precentral medial cortex.

It appears likely that this stress-induced downregulation of metabolic brain activity may mediate experience- (in this case stress-) induced synaptic reorganization, especially during time windows of elevated synaptic plasticity. In line with this idea, we found significantly elevated densities of dendritic spines in the anterior cingulate cortex of 21-day- and 45-day-old degus that were repeatedly exposed to daily parental separation during the first three weeks of life (Helmeke et al., 2001a; b; Poeggel et al., 2003). We also found significantly elevated spine densities in hippocampal CA1 pyramidal neurons, whereas significantly reduced spine densities were observed in granule cell dendrites in the dentate gyrus and on apical dendrites of large pyramidal-shaped neurons in the medial nucleus of the amygdala (Poeggel et al., 2003). Dendritic spines are the postsynaptic (i,e. "receiving") part of a synapse that receives predominantly excitatory signals.

Studies in laboratory rats that experienced maternal separation between postnatal days 14-16 confirmed the findings in degus, and in addition revealed a possible link between the synaptic changes and endocrine function (Bock et al., 2005). When rat pups were exposed to maternal separation during the stress hyporesponsive period of the hypothalamic-pituitary-adrenal (HPA) axis (i.e. when stress does not elevate corticosterone levels in the blood), no synaptic changes were observed in the anterior cingulate cortex. Thus, the magnitude and direction of stress-induced synaptic changes appear not only to depend on the brain region and cell type, but also on the developmental time window during which stress is induced. Rat pups that experienced maternal separation between postnatal days three to five ended up with fewer dendritic spines in the anterior cingulate cortex, as opposed to pups that experienced the same degree of maternal separation at days 14-16 (Bock et al., 2005). Furthermore, rats that were exposed to stress *in utero* during the last week of gestation (i.e. the pregnant dams were exposed to stress) showed decreased spine densities and reduced dendritic length in the anterior cingulate and orbitofrontal cortex (Murmu et al., 2006, see also Figure 6).

Similar alterations of neuronal connectivity have also been described in rats and monkeys after early social deprivation. For example, in early deprived rats reduced numbers of perforated synapses and synaptic contact zones in the amygdala have been found (Ichikawa et al., 1993). Monkeys raised in isolation diplayed reduced dendritic complexities and synaptic densities in the motor cortex (Struble & Riesen, 1978). In rats, it was shown that maternal care can directly influence the maturation of nerve cells and their connectivities (Liu et al., 2000). A recent study also revealed that maternal behavior can stimulate synaptogenesis and simultaneously prevent neuronal cell death (apoptosis) in the hippocampus (Weaver et al., 2002).

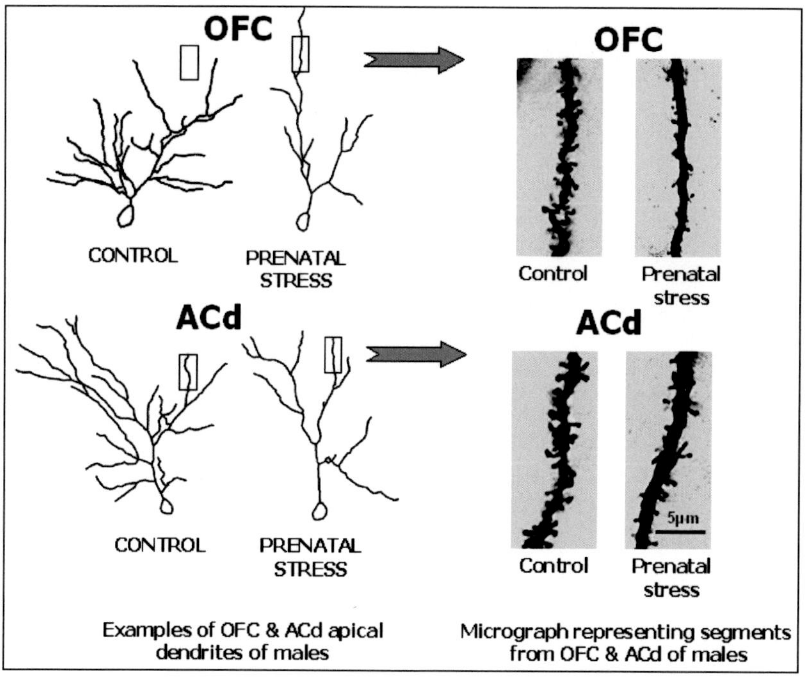

Figure 6. Illustration of dendritic (left) and synaptic /right) changes in male rats after exposure to prenatal stress. Stress in utero leads to a dramatic reduction of neuronal networks in prefrontal cortical regions, i.e. in areas that are essential for higher cognitive and emotional functions. In both prefrontal regions, the orbitofrontal (OFC) and the anterior cingulate (ACd) cortex pyramidal neurons in layer II/III display significantly reduced dendritic extensions (left). In addition, the prenatally stressed animals showed reduced densities of spine synapses (right).

Alterations of synaptic connectivity induced by neonatal and juvenile exposure to stress and social deprivation also result in neurochemical changes in the limbic system, i.e. changes in the balance of neurotransmitters, in particular of catecholaminergic systems, which play a major role in the modulation of emotional responses. A number of studies provide evidence that the activation of the HPA-axis, and therefore an increase in the level of stress hormones, directly modulates monoaminergic systems (Fuchs & Fluegge, 2002). For example, it has been shown that rats raised in social isolation display dramatically decreased dopamine metabolites (Miura et al., 2002). This is in line with anatomical findings in degus and gerbils that showed reduced densities of dopaminergic fiber innervation in some subregions of the prefrontal cortex for animals raised in socially deprived conditions (Winterfeld et al., 1998, Braun et al., 2000). Socially deprived degus develop a dysbalance of dopaminergic and serotonergic fiber systems in the orbitofrontal cortex and several limbic brain areas such as nucleus accumbens, hippocampus and amygdala (Poeggel et al., 2003; Gos et al., 2006). Similar findings have been reported in monkeys: Maternal deprivation has been shown to result in a significantly reduced dopaminergic innervation of the striatum and the nucleus accumbens (Martin et al., 1991).

Can Brain Dysfunctions be Prevented and are they Reversible?

One of the questions derived from clinical as well as from experimental data is: are the neurobiological changes that have occurred in childhood really irreversible, or is it possible to at least partly reverse or restore them through learning, i.e. induced by psycho- or pharmacotherapy. If this is possible, to what extent can brain function be normalized in adulthood?

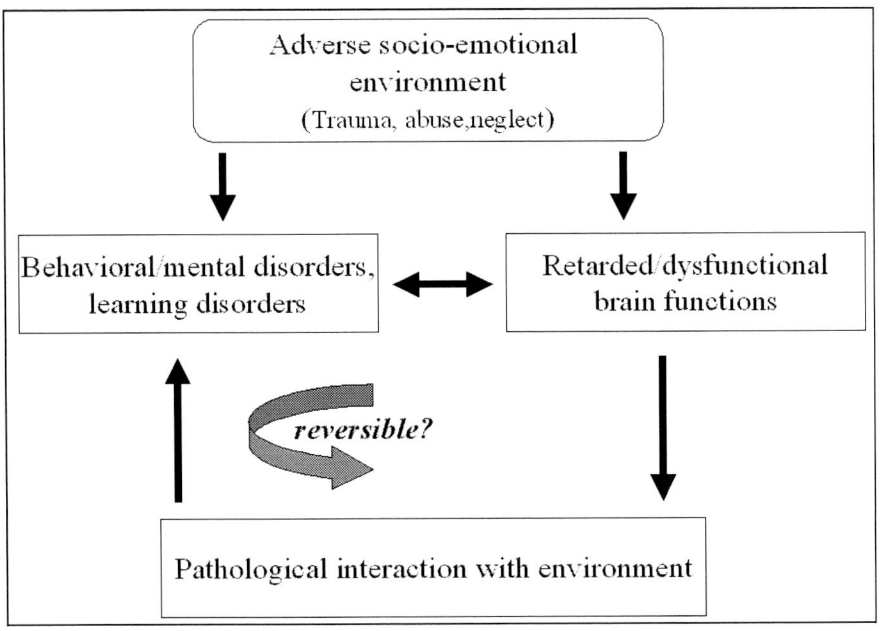

Figure 7. Role of adverse childhood environment on brain development.

Research in animal models demonstrated that being exposed to intensive maternal care downregulates endocrine functions and the behavioral responses to stress in adulthood. Extended periods of separation from the mother causes the reduction of growth hormones and brain derived neuronal growth factors, and it elevates adrenal glucocorticoids, adrenocorticotropic hormone (ACTH) and corticosterone response towards stress. These separation-induced changes can be prevented by stroking the young animals with soft hair brushes while they are separated from the mother, which gives an "imitation" of the tactile stimulation that is normally provided by the mother when she is licking, grooming and nursing her pups (Schanberg & Field, 1987; van Oers et al., 1998).

However, it is yet unclear whether normalizing abnormal behaviors by psycho- or pharmacotherapy is transient or whether it includes long-term "correction" of dysfunctional limbic circuits in the brain. There is evidence that treatment with antidepressants can stimulate neurogenesis in the hippocampus, and thereby might compensate for the stress-induced dendritic atrophy of hippocampal neurons (Harvey et al., 2003). However, this interpretation remains speculative. In order to evaluate if and to what extent developmentally-induced limbic dysfunctions can be corrected in the adult brain, the following question has to be addressed: What is the plasticity potential of the adult brain and its neurons, which can be activated by (re-)learning during therapy?

In general, the potential of neurons to undergo structural changes gradually declines with increasing age. On the one hand, eliminating supernumerary, redundant synaptic contacts during normal aging results in a more selective neuronal network that works with much more precision and efficacy. On the other hand, the progressive loss of synapses most likely reduces the network´s flexibility that is required for new learning. With increasing age this synaptic loss can reach extreme, in some cases pathological, forms (e.g. in Alzheimer´s dementia, alcoholism, etc.) and can dramatically impair learning capacity. Factors such as acute or chronic stress and changes of the hormonal status, for example during puberty or during aging (McEwen & Woolley, 1994, Murphy & Segal, 1996) contribute to normal as well as pathological synaptic elimination (Magarinos et al., 1996; McEwen & Magarinos, 1997). However, it is important to note that, particularly in humans, the adult brain retains its plasticity throughout life. From a developmental point of view the human brain retains life-long "immaturity". For instance, there is neurogenesis in the adult brain, and there is considerable synaptic formation and reorganization following brain infarcts, brain trauma and during and after epileptic seizures. This synaptic reorganization involves local sprouting of existing axons and synapses, and ingrowth of axons into cortical regions adjacent to the injured area. In this way, neighboring brain regions can at least partly take over the functions that have been lost due to the injury (Buonomano & Merzenich, 1998). Thus, the adult brain is still capable of growing new synaptic connections, but the main question is whether, and in which dimensions, this is used for learning and memory formation compared to the events in the juvenile, immature brain. Results from animal studies indicate that higher associative learning in the adult brain involves not so much a numerical change of synaptic connections, but rather more subtle mechanisms, such as weakening or strengthening of existing synapses (Lynch, 2004). Thus, at present the question of whether the reduced synaptic plasticity potential in the adult brain is sufficient to "repair" and correct dysfunctions that have occurred in childhood still remains unanswered and requires more detailed investigation (Braun & Bogerts, 2001; Kandel, 1999).

Born to Learn: Early Learning Shapes the Brain

Extrapolating and interpreting the results described in the previous sections of this chapter in "opposite" direction, leads to the powerful role of education on brain development. The results from neuroscience research demonstrate in an impressive way that education, particularly during the first six years of life (i.e. education which takes place in the family, day care and Kindergarten) is critical for shaping neuronal networks involved in emotion and cognition.

What is learning, and what happens in the brain during learning? From an evolutionary point of view, play is the oldest form of learning, and psychologists as well as biologists acknowledge that play and learning are inseparable. Plato appreciated that playful acts prepare the growing individual to acquire skills and behaviors needed in adulthood, and he recommended supplying children with toys and tools to enable them to practice. Animals also learn skills and behavior that are essential for survival in the wild through play, as anybody who has ever observed young cats, dogs or horses playing knows. The assumption that learning is almost a synonym for playing implies that learning is fun, and it is not necessarily associated with being "imprisoned" at school. Thus, preschool education should support and encourage in a playful way the innate curiosity and exploratory behavior of children. Starting at birth, children possess a natural drive to learn; they want to learn about everything in the world, and their insatiable curiosity and fascination for novel things and events is hard to stop. Their brains "search" for inspiration; they muse about the world, and they analyze and try to find explanations. This well-known "drive to learn" springs from brain biology, and neuroscientists try to understand the neurological basis for this restless search for knowledge that is so typical for the young, immature brain.

Experiments in animal models revealed that successful learning induces a strong feeling of reward in the brain that is mediated by the release of endogenous "drugs of reward". In particular, the monoaminergic neurotransmitters dopamine, serotonin and noradrenalin modulate arousal and mood. The elegant experiments of Henning Scheich and collaborators with Mongolian gerbils revealed that dopamine is released in the medial prefrontal cortex when, during training, the animals eventually found the solution to a task, i.e. when it "clicked" in their brains and they realized which behavioral strategy is correct (Stark et al., 2000; 2001; 2004). (I think that a sentence referring to "insight learning" (first discovered by Wolfgang Köhler during WWI with his chimpanzee colony on Tenerife), would be appropriate here.) It is known that in humans dopamine acts as a mood enhancer, as demonstrated in individuals who are (ab)using cocaine or amphetamines, both of which act through an increase of dopamine levels in the brain. Hence, the animal experiments clearly indicate that learning makes individuals happy by elevating mood enhancers such as dopamine. Moreover, the innate and insatiable drive of children to learn might reflect an "addiction to learning", i.e. the juvenile brain craves the "dopamine kick" because it can self administer this "home made" reward drug by successfully mastering a learning task.

These animal research results confirm the more or less intuitive knowledge of successful educators and teachers, namely that an emotional commitment is absolutely essential for higher associative learning. Furthermore, the animal experiments demonstrate that it is not necessarily an external reward ("carrot") which supports the learning success, but that the brain can reward itself. Moreover, these experiments revealed that mild stress, effort and labor ("stick"), during which stress hormones and neurotransmitters such as noradrenalin are released, are also essential for successful learning. It appears that the brain works at its best when it goes through a "roller coaster" of positive and slightly stressful emotions, and that the old principle of "carrot and stick" provides a good principle for "brain based learning".

Learning is not just “swallowing” and “digesting” information and facts, rather the acquisition and storage of knowledge has to be generated within the brain. Thus, teaching only works successfully if the scholar is willing to participate with his/her brain turned on! As mentioned earlier, higher associative learning is tightly linked to emotion, and this is due to properties of the neurobiological substrate, i.e., the limbic system. The limbic system, an evolutionarily old system in the brain, is critical for learning and memory formation and, in addition, its main function is emotional regulation. Thus, learning and the associated emotional components induce electrical, chemical (dopamine, endogenous opiates, stress hormones etc.) and - as a long-term consequence - also structural (i.e. rewiring) changes within the limbic neuronal networks. The child´s brain has an enormous, almost unlimited capacity to learn, and it is hard to overburden or overstrain it. It is much more jeopardized by not being challenged enough. Moreover, it is relatively easy to de-motivate the brain by monotone drilling (not to be confused with focused rehearsal and useful practicing), frequent experience of failure and disappointment, destructive and inconsequent criticism, punishment and humiliation.

Juvenile experience and learning leave “footprints” in the developing brain; they “imprint” on the neuronal networks, in particular of the limbic system which develops relatively late and slowly. The result of this experience- and learning-induced reorganization of synaptic networks during childhood is not so much the establishment of an encyclopedia of facts, but rather serves to establish a “grammar” of thinking, the development of thinking concepts and behavioral strategies. In parallel, a “grammar” or “language” of emotionality is created that is similar to the cognitive concepts set up in the brain networks. The establishment of cognitive as well as emotional capacities critically determines learning and emotionality throughout life, and failures in establishing a normally functioning “reward system” (= limbic system) will inevitably result in cognitive as well as emotional deficits. The observations in orphaned children and the results from various animal models impressively illustrate that in particular the severe, almost uncorrectable failures to develop the emotional “language”, resulting in “muted” emotional systems in the brain, are detrimental for learning as well as for social and emotional life.

Recent fMRI studies in humans confirmed findings from earlier EEG studies (Altenmueller et al., 2000) and revealed that learning appears to optimize the economy of energy consumption in the brain. For example, students who have been successfully trained to learn a simple acoustic discrimination task (i.e., to distinguish rising from falling frequency modulations) show an overall reduced metabolic activity in the auditory cortex compared to the wide spread activation seen prior to training (Brechmann & Scheich, 2004). Thus, the well-trained and experienced brain, which can retrieve learned information and also efficiently incorporate it into a new learning context, is less energy-consuming than a naïve, untrained brain, which has to activate broad areas while it is struggling to solve a learning task.

The Neurobiological Basis of Memory

What is memory and how do we remember? Behavioral as well as brain structure studies indicate that the retrieval of memories is not a passive phenomenon. Instead, it triggers a number of processes that either reinforce or alter stored information. The hypothesis that new memories consolidate slowly over time continues to guide memory research. It is assumed that new memories are initially 'labile' and sensitive to disruption before undergoing a series of processes (e.g., glutamate release, protein synthesis, neural growth and rearrangement) that render the memory representations progressively more stable, events that are referred to as "consolidation". Animal experiments have shown that retrieval of memory requires the activation of genes and protein synthesis, which is required for a second memory consolidation cascade termed "reconsolidation". Thus, memory retrieval is not at all the "downloading" of identical copies of stored information, but rather a reconstruction of stored templates. Whenever memory is retrieved and applied in a new context, details are slightly modified, and these modifications are stored. This mechanism of continuously changing memories might also underlie the well-known phenomenon of yarning cock-and-bull stories (i.e., fanciful and unbelievable tales). Thus, a new and fresh memory is initially unstable and fragile; it has to be stabilized by reconsolidation. The optimal time-frame for reconsolidation is within 24 hours, in order to successfully store a given information into long-term memory. During the unstable period of memory (short-term memory) the obtained information can relatively easily get lost, or more precisely, it will be "over-written" if reconsolidation does not take place within the optimal time frame. Recently, however, there has been support for the idea that stable memory representations can revert to a labile state on reactivation. In a way, this is not surprising. As already mentioned, retrieval is a dynamic process during which new information merges with and modifies the existing representation - memory should be seen as reconstructive, rather than a simple replaying of stored information. There is recent evidence that supposedly stable memory has become labile again after reactivation, i.e. via "reconsolidation". Accordingly, it can be assumed that consolidation, rather than being a one-time event, occurs repeatedly every time the representation is activated. This raises the following question: Does reconsolidation replace or "overwrite" the previously stable memory, or does it establish a new memory that coexists with the old? Whether reconsolidation is the creating of new memory or the modifying of an old, is this something other than the reconstruction of memories as they are retrieved?

What is forgetting, and what is forgotten? Memory retrieval (see above) does not only initiate reconsolidation, but also extinction of memory. Extinction is not the forgetting of the original memory trace, but rather reflects new learning. Animal experiments have revealed that reconsolidation acts to stabilize, whereas extinction tends to weaken the expression of the original memory. One important determinant of memory processing is the duration of a reminder: brief reminders lead to reconsolidation, whereas longer reminders result in memory extinction (Debiec et al., 2002; Eisenberg et al., 2003; Pedreira & Maldonado, 2003). One already mentioned mechanism of "forgetting", over-writing, has already been mentioned. An additional mechanism, which was postulated 100 years ago by Sigmund Freud, is the active blockade of memory retrieval (Andersen et al., 2004). The brain regions that influence "active" forgetting are the prefrontal cortex and the anterior cingulate cortex, i.e. regions that modulate the perception of emotions in others, as well as the regulation and control of one's own emotions. These brain regions appear to suppress the activity of the hippocampal formation, a region that is critical for memory retrieval, so this area becomes dysfunctional. Interestingly, premotor cortical regions, i.e. regions that control voluntary body movements are also involved in active forgetting. For example, if we decide to catch a cup that is falling down the table, we initiate quite complex arm and hand movements. If, however, when we start executing these movements we suddenly become aware that the cup contains hot coffee, our brains can immediately suppress further movements, which protects us from being hurt.

WHAT CAN NEUROSCIENCE CONTRIBUTE TO EDUCATION?

Similar to physicians, whom we expect to have detailed knowledge about the organs that they try to heal, parents, educators and teachers should have a realistic concept of brain functions as well as brain development, since the brain is the organ that they are changing every day.

Knowledge gained from neuroscience research reveals that learning and memory formation in the juvenile and adult brain are accompanied by physiological, biochemical, genetic-molecular and structural changes in the brain. Particularly in the juvenile brain the synaptic connections, which are activated during learning and during the formation, retrieval, consolidation and reconsolidation of memories, are stabilized within the neuronal networks and provide the basic "hardware" for future learning events throughout life. The previously-mentioned initial instability of "fresh" memories indicates that it is essential to rehearse and practice the contents of a school lesson within approximately 24 hours. Thus, doing "homework" makes a lot of sense. The vulnerability of unstable memories also makes clear that they can easily be overwritten or severely modified if, instead of doing their homework, the children watch TV or video games, which do not have any logical and meaningful connection to the content of the school lessons. The mechanism of active forgetting documents another essential, previously emphasized component of learning: the critical contribution of emotions for learning and memory formation and retrieval. Educators and teachers have to make sure that the content of their lessons are associated with positive emotions in order to prevent the previously described mechanisms of active blockade of emotionally negative memories that would occur if the child associates the learning situation at school with negative emotions.

The structural changes in the brain that accompany learning and memory formation are particularly pronounced in childhood. In other words, the juvenile, immature and still-developing brain undergoes much more dramatic changes while it learns than the adult, more stable brain. If the brain is "born to learn", why do we need teachers? Learning is a very individual inherent feature of the brain of humans and animals and it cannot be evoked from outside just by instruction and presentation of meaningless collections of facts. Our own experiments in young domestic chicks have shown that passive exposure to emotional stimuli, i.e. not allowing the animal to interact with the stimulus and express behavioral responses, cannot induce synaptic changes in the brain. On the other hand, learning can be evoked through stimulation from the environment, and this is precisely the working area and the expertise of professional educators. What stimulates learning and memory? From the neurobiological view learning is an active and a holistic process during which different brain regions are simultaneously activated, similar to the instruments of an orchestra. In order to co-activate remote brain regions, "gating" mechanisms are required, which functionally connect these brain modules. This is to ensure that attention can be focused on a specific task by suppressing irritating environmental stimuli that might disturb or block learning and the formation or retrieval of memory.

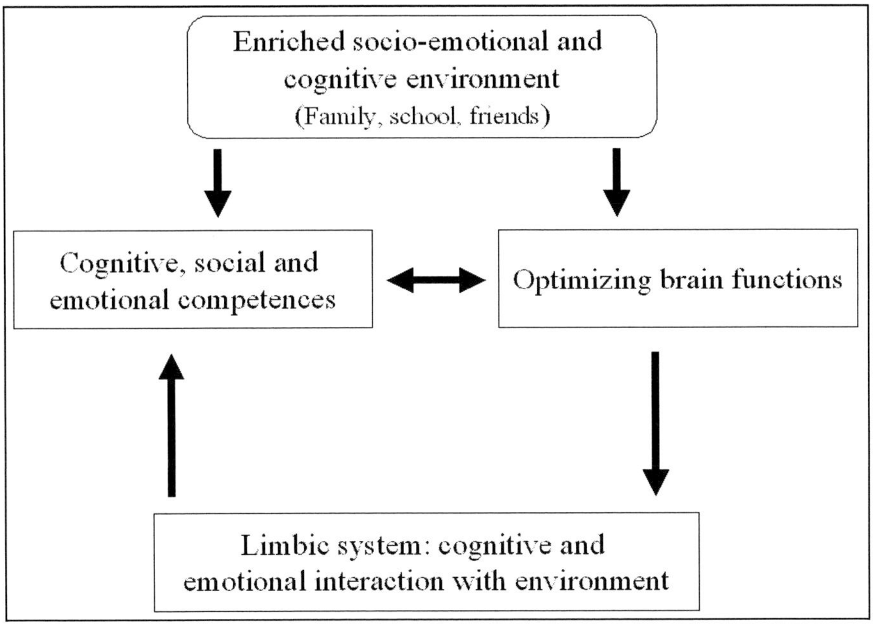

Figure 8. The role of childhood education on brain development.

Focused attention is an essential aspect that critically determines the efficiency of associative learning and memory processes. Selective attention is carried out by prefrontal cortical regions, and since it bears selective, experience-based evaluation criteria, it acts in a very individual way for each person. The frontal cortical regions modulate cognitive activity in the brain by filtering environmental stimuli through the enhancement or attenuation of specific stimuli. We all are aware of the phenomenon of being relatively immune to disrupting stimuli, such as human voices or background music, as long as we intensely focus our attention towards a task or a thought. Neurons in the prefrontal cortical regions, i.e. the brain areas that control attention and motivation, are highly interconnected with regions of the limbic system, the system for learning, emotion/reward and memory formation. As stated previously, learning is much more efficient if it is linked to emotions. A fan of horse racing does not have to be forced to read about breeding or winner statistics of race horses; all these facts (that are for "normal" people pretty meaningless and boring) will be immediately stored in his brain. School teachers know that presenting their lectures would be so much easier if the pupils are as enthusiastic about biology, chemistry or physics as they are excited about movie stars, rock singers or computer games. Thus, the greatest difficulty of teaching is not so much the didactics, but to focus the attention, to stimulate motivation and evoke enthusiasm in order to optimize learning and memory. This principle is found in all classic and new educational concepts, and it can neither be achieved by mere "cuddle pedagogy", i.e. using only rewarding stimuli, nor does it work through constant threatening and punishment.

The principle of evoking anxiety in pupils is a relatively simple but not too efficient way of getting pupils to learn, and it was used in old, and still – in a much milder form - in recent educational concepts, induced i.e. by examinations, school reports etc. On the one hand, mild stress and anxiety is very beneficial for learning, and in fact, recent research indicates that learning to avoid an unpleasant situation might be the most rewarding type of learning. For example, when a child is afraid of being humiliated in front of her classmates or if she does not want to disappoint her favorite teacher, this mild stress can be beneficial for learning performance. Animal experiments (Stark et al., 2004; Izquierdo et al., 2006; Schable et al., 2007) revealed that novel behavioral strategies for avoiding punishment or an unpleasant event are very rapidly learned. Results in humans as well as in animals have shown that the optimal teaching strategy is the "carrot and stick" principle. Consequent, individually dosed, predictable reward and reinforcement that is paired with constructive, comprehensible criticism and mild "punishment" appears to optimize brain function for learning. Too much reward, or learning tasks that are too simple, are "boring" for the brain and make it "lazy", as can be seen in children of high intelligence, who often fail to perform at their best in conventional schools. Similarly, the constant fear of criticism and humiliation by the teacher and the classmates, and an endless series of experiencing failures leading to constant stress levels, also strongly impair learning. Moreover, if learning at school is associated with negative, aversive emotions, it is detrimental for learning success and might induce the previously-mentioned blockade of memory formation and active forgetting, or it might induce selective extinction of the contents of the school lessons. In the long run, such adverse learning conditions can chronically harm the brain in the sense that it loses its innate curiosity and "addiction" to learn and eventually gives up. It is difficult and takes a lot of effort and patience to rekindle the flame of enthusiasm and joy for learning once it is extinguished, especially because in most cases children's self esteem and confidence in their abilities, which determine their motivation to learn, has also been destroyed.

Most teachers, however, work with the principle of positive motivation and reinforcement, to kindle the children′s interest in learning and to keep and further activate their joy to learn. Although the brain is born with an "addiction to learn", no child is born with a natural interest for all themes and topics. How can we motivate them, and how can we stimulate their interest for topics in which they are initially not naturally interested? It is a good idea to remember ourselves being as children at school. What evoked our interest in a certain subject? In many cases we might not remember a specific "trigger" and believe that we were always "naturally" interested in a certain topic or subject. Sometimes, however, we do remember specific situations, persons or books which evoked or stimulated our interest and enthusiasm for a theme. Such stimulating experiences and situations should be provided at school. Teachers have to create and present their lectures to be emotionally stimulating. They have to transmit their own enthusiasm in the subject, and the spark of enthusiasm has to jump across to the students. Children have very sensitive antennae when it comes to distinguishing between true and false enthusiasm, and it is difficult, if not impossible, to "emotionally cheat" children. Children should feel that they are valued and accepted as people, so they will realize that their individual learning success is important for the teacher.

With some pupils, teachers are lucky in the sense that they already have a positive and enthusiastic attitude towards learning. The learning drive of these children most likely has been reinforced by previous learning experiences, so they know that to learn something new, and to solve questions and problems, - in particular when they are difficult and take some time to find the solution -, "tickles" the brain. It "turns them on" (mediated by the "dopamine kick"). The majority of pupils, however, need some inspiration and animation in order to catch their attention and kindle their interest. Teachers have to show these children that working on difficult tasks and solving complicated problems creates a great feeling of reward, the "kick". This also includes the reinforcement of the child´s self confidence and he/she has to realize his/her competence to successfully work on difficult subjects. When this "high performance training" starts early in childhood (in fact it should be started at birth!) the neuronal networks, i.e. the child´s "learning machine" can be optimized for learning, and in addition we can prevent the development of reluctance or aversion against learning.

Although there are general neurobiological principles for learning and memory functions that apply to all mammalian brains (including ours), the details of the mechanisms underlying brain development, learning and memory are far from being completely understood. When the knowledge available from brain research is applied to education, it is important to be aware of the fact that brain functions are very individual and differ for each person. All brains use identical neuronal principles, which are outlined in this article, when they perceive and process information from the environment, when they learn and when they memorize. Nevertheless, each individual person learns in an individual way, according to his or her brain wiring, and when two people encounter an identical situation they may mentally handle it quite differently. Each person applies slightly different learning strategies. This is, among other factors, also due to the fact that the selection of presented information is not under complete conscious control, but is rather dependent on a personal cognitive and emotional evaluation, and this highly individual judgment of meaningfulness is based on previous experience. Brain-based individuality of learning requires individual teaching concepts, and there will be no generally applicable "recipes" or "prescriptions" for teaching concepts at school.

References

Altenmueller, E., Gruhn, W., Parlitz, D., & Liebert, G. (2000). The impact of music education on brain networks: evidence from EEG-studies. *International Journal of Music Education, 35*, 47 – 53.

Anderson, M.C., Ochsner, K.N., Kuh,l B., Cooper, J., Robertson, E., Gabrieli, S.W., Glover, G.H., & Gabrieli J.D. (2004). Neural systems underlying the suppression of unwanted memories. *Science, 303(5655),* 232-235.

Bock, J., Gruss, M., Becker, S., & Braun, K. (2005) Experience-induced changes of dendritic spine densities in the prefrontal and sensory cortex: Correlation with developmental time windows. *Cerebral Cortex, 15,* 802-808.

Braun, K., & Bogerts, B. (2001). Experience guided neuronal plasticity. Significance for pathogenesis and therapy of psychiatric diseases. *Nervenarzt, 72*, 3-10.

Braun, K., Lange, E., Metzger, M., & Poeggel, G. (2000). Maternal separation followed by early social deprivation affects the development of monoaminergic fiber systems in the medial prefrontal cortex of Octodon degus. *Neuroscience, 95,* 309-318.

Braun, K., & Bock, J. (2003). Early traumatic experience alters metabolic brain activity in thalamic, hypothalamic and prefrontal cortical brain areas of *Octodon degus. Developmental Psychobiology 43/3,* 248.

Brechmann, A., & Scheich, H. (2004). Hemispheric shifts of sound representation in auditory cortex with conceptual listening. *Cerebral Cortex, 15(5),* 578-587

Brower, M.C., & Price, B.H. (2001). Neuropsychiatry of frontal lobe dysfunction in violent and criminal behaviour: a critical review. *Journal of Neurology, Neurosurgery & Psychiatry, 71,* 720-726.

Bryan, G.K., & Riesen, A.H. (1989). Deprived somatosensory-motor experience in stumptailed monkey neocortex: dendritic spine density and dendritic branching of layer IIIB pyramidal cells. *The Journal of Comparative Neurology, 286,* 208-217.

Buonomano, D.V., & Merzenich, M.M. (1998). Cortical plasticity: from synapses to maps. *Annual Reviews of Neuroscience, 21,* 149-186.

Canetti, L., Bachar, E., Bonne, O., Agid, O., Lerer, B., Kaplan De-Nour, A., & Shalev, A.Y. (2000). The impact of parental death versus separation from parents on the mental health of Israeli adolescents. *Comprehensive Psychiatry, 41,* 360-368.

Changeux, J.P., & Danchin, A. (1976). Selective stabilisation of developing synapses as a mechanism for the specification of neuronal networks. *Nature, 264,* 705-712.

Chugani, H.T., Behen, M.E., Muzik, O., Juhasz, C., Nagy, F., & Chugani D.C. (2001). Local brain functional activity following early deprivation: a study of postinstitutionalized Romanian orphans. *Neuroimage, 14,* 1290-1301.

Comery, T.A., Shah, R., & Greenough, W.T. (1995). Differential rearing alters spine density on medium-sized spiny neurons in the rat corpus striatum: evidence for association of morphological plasticity with early response gene expression. *Neurobiology of Learning and Memory, 63,* 217-219.

Debiec, J., LeDoux, J.E., & Nader, K. (2002). Cellular and systems reconsolidation in the hippocampus. *Neuron, 36,* 527–538

DeCasper, A.J., & Fifer, W.P. (1980). Of human bonding: newborns prefer their mother's voices. *Science, 208,* 1174-76.

Eisenberg, M., Kobilo, T., Berman, D.E., & Dudai, Y. (2003). Stability of retrieved memory: inverse correlation with trace dominance. *Science, 301,* 1102–1104.

Emde, R.N., Polak, P.R., & Spitz, R.A. (1965). Anaclitic depression in an infant raised in an institution. *Journal of the American Academy of Child Psychiatry, 4,* 545-553.

Fuchs, E., & Flugge, G. (2002). Social stress in tree shrews: effects on physiology, brain function, and behavior of subordinate individuals. *Pharmacology, Biochemistry, and Behavior, 73,* 247-258.

Goodman, C.S., Schatz, C.J. (1993). Developmental mechanisms that generate precise patterns of neuronal connectivity. *Cell, 72 Suppl,* 77-98.

Gos, T., Becker, K.; Bock, J., Malecki, U., Helmeke, C., Poeggel, G., Bogerts, B., & Braun, K. (2006). Early neonatal and postweaning social emotional deprivation interferes with the maturation of serotonergic and tyrosine hydroxylase- immunoreactive afferent fiber systems in the rodent nucleus accumbens, hippocampus and amygdala. *Neuroscience, 140,* 811-821.

Harlow, H.F., & Harlow, M.K. (1962). Social deprivation in monkeys. *Scientific American, 207,* 137-146.

Harvey, B.H., McEwen, B.S., & Stein, D.J. (2003). Neurobiology of antidepressant withdrawal: implications for the longitudinal outcome of depression. *Biological Psychiatry, 54,* 1105-1117.

Helmeke, C., Ovtscharoff jr, W., Poeggel, G., & Braun, K. (2001a). Juvenile emotional experience alters synaptic composition in the anterior cingulate cortex. *Cerebral Cortex, 11,* 717-727.

Helmeke, C., Poeggel, G., & Braun, K. (2001b). Differential emotional experience induces elevated spine densities on basal dendrites of pyramidal neurons in the anterior cingulate cortex. *Neuroscience, 104,* 927-931.

Huttenlocher, P.R., & Dabholkar, A.S. (1997). Regional differences in synaptogenesis in human cerebral cortex. *The Journal of Comparative Neurology, 387,* 167-178.

Huttenlocher, P.R. (1979. Synaptic density in human frontal cortex - developmental changes and effects of aging. *Brain Research, 163,* 195-205.

Ichikawa, M., Matsuoka, M., & Mori, Y. (1993). Effect of differential rearing on synapses and soma size in rat medial amygdaloid nucleus. *Synapse, 13,* 50-56.

Izquierdo, I., Bevilaqua, L.R., Rossato, J.I., Bonini, J.S., Medina, J.H., & Cammarota, M. (2006) Different molecular cascades in different sites of the brain control memory consolidation. *Trends in Neurosciences, 29,* 496-505.

Kandel, E.R. (1999). Biology and the future of psychoanalysis: a new intellectual framework for psychiatry revisited. *The American Journal of Psychiatry, 156,* 505-524.

Liu, D., Diorio, J., Day, J.C., Francis, D.D., & Meaney, M.J. (2000) Maternal care, hippocampal synaptogenesis and cognitive development in rats. *Nature Neuroscience, 3,* 799-806.

Lynch, M.A. (2004). Long-term potentiation and memory. *Physiol Rev, 84,* 87-136.

Magarinos, A.M., McEwen, B.S., Flugge, G., & Fuchs, E. (1996). Chronic psychosocial stress causes apical dendritic atrophy of hippocampal CA3 pyramidal neurons in subordinate tree shrews. *The Journal of Neuroscience, 16, 3534-3540.*

Manoach, D.S. (2003). Prefrontal cortex dysfunction during working memory performance in schizophrenia: reconciling discrepant findings. *Schizophrenia Research, 60,* 285-298.

Martin, L.J., Spicer, D.M., Lewis, M.H., Gluck, J.P., & Cork, L.C. (1991). Social deprivation of infant rhesus monkeys alters the chemoarchitecture of the brain: I. Subcortical regions. *The Journal of Neuroscience, 11,* 3344-3358.

McEwen, B.S., & Magarinos, A.M. (1997). Stress effects on morphology and function of the hippocampus. *Annals of the New York Academy of Sciences, 821,* 271-284.

McEwen, B.S., & Woolley, C.S. (1994). Estradiol and progesterone regulate neuronal structure and synaptic connectivity in adult as well as developing brain. *Experimental Gerontology, 29,* 431-436.

Miura, H., Qiao, H., & Ohta, T. (2002). Influence of aging and social isolation on changes in brain monoamine turnover and biosynthesis of rats elicited by novelty stress. *Synapse, 46,* 116-124.

Murmu, M.S., Salomon, S., Biala, Y., Weinstock, M., Braun, K., & Bock, J. (2006). Changes of spine density and dendritic complexity in the prefrontal cortex in offspring of mothers exposed to stress during pregnancy. *The European Journal of Neuroscience, 24,* 1477-1487.

Murphy, D.D., & Segal M.(1996). Regulation of dendritic spine density in cultured rat hippocampal neurons by steroid hormones. *The Journal of Neuroscience, 16,* 4059-4068.

O'Connor, T.G., & Rutter, M. (2000). Attachment disorder behavior following early severe deprivation: extension and longitudinal follow-up. English and Romanian Adoptees Study Team. *Journal of the American Academy of Child and Adolescent Psychiatry, 39,* 703-712.

Pedreira, M.E., & Maldonado, H. (2003). Protein synthesis subserves reconsolidation or extinction depending on reminder duration. *Neuron, 38,* 863–869.

Poeggel, G., Helmeke, C., Abraham, A., Schwabe, T., Friedrich, P., & Braun K. (2003). Juvenile emotional experience alters synaptic composition in the rodent cortex, hippocampus, and lateral amygdala. *Proceedings of the National Academy of Sciences of the United States of America, 100,* 16137-16142.

Poeggel, G., Nowicki, L., & Braun, K. (2003). Early social deprivation alters monoaminergic afferents in the orbital prefrontal cortex of Octodon degus. *Neuroscience, 116,* 617-620.

Raine, A., Buchsbaum, M., & LaCasse, L. (1997). Brain abnormalities in murderers indicated by positron emission tomography. *Biological Psychiatry, 42,* 495-508.

Rakic, P., Bourgeois, J.P., & Goldman-Rakic, P.S. (1994). Synaptic development of the cerebral cortex: implications for learning, memory, and mental illness. *Progress in Brain Research, 102,* 227-243.

Rosenzweig, M.R., & Bennett, E.L. (1996). Psychobiology of plasticity: effects of training and experience on brain and behavior. *Behavioural Brain Research, 78,* 57-65.

Rubia, K., Overmeyer, S., Taylor, E., Brammer, M., Williams, S.C., Simmons, A., & Bullmore, E.T. (1999). Hypofrontality in attention deficit hyperactivity disorder during higher-order motor control: a study with functional MRI. *The American Journal of Psychiatry, 156,* 891-896.

Rutter, M., Kreppner, J.M., & O'Connor, T.G. English and Romanian Adoptees StudyTeam (2001). Specificity and heterogeneity in children's responses to profound institutional privation. *British Journal of Psychiatry, 179,* 97-103.

Schable, S., Poeggel, G., Braun, K., & Gruss, M. (2007). Long-term consequences of early experience on adult avoidance learning in female rats: role of the dopaminergic system. *Neurobiology of Learning and Memory, 87,* 109-122.

Schanberg, S.M., & Field, T.M. (1987). Sensory deprivation stress and supplemental stimulation in the rat pup and preterm human neonate. *Child Development, 58,* 1431-1447.

Scheibel, M.E., Lindsay, R.D., Tomiyasu, U., & Scheibel, A.B. (1975). Progressive dendritic changes in aging human cortex. *Experimental Neurology, 47,* 392-403.

Scheibel, M.E., Lindsay, R.D., Tomiyasu, U., Scheibel, A.B. (1976). Progressive dendritic changes in the aging human limbic system. *Experimental Neurology, 53,* 420-430.

Singer, W. (1995). Development and plasticity of cortical processing architectures. *Science, 270(5237),* 758-764.

Skeels, H.M. (1966). Adult status of children with contrasting early life experiences: a follow-up study. *Monographs of the Society for Research in Child Development, 105,* 31, 1-65.

Spitz, R.A. (1945). Hospitalism. An inquiry into the genesis of psychiatric conditions in early childhood. *Psychoanalytic Study of the Child, 1,* 53-74.

Stark, H., Bischof, A., Wagner, T., & Scheich, H. (2000). Stages of avoidance strategy formation in gerbils are correlated with dopaminergic transmission activity. *European Journal of Pharmacology, 405(1-3),* 263-275.

Stark, H., Bischof, A., Wagner, T., & Scheich, H. (2001). Activation of the dopaminergic system of medial prefrontal cortex of gerbils during formation of relevant associations for the avoidance strategy in the shuttle-box. *Progress in Neuro-Psychopharmacology & Biological Psychiatry, 25,* 409-426.

Stark, H., Rothe, T., Wagner, T., & Scheich H. (2004). Learning a new behavioral strategy in the shuttle-box increases prefrontal dopamine. *Neuroscience, 126(1),* 21-29.

Struble, R.G., & Riesen, A.H. (1978). Changes in cortical dendritic branching subsequent to partial social isolation in stumptailed monkeys. *Developmental Psychobiology, 11,* 479-486.

Suomi, S.J. (1997) Early determinants of behaviour: evidence from primate studies. *British Medical Bulletin, 53,* 170-184.

Suomi, S.J. (1991). Early stress and adult emotional reactivity in rhesus monkeys. *Ciba Foundation Symposium, 156,* 171-188.

Turner, A.M., & Greenough, W.T. (1985). Differential rearing effects on rat visual cortex synapses. I. Synaptic and neuronal density and synapses per neuron. *Brain Research, 329,* 195-203.

van Oers, H.J., de Kloet, E.R., Whelan, T., & Levine, S. (1998). Maternal deprivation effect on the infant's neural stress markers is reversed by tactile stimulation and feeding but not by suppressing corticosterone. *The Journal of Neuroscience, 18,* 10171-10179.

Weaver, I.C., Grant, R.J., & Meaney, M.J. (2002). Maternal behavior regulates long-term hippocampal expression of BAX and apoptosis in the offspring. *Journal of Neurochemistry, 82,* 998-1002.

Winterfeld, K.T., Teuchert-Noodt, G., Dawirs, R.R. (1998). Social environment alters both ontogeny of dopamine innervation of the medial prefrontal cortex and maturation of working memory in gerbils (Meriones unguiculatus). *Journal of Neuroscience Research, 52,* 201-209.

In: Neurosciences in Music Pedagogy
Editors: W. Gruhn and F. Rauscher, pp. 53-99
ISBN: 978-1-60021-834-7

Chapter III

Music Learning in Childhood Early Developments of a Musical Brain and Body

John W. Flohr and Colwyn Trevarthen

"My aim is to show, although this is not generally attended to, that the roots of all sciences and arts in every instance arise as early as in the tender age, and that on these foundations it is neither impossible nor difficult for the whole superstructure to be laid; provided always that we act reasonably as with a reasonable creature."

(John Amos Comenius (1592-1671) The School of Infancy.
Translated by D. Benham, London, 1858)

Abstract

This chapter presents a psychobiological model of early musical behaviours or ways of moving musically – one that accepts music as a natural behaviour essential to being human. At least in infancy and preschool years, an innate *communicative musicality* is inseparable from other ways children demonstrate a lively self-expression while acting in intensely sociable ways. New evidence on the remarkable musical awareness in infants has important implications for teaching of musical knowledge and skills that are specific to a particular culture. We record how a preschool child's spontaneous vocal performances and enjoyment of musical sound is interwoven with growth of all expressive movement and gesture, with the emotions, and with cognitive processes of exploration, discovery and relating on which every kind of cultural learning depends. We give special value to the vital sense of individuality and pride in recognised achievement that can grow in interactions with appreciative caregivers, peers and teachers. In accord with the social cultural theories of learning, we view education, including musical education, as a part of natural human social life, a life that has evolved to make and use culture as a community of meaning.

First, to introduce the topic, we contrast "teacher-centred" or instructive training of preset skills, in perception and performance of established forms of music and musical literacy, with approaches called "child-centred" that claim to recognise, and give primary importance to, the young child's spontaneous pleasure in creating and sharing song and dance. In support of an approach that recognises the musicality children bring to music learning, we then review research with infants that has provided evidence in the last few decades for a theory of the innate foundations of communicative musicality. An age-related program of development has been found that reveals intrinsic motives that prompt the child to act in increasingly sophisticated ways, and to learn cultural skills with an increasingly competent body and mind, by collaboration with what other people do and by attending to the conventional organisation of what they know. Third, we look at modern science of movement to gain a clearer picture of what it is the mind of a child does when he or she shows intentions, interests and feelings by rhythmic and expressively modulated activity of the body, and a willingness to learn new motor skills. This leads us to consider how the brain is formed to communicate both cognitive and emotional states by controlling the time and effort of body movement, and how it adapts to cultural forms of art and language.

Musicality is identified with the capacities of *human brain motives* to (a) regulate the rhythmic moving of an exceptionally complex bipedal body with intricately mobile limbs as one gracefully coherent system, and (b) to be moved to sympathetic activity by other persons' intentions and emotions expressed in their gestures, and thus (c) to collaborate in learning and the social creation of meaningful activities of great complexity. We draw support from new evidence on the power and adaptive flexibility of the developing human brain to mirror the motive impulses of other people, to pick up and engage with their purposes and emotions, and to seek to be transformed by new ways of perceiving and new ways of acting. It is clear that the aptitude for learning a particular tradition of music built by previous generations depends upon childish motives and emotions, and on the influence of sympathetic communication with other persons on both cognitive growth and motor skills learning in brain tissues when they are at their most impressionable.

In the last section we seek to identify principles for a developmentally appropriate music education by comparing selected music education methods. Adaptation to musical culture appears to richly exercise all the special motivating and learning processes of the young human brain in its regulation of a body alive in the world of sound and human company. Music and music education serve a role in the promotion of the individual well-being, confidence, and social adaptation of the young child, and it can open the way to a lifelong enjoyment of learning – in making, responding to and sharing the inventions of music.

Introduction: Cultivating Musical Beginnings

Scholastic work on the culture of education (Bruner, 1996) has a long history in our complex and highly technical civilisation, and there are many theories of how a young pupil can learn to play music, and how children should be taught to perceive and understand music (Flohr, 2004). Usually these theories begin by a systematisation of the material to be learned, a classification of the established structural elements, symbols, codes of practice and texts of the tradition. In this chapter, however, we do not intend to discuss how a general learning theory might apply to the teaching of advanced music cognition or performance skills. We propose that musical development, especially in early years, can best be fostered by supporting and encouraging the spontaneous vitality and inventiveness of human movement and gesture, by recognising children's rhythmic expression of motives and emotion, and their communication of affections, thoughts, ideas and cooperative activities in singing ways. Music can be taught in such a way that is supports how every young child is motivated, from within, to form collaborative and creative relationships in moving, and to pick up new ideas and elaborate rituals of performance from other people.

Schooling is by no means universal–children everywhere learn much by sharing activities and collaborative tasks, by *intent participation*, and in some communities there are no schools (Rogoff et al., 2003). Focusing on infancy and preschool years, we view formal education in any musical tradition as a cultivated product of natural human social life and the active creation and using of meaning by *mimesis*, that is, the social presentation of ideas, purposes and feelings in patterns of whole body movement, by intuitive analogy (Donald, 2001). Thus we align our position with the social-cultural theories of learning; for example, Comenius (1642/1969), Rousseau (1762), Malinowsky (1923), Dewey (1938/1963), Vygotsky (1978), and Bruner (1996). We also draw support from recent developments in the theory of how the human culture-making mind evolved through elaboration of expressive ways of moving that are exhibited from birth, and that receive strong intuitive support from parents and other companions of the young child (Dissanayake, 2000a, b; Donald, 2001; Freeman, 2000; Mithen, 2005).

We are led to contrast two strategies for the teaching of music. One gives primary importance to the form of music that has been assigned to be learned, from outside–a new skill to be mastered according to externally defined criteria and rules; the second approach consciously appreciates and shares the spontaneous, creative, and perhaps anarchic, musicality that the child already possesses, inside, as motives for living and learning with companions. The teaching is either more a putting in from without, with distant curricular ends in view, or more propelled from within the expectations of teacher and child together, by the enjoyment of discovery of new meaning for moving in the present, a meaning that can become a memorable guide to future adventures.

The first *teacher-centered* approach defines the music in its conventional, cultural, largely artificial textual form. It specifies a process of learning, the structures to be perceived and measurable levels of execution to be achieved. It is led by prescribed concepts and categories, identifies the notes and follows the score from the teacher's point of view. There is, usually, critical appraisal and formal assessment by an educated listener who gives importance to the level of performance and the students' ability to discriminate standard and well-defined musical elements. The second *child-centered* approach is intentionally open to the child's fluid experience as a naturally creative musical being, to the flow of experience and emotion of one who, while certainly naive in understanding of music, is able to try and to express messages that have the rhythms and graceful regulation of song or dance released in a native state – a state that is at first free of any cultural rules or language, but ready to learn new ideas and stories by active experimenting, and from a teacher of any age.

While the two approaches are very different in their basic assumptions about learning, in practice music teachers normally apply both in varied combinations, with different emphasis. For example, a teacher utilizing the Dalcroze or Orff method of teaching music will often emphasize the child-centred approach, stressing the value of improvisation and creative experiences. The same teacher may also operate a teacher-centred, instructive method for musical skill training. Any experienced teacher of young children who prefers to instruct by the book, will appreciate the importance of communicating well with the attentiveness and willingness to practice of children, disciplining by engagement, communicating by what Fred Erickson calls "going for the zone", meaning the Vygotskian zone of proximal development of the child (Erickson, 1996). Few teachers will practice only a teacher-centred or child-centred approach. (For a comprehensive review of the range of music teaching methods for early childhood see Flohr, 2004).

We propose that, even for an adult, learning to be a good musician (mastering the music with discipline and how to perform pieces composed by other musicians for the appreciation of listeners of the tradition) requires full encouragement and use of the same inherent musicality that children demonstrate from birth, as well as years of disciplined practice. It needs, besides a willing effort to learn and to relish the work of learning (Csikszentmihalyi & Schiefele, 1992), the expressive manifestation of pleasure in telling one's own story in sound, making patterns that are enchanting, exciting, enthralling and emotive or moving, for the self and for others. Musical learning certainly begins best that way (Custodero, 2002; Custodero & Johnson-Green, 2003; Fröhlich, 2002, in preparation; Moorehead & Pond, 1978). A teacher must not merely recognize and respect the child's inventiveness of musical expression, but must be prepared to share it – to be as creative or playful as the child while stretching the aims and expectations in constructive ways, teaching an established knowledge that confirms intuitions and motives while enriching them and making them part of a large community of art and knowledge.

It has been demonstrated that the acquisition of high musical skill by an older child, over about 7 or 8 years, requires both many hours of practice and a close and sustained attachment to a responsive and appreciative teacher (Sloboda & Davidson, 1996). There is much evidence that the motivation, imagination and memory of a child, and learning ability, are changing greatly after this age, a time when formal schooling is beginning, or has already begun, in industrialised societies (Donaldson, 1992). Our task is to explore theories of motivation in younger children, before this development of the human mind, for a life of music making in the company of other people, to discover how the learning of intricate traditional forms of musical work and invention is prepared for in childish human nature, in the ways all of us act and experience our actions in sound from birth.

This, we know, is not a new direction of inquiry, but it is a lively and challenging one, especially so in the last few decades as the intrinsic musicality of infants has been demonstrated and the intuitive musicality in responses of parents has been recognised as an essential support for language learning (Malloch, 1999; Papoušek & Papoušek, 1981; Trehub, 2004; Trevarthen, 1999). We find it is necessary to further explore children's musicality and the teaching of music following ideas of Emile Jaques-Dalcroze (1921), Moorehead & Pond (1978), and others concerning the importance of using all resources of the body, of the brain and of the environment, and especially the pleasure of creating with willing companions.

We will attempt to relate music learning to the unique adaptations of the human body and brain for an active and creative life in communication with other people. It is interesting to observe how brain scientists with access to new methods of functional brain imaging are often encouraged by psychologists to measure differences in neural tissue volume or the distribution and strength of neuronal activity that correlate with narrowly defined levels of musical skill, comparing experienced and inexperienced subjects to distinguish possible innate differences from effects of the practice to develop that skill. They ask if the results of conventional tests of musical discrimination, for rhythm, pitch melody or harmony, or of performance are related to the growth of hypothetical cognitive organs of *musical intelligence*, or if they are entirely products of stimulation. Great importance is given to the *plasticity* of regions of the brain that appear to be imprinted with stimulus information and enlarged or structured so they can function as repositories for representations of practiced actions or often-experienced cognitive categories.

But the neuroscientists are increasingly aware that they will have to study the functions of the *embodied and active mind embedded in its world* (Clark, 1997; Varela, Thompson & Rosch, 1991), with sensitivity to the regulation of rhythms and temporal phases of motor activity and attention, to emotions, and to the extraordinary and still little understood capacities of a human brain in one person for mirroring or sympathising with the intentions, interests and feelings of other brains in other persons (e. g. Adolphs, 2003; Decety & Chaminade, 2003; Gallese, 2003; Schilbach et al., 2006; Thompson, 2001).

Until recently few researchers have investigated the movements, or the *motivations* for moving and perceiving, that lead to or transmit musical experiences in real time, and it is technically difficult to do so, though new technology for tracking brain activity with high temporal resolution is changing that (Ioannides, 2001; Turner & Ioannides, in preparation). At the same time evidence is accumulating that what most readily transfers between human brains are the dynamics of intentions and emotions implicit in other persons' forms of movement (Gallese, 2003; Schilbach et al, 2006). When the dynamic, body-moving, emotional and interpersonal emotive aspects of musical enjoyment and training are considered, and not just the perceptual, cognitive and memory functions, the answers from brain research may be very different from those presently available.

We come to the conclusion that to practice and learn singing or instrumental music performance is to cultivate both the body and brain in ways that human bodies and brains are evolved to be cultivated. Brains are evolved to be changed by use while they are in control of movements. Music education is attained through personal effort at making actions of the body efficient and satisfying, to benefit the vital emotional regulations of a conscious, intentional self. Music education helps to share life, actions and experiences in sound that can become a home for a community of meaning where the individual is a recognised and appreciated companion with a musical individuality, no matter what level of skill in musicianship he or she has attained. To be helped to experience and enjoy music in this way is a natural right of every child (MENC, 1991). We have to keep intact the natural vitality of singing and dancing when we work as music educators, for this is the source from which all the aesthetic and social powers of musical culture flow.

THE INNATE MUSICALITY OF INFANTS AND YOUNG CHILDREN

It takes years of self-motivated work and guidance from expert teachers to train a skilled singer or instrumental musician, and a lifetime to know and understand or deeply appreciate a musical tradition. No infant or toddler is born a musician, although their musical expressions can be interpreted that way (McPherson, 2006). They do not have the vocal organs or control of the voice to be a developed singer, nor do they have the oral and manual dexterity to play an instrument properly to make what an adult would recognize as music. Nevertheless, the foundations for making and sharing musical sounds and dancing movements of the body are there long before a baby can stand, and babies love simple music and musical games from birth. Infants have considerable capacities for appreciating elements of musical sound, and quite remarkable abilities to time their limited vocal repertoire in rhythmic engagements with an adult who is responding musically. This is the paradox of infant *communicative musicality*. By means of detailed musical acoustic analysis, Stephen Malloch has identified expressive parameters, of *pulse*, *quality* and *narrative* in the earliest vocal dialogues between parents and infants (Malloch, 1999; Trevarthen & Malloch, 2002). How should we interpret these behaviours that appear to anticipate the performance and appreciation of musical art? Why do human infants have sensitive talents for appreciating musical sounds when even a modest proficiency in music is, at best, several years away? Clearly infant musicality must express and serve some other fundamental developmental need of every human being. Perhaps it is a need to express intentions by moving the voice and limbs in expressive ways, to communicate. Maybe music is a cultivated product of motives of all humans for sharing what goes on in their minds and bodies.

Research of the last few decades has transformed scientific recognition of the infant mind, and especially the capacity a baby has from birth to engage an affectionate adult in communication, using all the senses to appreciate the qualities of human expressive movement (Stern, 2000; Trevarthen, 1998). Infants just a few weeks old express themselves with face, hands and voice, and all their body, when they communicate in response to certain of the lyrical, rhythmic and repetitive patterns of parental talk. These are the very patterns of expression that distinguish the sounds of music and poetry (Miall & Dissanayake, 2003). Infants also discriminate and are attracted to subtle features of the human sounds of song or musical performance, showing certain preferences for pitch level, rhythm and harmony that are similar to those of adult listeners (Trainor & Zacharias, 1997; Trehub, 2004). These delicate auditory sensibilities are met by intuitive forms of parental communication that engage immediately with the infants' expressions, joining adult and infant in the intimate art of shared human vitality, beginning, many researchers now believe, the teaching of culture specific rituals, ideas, and the meanings of things that people make and use, including the language they talk (Fernald, 1992; Freeman, 2000; Halliday, 1979; Papoušek, H., 1996; Papoušek, M., 1996; Papoušek, Papoušek & Symmes, 1991; Trehub, 2002). The abilities of human infants exceed even the remarkable skills that some birds and mammals show for emotional expression and social communication, and they are far more adaptable to the habits and inventions of the community (Wallin, Merker & Brown, 2000; Eckerdal & Merker, in preparation).

Proto-Musical Responses of Foetuses and Newborns

Infants are born with sensitivity for the rhythmic expressive movements of other people. They attend with eyes, ears and touch, and show responses with a matching sense of time in movement. Affectionate adults are attracted to the infant's interest and modulate their talk and gestures to engage in games of address and reply or assertion and apprehension of intentional looks, vocalisations and hand movements and touches. Even breast-feeding becomes a conversation as the infant makes rhythmic bursts of nipple-stimulating non-nutritive sucking in interaction with the mother's talking, stroking or jiggling. This kind of sucking seeks responses and learns from their contingency of timing. Newborns call for attention with pulses of breath, exercising their immature and high placed vocal apparatus with powerful cries and rhythmic sobbing.

In the month and a half before a full term birth (before 30 weeks gestation), with senses adapted to an aqueous environment, the foetus listens to the mother's voice, learns its distinctive timing and tonality, moves in responsive contact with her body, sometimes reacting with bursts of rhythmic movement to the pulse of sounds outside her, including the sounds of music (DeCasper & Spence, 1986; Fifer & Moon, 1995; Lecanuet, 1996; Trevarthen et al., 2006). In fact, the child's hearing develops in the third trimester usually by the 26th week (Flohr, 2004, p. 20). The foetus drinks amniotic fluid by swallowing, and the lungs also expand and contract breathing the liquid. Changes of heart rate and foetal breathing, and of gestures and expressions seen with ultrasound, give evidence of auditory awareness. Although there is little scope for the mother to be consciously in dialogue with the consciousness and movements of the human being inside her, the foetus can learn the expressive rhythm and intonation of her voice so that the signature of her speech will be known and sought for after birth.

The potential of this foetal sensitivity for musical sounds in the last trimester of gestation is demonstrated by the effective use of music as therapy in the Neonatal Infant Care Unit (NICU) (Standley, 2002). They are confirmed by the remarkable capacity of one prematurely born infant to take part in a delicately timed vocal exchange. Spectrographic representation of a vocal dialogue between a baby girl at 32 weeks gestational age and her father who was kangarooing her against his chest under his shirt and imitating her simple calls indicated that the infant and adult could precisely match one another's timing for both syllabic (0.7 seconds) and phrase (4 seconds) units of time, and the infant could improvise the phrase intervals to call the father repeatedly when he failed to respond (Malloch et al., 1997; Malloch, 1999; Trevarthen, Kokkinaki & Fiamenghi, 1999). These times match the durations of utterance and phrase found in spontaneous vocal productions of infants (Lynch et al., 1995), as well as in mutually regulated conversational exchanges with parents that develop in early months. It should be noted that the father of this premature infant closely matched her performance by imitating the pitch, timbre and duration of her sounds. He and she were clearly establishing mutual sympathy and reciprocal interest by matching vocal expression with comparable auditory awareness, and essentially the same brain-generated sense of time. Very similar timing in vocal exchanges with premature infants have been recorded in a large NICU in Tokyo (Watanabe, et al., 2006).

This kind of sympathy, extending to a wide range of visible and audible expressions involving movements of eyes, face, mouth and hands, has been demonstrated in studies of newborns imitating adults. Infants imitate a surprising range of odd human expressions in the first days or hours after birth including large tongue protrusion, a wide open mouth, raised index finger and other finger extensions, tightly closing eyes, looking up high, simple vocal calls (Field et al., 1982; Heimann, 1998; Kugiumutzakis, 1999; Maratos, 1982; Meltzoff & Moore, 1977, 1998). These imitations can become a well-regulated turn-taking dialogue if the adult is appropriately respectful of the infant's initiatives, for example, waiting for and accepting the infant's repetition of an imitation as an invitation for *reciprocal imitation* (Nagy & Molnár, 2004). Emese Nagy has shown that as the infant changes between imitating and giving back the audible or visible expressions that the adult has presented as models, there are changes in his or her heartbeat, which are indicative of alternating brain states of effort or attention. With *imitation* the heart of the infant *accelerates*, and with *provocation or questioning* of the adult, the heart *decelerates* (Nagy & Molnár, 2004).

Neonatal vocal imitation is limited, but within a few weeks the baby can match simple vowel sounds of differing timbre as well as the rhythm of a short group of repeated sounds. Research on imitations of infants from the first weeks has found that imitative exchange with adults (*infant mimesis*) is characterised by expressions of emotions of *interest* and *pleasure* (Kugiumutzakis et al., 2005). The infant's behaviours are not simply automatic mirroring of the shape of a movement, but a means of establishing sympathetic motives, or intentions to communicate (Kugiumutzakis, 1998; Trevarthen, Kokkinaki & Fiamenghi, 1999; Trevarthen, 2006).

Newborns demonstrate, as we have seen, a precise sense of *timing*, integrating together all forms of movement, one that matches the natural cadences of adult movements (Trevarthen, 1986, 1999). The graceful efficiency of the baby's complex, flowing postures and gestures, the movements of hands that feel objects they touch, the precision of the coupled saccadic eye movements that search the newly seen world, and life-sustaining movements of feeding, breathing and vocalising indicates they possess an integrated prospective guidance of their actions (Adamson-Macedo, 2004; Craig & Lee, 1999). They display the beginnings of a sense for *emotional episodes* or *emotional narratives* in the expressions of other persons, anticipating cycles or flow of energy or mood (Stern, 1992, 1999). These emotions, regulated by inborn *affective systems* of the brain, are adapted to control the seeking of experience, systems that provide defence against stress, pain or harm, and that make appeals for parental care (Panksepp & Bernatzky, 2002; Trevarthen et al., 2006). Their purposeful movements and emotional expressions indicate that the newborns are conscious and intending human creatures, who can be aware of other humans as different from and sympathetic with themselves (Trevarthen & Reddy, 2006).

For their part, mothers and fathers seeking to communicate with their babies, in any language, greet them with melodious vocal sounds and talk and move rhythmically as if dancing and singing (Fernald, et al., 1989; Papoušek, M.,1996). Their infants attend to this *parentese* and, with their limited vocal repertoire, often coo in precisely complementary ways, accompanying their sounds with expressive facial mimics and gestures of the hands.

The Musicality of Proto-Conversations

Rapid improvement in visual perception in early weeks after birth with developments in the vocal system, increase the variety precision of exchanges between infant and parent. In the second month enjoyable *protoconversations* involve intent mutual gaze, and transactions of face expression, vocalization, hand gestures and touching (Bateson, 1975). A parent holding a two-month-old adjusts posture and movement to favour face-to-face communication, because the baby is so obviously interested to see and hear them. The infant's intent gaze evokes periodic and carefully modulated responses. Acoustic analysis of the vocalisations passed back and forth has revealed the complex and precisely regulated rhythmic and expressive collaboration of sound making (Malloch, 1999). The baby has a good sense of conversational timing, and this contributes significantly to the attachment the parent feels for the baby, fostering care and enjoyment of the baby, which benefits development (Beebe et al., 1985; Jaffe et al., 2001). If the parent is depressed and fails to be responsive with vocal and other means of expression, this discourages the infant, who becomes withdrawn and difficult to console (Murray & Cooper, 1997; Robb, 1999). Experiments have shown that this reaction is due to a high sensitivity for the immediate contingency and emotional appropriateness of parental responses (Murray & Trevarthen, 1985; Nadel et al., 1999; Trevarthen, 1975; Tronick et al., 1978).

From three months, as the body gains more strength to move the head free of support and the baby can extend and direct the arms and hands to track and grasp objects (von Hofsten, 2004), the rituals of gesture play and baby songs are learned and anticipated by infant and parent with increasing sociable joy or pride (Trevarthen, 2002). By six months the infant practices melodious improvisations of voice and gesture, becoming both a composer and a collaborative performer (Littleton, 2002; Papoušek, M., 1996). The infant's expressive sense is gaining an aesthetic/moral emotionality that identifies and regulates attachment and companionship and the formation of the child's identity as a performer with a repertoire that familiar company will appreciate with pleasure (Reddy, 2003; Trevarthen, 1990). Parents perceive that clear emotions are expressed in the effort and tonality of extended vocalisations of babies, and they use song and body movements to excite play or modulate distress (M. Papoušek, 1996; Trainor, 1996, 2002; Trehub et al., 1993, 1997). This endows the child with a general sociability so that groups of infants under 9 months old, with no adult present, can set up intricate negotiations of interest and mood, and they may improvise together displays of vocalisation and body dancing (Bradley, in preparation; Selby & Bradley, 2003). The baby is beginning to show clear entrainment to the rhythms of music, bouncing the body in imitation, and expresses delight when recognising the beat and melody of a lively tune (Littleton, 2002; Mazokopaki & Trevarthen, in prep.).

After 3 months, infants are willing subjects in laboratory tests of their interest in and consciousness of human made sounds, including musical sounds and the sounds of speaking (Kuhl, 1985; Trainor, 1996; Trehub, 1990). They prove to be perceptive of far more features of the musicality of sounds than they can control in their own vocal productions or by manipulation of resonant objects as instruments. As was true for their awareness before birth, proficiency of *listening* is far in advance of versatility in *performance*, serving as a way to establish synchronisation and alternation with the much more elaborate productions of their older human companions (Eckerdal & Merker, in preparation).

Collaborative Improvised Musical Games and Traditional Nursery Songs

As the baby grows in powers of expression, affectionate parents generate expressive movements, vocalisations and talk that respond to and regulate the infant's playful motives. As the baby becomes livelier and more distractible by things to look at, grasp and manipulate, caregivers recall traditional baby songs or improvise song-like vocal play with dancing movements either to engage a playful infant or to pacify a restless or distressed one.

There are differences in non-European cultures, especially clear when comparisons are made to communities where schooling and literacy do not exist, and in which different value is placed on talking to infants or on giving them physical exercises to enhance their postural development and locomotion (Dissanayake, 2000a; Takada, 2005). Nevertheless, universal features of timing and modulation in infant directed speech, baby songs and the movements of gesture games and baby dancing confirm the intrinsic, culture-independent motive processes of both infants and adults, and a fundamental rhythmic musicality, which has been inferred from the communicative behaviours of newborn infants (Trevarthen, 1999).

A film of a blind 5-month-old Swedish baby moving her left hand while her mother sings two baby songs composed by Alice Tegnér reveals an amazing, completely intuitive, complexity of gestural imagination as the little girl, who has never seen any human hand, listens to her mother's singing (Tegnér, 1995). She dances or moves expressively to the melody of the song, matching subtleties of rhythm and tone of voice with waving of the arm, lifts and rotations of the wrist and spreading and pointing of the fingers. Data from microanalysis of the film compared to a spectrograph of the mother's voice indicated that the infant was *anticipating* her mother's vocal gestures by an interval of 300 milliseconds. She knew the songs and could lead her mother in their duet of expressive moving. Moreover, the analysis reveals that her active participation is intermittent. At times she seemed to pause to just listen, and occasionally she appeared to reflect on or think about a previous phrase, gesturing to herself, out of time with the singing at that moment.

At this level of human sympathetic engagement, where the referential content of any language, including the text of the song, is of no importance and where cultural rituals of gesture and voice are just beginning to be emphasised and learned, human musicality has to be seen as a part of a much more general *innate creative impulse* driving actions and awareness, one that is seeking to share with other persons the experience of moving. The blind baby was not taught to move to the music.

Research on baby songs, vocal chanting and dancing games or other ritualised forms of moving and touching has found that every human society uses them, in some form, to stimulate play and joint participation after the infants are about three months old. After three months the infants are highly attentive to rhythms and rhyming of movements and vocalisation, and to the emotional quality or aesthetic style of the adult's *performance*. They are also more agile in their gestures and vocal output than before, as well as more alert visually. A song or action game made with joy and affection elicits happy engagement, and the infant learns a repeated enjoyable ritual, shows pleasure at hearing the beginning of a familiar melody, and can demonstrate anticipation of ends of phrases or the climax and ending of a song by vocalising in tune (i.e. matching pitch), or by gesturing in time. The baby is then apparently engaged by the *emotional envelope* of the presentation, or its narrative (Malloch, 1999; Stern, 1992, 1999; Trevarthen, 1999).

What is *innate* here is an avidity *to learn others' ways of acting and thinking by moving expressively*, and the parent responds intuitively to *teach by moving expressively in deliberate ways*. By six months simple culture specific ways of moving in performances, such as the actions of a hand-clapping song, are taken up and used or 'shown off' by infants in games with their familiar companions (Trevarthen, 2002). These learned behaviours are associated with strong expressions of *shared enjoyment, or pride*, which recall the emotion of pleasure that is found to be associated with even neonatal imitations. A six-month-old, now able to sit up and move arms and hands effectively, has limited vocal capacity compared to a one-year-old, but can call, gurgle, giggle and squeal, and can assume a special voice, matching the rhythm of vocal expression with waving or bouncing gestures (Reese, 1998; Littleton, 2002). He or she (girls being a couple of weeks in advance at this age) may eagerly take part within a skilful performance of a range of songs and action games, inserting small contributions at appropriate times and with appropriate pitch and intonation (Trevarthen, 1999). The rhythms of babbling which appear around 6 months match those of banging objects with the hands, and though the age of this development, like others, varies considerably between babies, the two kinds of rhythmic activity develop together for each individual. Vocal sounds are varied with communicative intent before the development of protolanguage at around nine months (Halliday, 1979).

Musical games that have been shared are both recognised with pleasure and their repetition actively solicited. The infant is clearly interested in becoming an active agent in family rituals of play that others enjoy. Soon after six months a baby can recognise his or her name and certain other frequent utterances made by familiar people, can imitate speech sounds, and sometimes emits what seem like poetic or musical phrases (called *jargon*) (Locke, 1993; Jusczyk & Krumhansl, 1993). Gestures with fingers and toes are becoming more elaborate and expressive as hands and feet sensitively explore what they contact.

It is noticeable that a happy 6 to 8 month old baby is, becoming more of a show off or a clown (Trevarthen, 1990). This behaviour indicates a growing consciousness of how other persons appreciate and react to what is displayed for their attention. It has been taken as evidence of a new self-awareness, but it is manifestly, at the same time, a heightened *other awareness* (Reddy, 2003). Now the identity of the Other becomes of critical significance and a factor in the establishment of a Self identity, including a musical identity for the infant (Trevarthen, 2002). Confronted by a stranger, a baby who shows great pride in demonstrating how to sing or how to show "Clap-a-Handies" in the approved manner to a family member, is confused and ashamed when in the face of an uncomprehending stranger, who clumsily fails to get any effort the baby makes to show off (Trevarthen, 1990; 2002).

By the middle of the first year at least, the sociability of the baby is strong enough, and willing enough, to generate complex, emotionally regulated, interactions between peers. Groups of two or three babies with no adults present can communicate and do so in subtle ways, showing that they can create a form of drama among them, inventing new mannerisms and ritualistic games to gain group acceptance or to protest exclusion (Selby & Bradley, 2003; Trevarthen, Kokkinaki & Fiamenghi, 1999). These games can include group vocal performances (Bradley, in preparation).

Infants' Musical Intentions

The temporal arts have their innate roots in language-free intimacy, where the experience of an embodied Self, moving and sensing the body in time and with rich emotional colour, evokes intuitive sympathetic response from affectionate others, those who assume parenting responsibilities for the baby agent/performer (Dissanayake 2002 a,b). We are persuaded by the precocious talents of infants as performers, and the value given to these behaviours by parents, that human infants are born seeking to engage with the intentions of other people in rhythmic musical ways. Babies sense that acting a part and playing an active roll in a ritual of shared movements is how to be part of the community, as anthropologists have observed to be the case across the world (Blacking, 1976, 1988; Dissanayake, 1988; 2000a; Takada, 2005; Turner, 1974).

The agency of the infant's Self includes active postural movements and gestures with orchestrated rhythms that define and explore awareness, first of the *body* itself, felt in its flesh and joints *proprioceptively*, then of the *world* outside that the body comes in contact with and appreciates *exproprioceptively*, by touch and hearing before birth, then by sight. This growing consciousness is guided emotionally. The flow of enterprise gives the efforts of moving evaluations of pleasure and elation, or tension, urgency, and fear (Csikszentmihalyi & Csikszentmihalyi, 1988; Custodero, 1998). In normal circumstances of human company, all these states and moods assessing the fate of the Self-in-action can be picked up and aided/abetted or opposed/rejected/neglected by Others.

For musical development it is the same. The innate musicality can be nurtured by intuitive and joyful mentoring, or it can be left to satisfy itself as it will do when the child is left alone to exercise inquisitive consciousness as best it can, or perhaps wither in competition with other ways of acting sociably. From the earliest stages of development, and clearly already by the foetal stage, a human body and brain are adapted for communication of enthusiasms and emotional experiences. The regulatory systems of the inner visceral Self are greatly elaborated for Self-Other regulations by a still mysterious capacity for sympathising with expressions of self-regulation (Porges, 1997; Trevarthen, 2001; Trevarthen et al., 2006).

The potential for development of cognitive powers and motor skills is greater by sharing with a more or differently experienced Other or group, and that is a primary way elaborated cultural practices and skills of awareness and exertion can be passed on in the community. This is the developmental use of human attachment (Bowlby, 1988) and why attachment figures matter. Attachments of the child to Others are from the start not just protective contracts, but teaching/learning relationships – *companionships* in which the purposes of acting are elaborated through expressive movements that play out the dynamics of consciousness and intentionality, and their emotions (Trevarthen, 2005; Trevarthen & Reddy, 2006). Humans learn culture in all its forms by mimetic sensitivity for intentional images expressed in movement projects of other persons (Donald, 2001; Gallese & Lakoff, 2005).

Musicality of voice and gesture is a crucial part of this evolved gift for showing and sharing endlessly elaborate human intentions (Cross & Moreley, in preparation; Mithen, 2005). It is actively learning simple cultural conventions in infancy.

Making Musical Meanings Beyond Year One: Children's Musical Culture

Infants make big advances in the second year, achieving new mobility, making new friends, creating new meanings imaginatively with them in increasingly elaborate play, and understanding language. They are surer of themselves in the world, and may assert their felt rights forcefully. Given cheerful acceptance of their musicality and shared discovery of new favourite formulae, they sing, for their own repeated pleasure or as part of delightful social encounters. The teaching, training or scaffolding of cultural learning by adults is stretched and transformed by intrinsic developments in the child. Adults shape what is *wanting* to be shaped. But different ideas about what is good for children and what they should learn in different families or cultures can have a large effect on how the innate motives to communicate and share purposes and experiences develop into knowledge and skills. In an environment that values song and music of a particular tradition, and that shares it with the toddler generously, the beginnings of real musicianship may become apparent. Nevertheless, there are competing forms of knowledge and skill, and musicality may be left little cultivated. In this connection it is interesting that children may enjoy the creation of musical forms of play just for their own amusement, or for sharing with other young children. This leads to the phenomenon of *children's musical culture*, a way of exercising musicality and creating new forms of musical play that may not be much affected by the ideas and practices of the adult world (Moorehead & Pond, 1978; Bjørkvold, 1992).

A toddler has new powers for understanding the purposes of other persons' ritualistic actions, and can, for example, even acquire the artificial strict time of an instrumental performance with percussive instruments (Eckerdal & Merker, in preparation). Toddlers, especially in groups with their peers, expand their embodied expression, testing the freedom to dance and jump in richly musical ways (Custodero, 2002). They transact imitations together, not just to do the same thing, with the same items of dress or with the same instruments or toys, but for the pleasure of negotiating and changing roles of leader and follower, which they find funny (Nadel & Pezé, 1993). Their imitations are a currency of friendship, like the play fighting of animals (Bateson, 1956; Bekoff & Byers, 1998), which appears to be associated with expressions of emotion joy (Panksepp & Burgdorf, 2003).

All who have carefully followed the progress of music making from infancy have been impressed with the importance of genuine companionship from adults or older children for encouragement of musical inventiveness and the development of the singing voice or the ability to manipulate rhythms, tones and melodies of human made sound. The extraordinary capacities of very young babies for listening with musical discrimination (Flohr et al., 2000) become harnessed to the task of guiding a toddler's newly acquired movements for making and controlling such sounds, and for sensing better in their bodies how it would feel to make music and dance like big people.

The Norwegian musicologist Jon-Roar Bjørkvold, in his book *The Muse Within* (Bjørkvold, 1992) used his training to observe closely the musical games of children in Oslo, Moscow and St Petersberg, and Los Angeles. The three countries approached the education of children in very different ways and still do. He asked, "To what extent is there a common child culture irrespective of these enormous cultural, social and political differences?" He took a forthright position: "Our Western conception of music includes an aesthetic dimension that controls our selective perception; it functions as a kind of subjective filtering mechanism. This aesthetic dimension, which among other things includes norms regarding pitch and tonality, is totally foreign to child culture. Nor does this concept of music, rooted as it is in the classical tradition, capture the sense of functional and expressive unity that is the warp and woof of the spontaneous singing of children" (Bjørkvold, 1992, p. 47). He noted that the Western concept of music separated from other dynamic expressions of human culture is not understood, or recognised in the languages, of many parts of the world, notably in Africa where music permeates all life activities. He insists, "It is critically important for children to master spontaneous singing, for it is part of the common code of child culture that gives them a special key to expression and human growth. This kind of singing, which springs forth without adult encouragement and is therefore spontaneous, is governed, like all other facets of play, by child culture's norms and rules in both use and form" (Bjørkvold, 1992, p. 63).

Bjorkvold charted three ways infants and young children use the voice with a singing kind of expression. *Fluid/amorphous songs* "evolve in a completely natural way from the infant's babbling as part of its first playful experiments with voice and sound. This type of song, with its fanciful glissandi, micro-intervals, and free rhythms, is quite different from what we adults traditionally identify as song" (Bjørkvold, 1992, p. 65). *Song formulas*, which as teasing songs, for example, assume symbolic forms are for communicating with other children and they flourish after the child begins to play with peers, typically at two or three years. Bradley (in preparation) has found something like a primitive form of this invented between infants under one year when they are left on their own in trios. Elements of musically more complex *standard songs* are picked up from play with adults and hearing them sing very early and soon are adapted to fit what the infant is doing, so they can appear surprisingly early.

"Children do not burn their bridges as they enter new territories in their ongoing conquest of the world. Their bold exploration of the new presupposes a secure footing in the old and familiar. The spontaneous singing of a six-year old, therefore, includes all three song types. Only an expressive vocabulary that includes all three song types...is adequate to the whole range of children's emotional, communicational, and social needs" (Bjørkvold, 1992, p. 68). The relation of forms of child musicality to communication with language is clear. "Modern language research has identified the following as the main functions of human speech generally: creating contact, communicating information, and making identity. These three functional areas...are clearly evident in the spontaneous singing of children" (Bjørkvold, 1992, p. 80). "The significance of such singing for socialization, identity, cognitive development, communicative ability and general human growth holds not only for children throughout the world, but for all oral and muse-ical folk cultures. Thus, we are dealing here with universals that are not only transitional, but *transgenerational*, for folk cultures include people of all ages" (Bjørkvold, 1992, p. 81).

Finally, Bjørkvold's observations taught him, as a musician and teacher of music, that, "A child who is asked to play a printed score must turn his attention from the primary experience of making music (spontaneous singing within the child culture, for example) to a kind of secondary music making in accordance with the notes on the page. For many children, the result is that their ability to make music in the primary sense withers and dies. Children studying Suzuki violin often encounter reluctance and difficulty with reading music after spending their early years learning their music by rote and without music. Nothing comparable occurs in the case of language: children's encounter with the alphabet does not rob them of their oral competence (though it can influence to some extend the way they speak the language). Their oral musical competence, however, "can be irretrievably lost as a result of premature preoccupation with written music" (Bjørkvold, 1992, p. 188).

Mechthild Papoušek (1996) describes, with detailed acoustic graphs and musical analysis of parent-infant vocal exchanges, how early phonation and cooing, exploratory vocal play, babbling, variegated babbling and early words of the first year, may, with the engagement of parents, become singing of songs. Mothers and fathers accurately reflect the infant's expressions with the pitch, melody, duration and rhythm of their vocalisations, as well as extending and embellishing them. Universal temporal and expressive features of the infant's sounds, and parents' intuitive mimicry of them to express different rousing or soothing intentions in regulating the infant's arousal show that families from cultures with very different languages may have similar expressive forms or emotional impulses. Vocal development of young children in terms of physiological characteristics, vocalization/speech characteristics, and singing development is outlined in *Musical Lives of Young Children* (Rutkowski & Trollinger, 2004, p. 83).

Music teachers who are familiar with the creativity of toddlers' spontaneous musicality can perceive the bridge that must be crossed for a budding musician, for learning to read musical scores and master an approved art tradition of singing or playing. The way is easier and learning more sure if the teacher strives to be a companion in the enjoyment of moving with the subtleties of children's musical invention and celebration (Custodero, 1998, 2002; Custodero & Johnson-Green, 2003; Flohr, 2004; Hargreaves, 1996; Imberty, 1996; Littleton, 2002; Papoušek, M. 1996; Sloboda & Davidson, 1996; Trehub, 2002).

To understand this bridge of musicality between generations, and indeed between cultures, we need to have an idea of how a human body may be moved to musical expression, and how people of different ages and experience may move in musical sympathy.

Musicality and Moving in Early Communication

Principles of Movement Science

"Understanding movement is central to understanding development. Without movement we - by which I mean the animal kingdom - would not be able to eat, avoid harm, reproduce or communicate by sound, gesture or facial expression. We would not be able to perceive, since perception is an active process. Consequently we would not be able to think, because there would be nothing to think about. We would not even be able to breathe or pump nutrients around the body. In short we would be dead" (Lee, 2005).

In spite of their differences in size, mobility and experience, a newborn infant and an adult evidently share a matching time sense and the same affective values of moving. They engage one another by hearing, sight and touch and regulate an exchange of states of interest, intention and emotions with intuitive ease, exhibiting intricate *synrhythmic* activity (Trevarthen et al., 2006; Trevarthen & Reddy, 2006). How can we explain this? Apparently we need a scientific account of how the actions of their communication become coordinated in mutual experience. But first we have to understand how one animal, such as a person, can make their own movements effective as actions with particular goals.

Modern movement science begins with the work of a Russian physiologist, Nicholas Bernstein (1967), who discovered in the 1920s that animal movements are organised, not by assembly of reflex reactions to stimuli, as Pavlov claimed, but from planned use of the body as an instrument for relating to the world (Pickenhain, 1984). Their motor impulses will need perceptual guidance to fit the circumstances in the body and in the world (Gibson, 1966), but they can arise with some independence from sensory prompting, and can specify the goal to be achieved with the aid of perception. Bernstein's research also clarified that the brain has an integrated capacity for substituting one action for another to get a desired goal, a function called *motor equivalence*, also recognised by the American psychologist Karl Lashley (1917, 1951). This needs a *body image*, of what the body feels like in all sorts of movement.

Guiding a movement to it goal requires an efficient and automatic discrimination, in advance, of the different effects inside the body that have to be perceived and controlled when activating properly a very complex system of muscles and joints. Then the action has to be sent along the path to the desired external goal. David Lee (2005) calls this 'gap closure', and he has made a mathematical model, called *tau* theory, that accurately describes how all animals use their brains to calculate, in advance, how to use the body and forces from events in the world to make single actions efficient and effective as intended. Substituting one movement for another to get to a particular goal needs a translation between two movement plans. The prospects of moving for each have to be known and adapted to help reaching the same goal by different bodily means. Versatility of movement plans is required for different ways of moving about in a cluttered world, or for shifting strategies when dealing with an object that is being handled outside the body. All the different movements and all the different ways of sensing their effects have to be coupled in one sense of time, in the present and into the immediate future.

This describes the essential cerebral organisation of a Self, and its principles of operation are as ancient as animal life (Merker, 2005). Conscious awareness and the acquired cognitive processes that organise it are dependent on the unconscious creation of *time images for movements* (Jeannerod, 2006; Sperry, 1952) and of *how it feels to move* (Damasio, 1999). Both the growing cognitive powers of infants and their abilities to communicate are dependent on the initiation and regulation by movements by these principles of prospective control, which are innate (Trevarthen, 1984; von Hofsten, 2004).

When intentions to move are communicated, i.e. passed between subjects – such as in the case of the blind Swedish baby who, with her left hand, expressed the sound of her mother's singing, matching the rhythm and melody of the music made by the mothers vocal activity with dancing of her hand – there has to be a way for the imitator to be in the role of (intend like) the model agent. Matching standards of timing, shape of body and sense of energy for moving are necessary (Trevarthen, 1986).

Pavlov's dogs were immobilised and subjected to stimuli not of their choice. In fact, they were made to be inactive like human subjects usually are in modern brain scan research. Bernstein (1967), however, used film analysis to study humans performing in free and natural ways many tasks, from athletic locomotion to use of tools, and he made detailed biomechanical studies of how the body masses were displaced with rhythmic ease. Modern research uses motion capture technology to achieve higher resolution, but the method is the same.

Images of Others' Moving: The Rhythmic Cerebral Mechanism of Sympathy, Active from Birth

It is interesting that a revolution in recent years of understanding concerning how adult human brains are aware both of the feel of their own actions, and how they are conscious of the intentions, experiences of other persons doing things and of their emotions, has come by functional brain imaging in conditions where people observe films of other people that truly engage their active conscious interest in spite of the enforced immobility of the situation in the scanner. The data show that they are *sympathising* with the other people by *tacit moving*, i.e. by activating virtual motor images of their own intentional system (Adolphs, 2003; Damasio, 1999; Gallese, 2003; Gallese et al., 2004; Jeannerod, 2006; Schilbach, et al., 2006). For example, when a person is watching a performance, or just listening to the sound of someone playing music or singing, parts of their brain that would be active were they singing or performing become active, as if they were the agent. This is called 'mirroring', but is a process of shared intentions and feelings that must be the foundation for all of what is sometimes called imitative learning. It is clearly a complex psychological function and it may involve very widespread activity of the brain, including, for example the basal ganglia and the cerebellum outside the cerebral cortex. Its foundations are in the future-directed processes that generate and guide movements.

A newborn baby cries, suckles or moves its face and limbs with the prospective motor control of a single dynamic personality, using cerebral generation of time in moving, and the *tau* guide, function to achieve effective actions (Craig & Lee, 1999). Babies hear human sounds, maybe all sounds, not as pitches, harmonies and so on, how music psychologists categorise sounds of music, but as expressions of the effort, affection, weakness or caution of human moving to make the sounds, and immediately in relation to the motor patterns of their own *subjectivity* or self-awareness. They have musical sympathy or an *innate intersubjectivity for musicality in moving* (Trevarthen, 1999). They hear sequences of sounds made by other persons, not as notes in musical measures or as phonological elements in words, or as words in phrases, but as minded gestures – actions of moving agents who may sense and engage with their Self contingently and responsibly, motivated by a matching time in the mind for moving and sensing (Wittman & Pöppel, 1999).

The infant's manifest sense of rhythm, phrasing and melody is attracted to the sounds, sights and feel of human bodies, shaped like their own, and moving with expression like their own. Their moving matches that of their affectionately attentive parents and to some extent the music in their environment. They have inner visceral interoceptive and somatic proprioceptive senses of extended time that lead them to express and detect narrative forms of fluctuating excitement, effort, recognition and repose, and they anticipate experiences of recognisable quality at certain moments in these cycles of acting, exploring and finding. Their life is conducted in *body time*, first sensitive to the changes of needs through the day, not clock time. It seeks engagement with the natural vital inner rhythms of other persons that harmonize with and support their own. This engagement of rhythms is improvised between infant and adult in much the same way, and with remarkably similar timing, as the playing of two jazz musicians improvising a duet (Schögler, 1999).

Researchers find that the mind's inner sense of agency imagines cycles of effort that cause fluctuations in heart beat and breathing in cycles of about 30-40 seconds while a person is asleep when the special identifying and recognizing senses are shut down and the mind dreams as a free spirit pushed by autonomic cycles of agitation and repose, and sometimes recognizing remembered events as if they are real (Delamont, Julu, & Jamal, 1999). This period, of about half a minute, corresponds to the intuitive time of narratives in protoconversation with a two-month-old, or the stanzas of a baby song shared with an older baby (Trevarthen, 1999). In the process of conscious, waking action in the world, this inner *narrative time* of musicality becomes experienced time, full of references to things discovered and describable, and linked to imagined action schemata regulated by *dynamic emotions*. This is true for a baby and mother playing with sound (Stern, 1999), as for an adult experiencing or playing music (Imberty, 2000). As the baby becomes older the stories of remembered moments in life (Stern, 2004) become linked in a line mode of consciousness (Donaldson, 1992), and then a child can learn how to explain intentions, thoughts, feelings and experiences in talk.

Language, the most elaborate and productive achievement of collective human action is not just a structured system of phonological, lexical and syntactic or grammatical components or rules, innate or learned. Its cognition rides, as music does, on metaphors and parables of embodied, moving experience (Gallese & Lakoff, 2005; Lakoff & Johnson, 1980; Turner, 1996). The metaphors that regulate even our most complexly rational thoughts are representations in the brain of movements of the body, and especially movements of the hands (Goldin-Meadow & McNeill, 1999). Einstein reported that his mathematical inventions came to him as sensations of bodily movement. The emotional qualities are carried especially in the voice, which reports the vital movements or *animus* of breathing, and their elaborations for communicating with other subjects.

The brain of an infant that achieves this inter-modal or a-modal sensory appreciation of inner moving is responsible for coordinating and regulating their own movements and for ordering their own awareness in their own body, and for situating this consciousness in a world gradually becoming filled with categories of experience. By the end of infancy the infant will be guided in their awareness and intentions by a rapidly increasing set of discrete recollections or recognitions – of objects, events, places, structures, action schemata, uni-modal experiences, and words. But the primary rhythms of their waking actions and awareness, and of their expressions in communication, remains that which we have recognised in infants as the Intrinsic Motive Pulse (IMP), the temporal foundation of intuitive musicality and the foundation for relating sympathetically in time with the behaviours and expressions of other people (Trevarthen, 1999). The young human mind is carried to cognition by an embodied sense of time, by purpose and the changing quality of feelings in moving, estimating vitality affects in its self and detecting them in other vital creatures, especially those that meet them with affection and joy in human bodies (Stern, 1999; 2000). It is true that "music is the food of love", and much else concerning human shared experience besides.

Steps, with Growing Body and Brain, to more Vigorous and more Inventive Communication

Of course, there are great developments in the power of an infant and young child to move, in their cognitive interpretations of experience and therefore in their goals and interests, and in their appreciation of the meaning of other persons' actions. These developments result from growth of the body and brain, as well as learning, and the emergence of new ways of moving, and new ways of vocalising. At the same time, the basic timing and emotional evaluations of motor control, which enabled intersubjective sympathy with parents and others from birth, do not change. The child's innate musicality, their IMP, is merely exercised in different ways, using different body parts and more intelligent discriminations.

The newborn can make small coos in a pulse of shallow, rapid breathing, and can hold the breath for a few seconds or make very long and powerful cries punctuated by rhythmic sobbing, sustaining oscillations between inspiration and expiration over long intervals. In a few weeks coos are produced that may be articulated by glottal stops to make sounds which adults in different countries call ah goo, an gu, ah ger, etc., and after about 3 months babies laugh. The timing of infants' spontaneous vocalizations, for their own entertainment, indicates that they do not depend on hearing the similar adult timing of syllables and phrases, and they can be demonstrated in the same measures by infants who are deaf (Lynch et al., 1995). The precise engagements in dialogues of protoconversation with adults, who intuitively facilitate with well-timed infant directed speech and imitations of the infant's coos, indicate that infants sense how to engage and synchronise their breathing and phonation with the movements that make syllables and phrases. They have a narrative expectancy for the energy of vocalisation extending over tens of seconds.

The vocal system of an infant, the first musical instrument, undergoes rapid change in the first six months after birth, and thereafter the larynx descends further as the body grows, lowering the pitch range of the voice (Rutkowski & Trollinger, 2004). For the period of infancy the vocal cords have less control than they will after that, and this limits singing types of vocalizations to simple calls, which may be pitched to match the baby talk or singing sounds of a parent. After one year a child plays with singing sounds before developing, with changes in both the vocal system and the brain, the articulatory capacity that allows speaking clear and recognisable words (Ménard, Schwartz & Boë, 2004; Rutkowski & Trollinger,

2004). Development of the most precise articulations at the front of the mouth continues with the development of connected sentences of speech after the second year. Singing develops before language, as it may well have done in human evolution (Cross & Moreley, in preparation; Mithen, 2005). All of these changes in vocal production and fluency are linked with developments in deep subcortical parts of the human brain much transformed by evolution to make communication of both musical and linguistic meanings possible (Donald, 2001; Lieberman, 2000). These changes are also linked with development in the corticalised systems that mediate sympathetic awareness between human bodies and their actions (Gallese & Lakoff, 2005).

The second stage in speech development is likely of special importance in relationship to the babbling associated with musical development in that the "infants have gained control of producing normal phonation with a vocal tract that is postured in a variety of ways and thus capable of transmitting contrasts of vocalic quality." (Oller et al, 1997, p. 414). "Speech science research concerned with infant vocalizations and pre-language toddlers indicates that sounds used in vocal experimentation are initially reflexive, but via imitation and learning, the child gradually develops vocal patterns that are appropriate to his or her culture." (Trollinger, 2004, p. 220-221). "Being that pitched vocal sounds (generated by the larynx) in speech are produced the same way for singing in young children (via the raising and lowering of the larynx), this awareness may contribute directly to perceived and produced pitches in singing" (Trollinger, 2004, p. 221).

Moving, Being Moved and Musicality: Theory for Education of Musical Expression

The work of Emile Jaques-Dalcroze suggests how we should conceive music, emotion, and movement as inseparable. He believed that humans feel emotions from various sensations produced at different levels of intensity in muscular contraction and relaxation. Human emotions are translated into musical motion, and hearing music we sense those emotions in various parts of the body, and feel emotions corresponding to the various levels of muscular contraction and relaxation.

Recent neurological research confirms that the nervous system is richly integrated and that the brain functions as a dynamic system transferring information at great speed, faster than research techniques can track except locally in very limited regions, or for very short periods of time. This is certainly true for the processes responsible for awareness or production of musical communication (Turner & Ioannides, in preparation). Body and mind work in tight reciprocal coordination in the generation of movements and consciousness.

Over one hundred years ago Emile Jaques-Dalcroze (commonly referred to as Dalcroze) observed how movement stimulates music learners and helps them feel the music. In answer to the question of what gives music life and expression, he wrote, "Movement, rhythm." (Jaques-Dalcroze, 1921, p. 101). The sensation of contraction and relaxation in human emotion is reflected in the tension and release of the rhythmic movement of music in time and space. Both emotion and music come about in movement. Dalcroze spoke of how there is "a gesture for every sound and a sound for every gesture". We might add here that there is a gesture for every sound that corresponds to an emotion, and a sound for every gesture that evokes an emotion.

The Dalcroze method for teaching music emphasizes the central importance of rhythm in human life. Moving rhythmically is not about dancing or making pretty movements. Rhythm is the 'heartbeat' of music. The word rhythm is related to the Greek word *rhythmos* (measured flow or change, through time, as in a river flowing). When a parent and child engage in rhythmic music making the musical elements of pitch, rhythm, dynamic energy are in play. The dynamic energy of music comes from movement, and corresponds to the way in which feelings are generated through brain-generated time and space. Dalcroze defined rhythm as varieties of flow through time-space (Jaques-Dalcroze, 1921). His idea of "flow" as a quality of live movement gives the word the same sense as it has for Csikszentmihalyi in his descriptions of optimal experience (Csikszentmihalyi, 1990, 1997; Custodero, 2002).

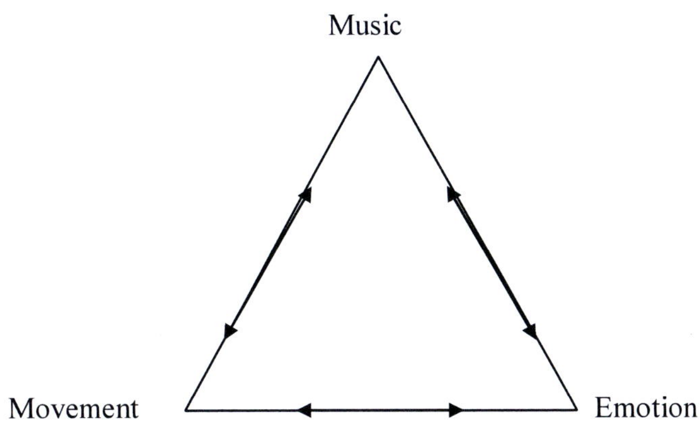

Figure 1. Interrelationship of Music, Movement, and Emotion.

Kinesthesia, from the Greek, was adopted by physiological science in the 19th century to designate awareness of movement. Howard Gardner identifies bodily kinesthetic as the sense that guides fine and gross movement, and that aids the perception of musical motion and controls the performing of music (Gardner, 1993, 1999). The Dalcroze teacher Robert Abramson calls kinesthesia or kinesthetic sense the missing link between sound, movement and feeling (Choksy et al., 2001), which accords with what is now known about the brain and human development. For example, Parsons found that the cerebellum, the organ most concerned with kinesthetic regulations, is activated in close coordination with areas of the cortex in music activity (Parsons & Fox, 1997). Hetland has used the term rhythm theory to describe Parson's module theory of music and spatial tasks (Hetland, 2000). The rhythmic elements of music (hence rhythm theory) controlled in the cerebellum and cerebral cortex are transformed with increasing ability following exercise in the performance of specific mental tasks. There can be no musicality without the innate source of time for rhythms of movement and an emotional appreciation of effort and grace in making movement. The perception of sounds made by the human body moving is also constrained by innate parameters or hearing that detect levels of effort in pitch, loudness and timbre or harmony (Flohr & Hodges, 2006).

A primary function of musical performance is to express emotions to others. All music begins with movement, and all music has a social function. Movement is excited by emotion and by the sharing of emotion. Music with good, well-regulated, rhythmic motion brings emotion to life in an audience. In his method Dalcroze categorizes the qualities of musical performance into three general types. *Arythmic* performance is spastic and offbeat. *Errythmic* performance has all the notes in the right place, but is dull and tedious, lacking nuance of weight, motion, and time. *Eurythmic* performance balances motion and rhythm and solves the problem of how to make music move so that the audience and performer are emotionally moved. To perform in a *eurythmic* way the musician must be able to control her/his tendency to perform in an *errythmic*, synchronized way, without expression. *Errythmic* performance – placing all the right notes in the right places but without rhythmic movement and hence without feeling – is analogous to a philosophy of young children's music development that does not taking into consideration the natural social interactions with caregivers and teachers, and the emotions they share. The infant's intrinsic musicality of movement and communication, in a mood of contended liveliness, has a natural regularity of impulse and sympathetic style that may be called 'eurythmic'.

MUSIC AND THE PSYCHOBIOLOGY OF A CHILD'S BRAIN

We commonly use 'the brain' as a metaphor for our theories of what governs our life and that of other people. This is increasingly so in a scientific culture that expects revelations of reality by intricate esoteric methods of investigation of nature. Unfortunately attention to detail and the complex methods used to gain information about the human organism–about genetics, nerve connections, brain functions and behaviour–means that an overall understanding of the brain-mind, one that relates to common sense of parents and teachers regarding how children have evolved to develop and what they need, the reality we live in, is confused. The pace of technological development in brain science is so fast that there is no time to correct errors of over simplification, and educational policies are often misguided by part truths. Moreover, as with all large, complex cultures, there is an emphasis in our education on what a child has to *learn* to become socially skilled. Just growing up in collaborative company is no longer enough. To explain how children learn music we must do our best to balance evidence for cerebral *plasticity* or changes in brain growth and function consequent upon special experience or practice, against a more general understanding of cerebral functional *elasticity*–the adaptive information–seeking actions of the developing child and his or her brain, and how these actions respond to encouragement and 'scaffolding' from teachers and lead to the changes in the brain that increase awareness and skill. We need a broad *psychobiological* perspective.

In 150 years of developmental science it has been revealed that brain and body grow together–maps of the body interwoven in nerve tissues to regulate action and to deliberately seek awareness that can guide moving, learning to perceive more to move more effectively. The body is mapped in the brain from embryo stages, and a foetus moves with coordinated integrity (Trevarthen, 2004). This moving is generated with rhythms that anticipate how the body will become effectively mobile, manipulative and expressive, and all the brain-generated moving and perceiving is coupled to non-nervous hormonal regulations of the body's vital inner functions. The dualistic idea of a separate mind and body, although still a prevalent belief, is no longer valid.

At a very early stage, just past mid gestation, the foetus has capacities for perceiving and learning sounds. The human brain has, at this stage, a cerebral cortex that is much larger in proportion to the body and the rest of the brain than in other mammals, including apes, and the size and growth cycle of the human neocortex correlates with the relatively long life of a human being. It is a tissue adapted to change in interaction with information from the body and from the outside world, especially the social world (Freeman, 2000). It augments motives that are generated in the brain as a whole with rules for action and images of experience stored in memory.

We have clear evidence that the brain tissues in and memory storing regions gain functional power and change in morphology as a result of stimulation and use (Johnson, 2005). But it is important to add that every change in cortical function has two causes–the incoming evidence of the environment from the senses and the activating or motivating input from core brain systems beneath the cortex that seek and validate this sensory evidence. The integrated subject or Self has to be 'interested', or the cortex will not learn. The developmental structures of the mind are self-regulatory as they learn (Ciccheti & Tucker, 1994).

In reviewing how the development of a young child's brain can be related to learning music we must, therefore, ask what is known about the *motivating/emotional mechanisms,* how they might facilitate adaptive change in the cortex due to practice, and especially how they respond to the motivations and emotions of other persons who would be teachers. We also must try to understand how any development in cortical function might be expressed in a more acute awareness of music, or new skill in the making of music. The natural foundation of music in song indicates that we should focus first on the brain/body mechanisms of breathing, vocalization and of hearing, and then we have to address the question of how the rhythmic and melodic expressions of the voice might be related to dancing of the body and the skilfully controlled gestures required to play a musical instrument according to the conventions of a particular musical culture. Here we can only do this in a summary fashion.

We limit our concern to the beginnings of musical awareness and musicality in moving before a child's brain is old enough to be the intelligent organ of a musician who knows how to recognise and name tones, scales, harmonies, phrases, melodies, and so on, and to read the notation of a musical score and turn it into beautiful sound by beating a drum or bell, moving a bow, plucking a string or fingering a keyboard. Recent developmental psychology research, summarised above, demonstrates that awareness of music has very strong origins in motives and mechanisms for communication that are active and important in a child's development from birth. Communicative musicality comes before language and serves as a support for learning how to use words. It also is a bridge over which other complex cultural skills, including being a singer or musician, can be transferred. Human brains in human bodies grow musically intelligent together, the elder ones adapted to join with the motives of the younger.

Embryogenic Motives and Awareness of Moving

Before the animal's brain is active as the coordinating centre of conscious mobile life, its intricate cell-linking networks are formed within many tissue arrays that chart the animal's body, with its symmetry and in alignment with the polarity of the animal's fate for swimming, walking or flying in the media and light, sounds, heat and shadows of the world and in resistance to the gravitational pull of the earth. The first neurons to carry messages that will mediate between these maps of the central nervous system and the body are those that excite muscles – the *motor neurons* that control moving. The integrative systems of the brain core are already prepared to give the whole a coherence in time and a polarised purposefulness – a rhythm of action and a directed future sense. These are the *affective or emotional neurons* that will also regulate energy in moving in balance with vital needs of the body and the risks and benefits of action (Trevarthen, 2004; Trevarthen et al., 2006; Panksepp & Trevarthen, in prep.). They include visceral efferent neurons that control the vital organs, including the heart and lungs, and that move the muscles of the eyes, face, mouth and throat, which become the main organs of communication. Hands are coupled to this expressive system by a unique development in humans. No other species can express so much of intentions and feelings by movements of the forelimbs.

Head, eyes and hands can listen, look and feel independently of the body as a whole, to select goals for future action, and they become of key importance in communication because they signal what a person is *going to do*, what interests them. Among the earliest perceptual systems to be active are those evolved to sense movement within the body (*kinesthetically* and *proprioceptively*) or in relation to the medium through which the animal is moving (*exproprioceptively*), and these include vestibular senses of displacement of the head, and hearing.

The movement controlling system of the brain rhythmic in its excitatory activity from the start, and it develops by responding to the sensory evidence of movement, selecting most effective regulations for holding the body in coherent activity, for engagement with the world and for maintaining the functions of the whole body in a vital state, protecting from injury and collecting the energy needed for action. Brain and body grow together as a 'going concern', anticipating adventures in action.

The human brain is built to move the human body. It learns from the body, and it teaches the body how to deal with forces that arise in itself when it is moving, and how to pick up information from the environment to guide moving to anticipated goals as one subject or Self. This Self becomes the source of intentions and of perceptions that change with experience, and its actions and expressive movements are the only signals that Other Selves can use to understand what is going on in its brain. All this adventurous, time-regulated activity, with its emotional passions, sets the foundations for the special human dynamics of music.

Tension and Tone in the Core of the Brain: The Start of Emotions for Knowledge and its Communication

In the motivating brain are formed *seeking* impulses that prompt the subject to gain new experiences. Out of these a unified consciousness is built up in which one time and space for many movements is represented, and the uses of many kinds of object are perceived. The neurochemical processes of *emotion* in the brain (Panksepp, 1998) are in communication with hormonal systems that link the body's vital needs with actions directed to satisfy or protect them. Emotions generated in the *periaqueductal grey* matter of the midbrain and integrated with experience in *limbic cortex* and *amygdala*, evaluate and regulate moving, and give values to places and objects the brain has learned to appreciate in relation to needs of the body. Thus every perceived thing or action that is remembered has an emotional 'colour' or quality as well as a cognitive or performative structure and form. The perception of the timbre of the voice or the musical sounds of instruments engages this core system of emotional values.

The neocortex assimilates a lifetime of information about how to move in and exploit the resources of surroundings. Its expectations are determined by the motivating systems of the brain stem. The limbic system, especially the *hippocampus*, grows as new motive and emotional controls are learned, activating the retentive tissues of the sensory and motor systems of the neocortex and mediating memory.

As intelligence and behaviour of animals selves have increased in complexity and daring, their brains have evolved systems of imagination, thought and emotion that seek to communicate and grow with other brains, creating adaptive *social intelligence*. The only way for functions of one brain to influence and collaborate with another is through body talk– using the body so other bodies can divine messages about intentions, attentions and feelings. The brain-to-brain activity mediated by expressive body movements started to evolve with the earliest animals that had brains–even when the brain was no more than a nerve net distributed throughout body of a jellyfish, interconnecting its organs of mobility and sensitivity to stimuli.

Human brains link human bodies with uniquely elaborate complexity of movements and senses. A long gestation permits the elaboration, in a protected environment, of a mutual *amphoteronomic* regulation of vital functions, preparing for a life with human sense of time and human emotions. After birth, the mother and other caregivers and companions, provide the infant with psychological company and guide development of new awareness and new ways of acting. They meet with mutual understanding or sympathy, sharing the same parameters of human *synrythmia* (Trevarthen et al, 2006). All this collaborative brain-body living depends on sharing of the *time and expressive form and style of movements.*

Time in the Mind: The Growth of Sympathetic Rhythms and Musical Imagination Before and After Birth

The brain is, in a sense, musical before birth, formed to put rhythmic impulses into movements of the body, and for using the body to communicate these impulses and to sense them in others. The first muscle activity the central nervous system excites in the embryo has rhythmical timing, before the body sends any sensations to drive it. The senses of touch, vibration and hearing come early in development. The foetus hears musical elements in the voice of the mother months before birth (usually around the 26th week), and sometimes it reacts to the rhythm of music, kicking with its legs. It has well-formed eyes, face, mouth, lips and tongue, vestibular system and cochlear, as well as motor organs for breath and voice and hands for gesture, inactive in their confinement, waiting for the freedom to move that will be granted at birth. When the foetus starts to *hear* and learn his or her mother's voice, the neocortex is just beginning to form. Its primary networks for for *seeing* undergo growth and organizational changes in the first 6 months after birth. The networks for hearing, which begin to be effective before birth, have a second burst of development after infancy, when the sounds of objects being manipulated and of the speech of other persons become important.

Through childhood to adolescence many formulae for motor skills and patterns for identifying sights, sounds and touch experiences that relate to the execution of new skills are incorporated in neocortical maps representing the eyes, vocal system and hands. But recollection of these experiences depends upon how the whole body is motivated to move and on the affective assessment of circumstances. These subjective functions of the mind are regulated to a large extent outside the neocortex, in the more irregularly textured archeocortex and limbic cortex, which function with the subcortical structures of the brain stem (MacLean, 1990).

Regulations of the rhythms of life have evolved in and grow in the neuro-chemically complex *interneuronal core* of the spinal cord and brain stem, the *basal ganglia*, the *cerebellum* and *limbic system* which are coupled by way of the *hypothalamus* to the pituitary and adrenal glands that secrete hormonal messages that circulate in the blood to all organs of the body. The cerebellum, as prime coordinator of the intricacies of motor activity of all the body and precise *timing* of movements, with more neurons than the cerebral cortex, develops throughout life, but most conspicuously through early childhood to adolescence, as the body grows in size and power and skills are practiced and retained. It is greatly involved in both the hearing and performance of music.

The cerebral *neocortex* has evolved as a supremely adaptive tissue to record and organise impressions from the senses and to give refinement to commands the muscles, always under the direction of impulses from the subcortical brain. This dependency of neocortex on the subcortex is clear in the way the cortex forms in the foetus (Trevarthen, 2004), and how its circuits change in readiness for, and response to, the world after birth. New connections and transmission complexes are set up in the developing brain under the dual influence of environmental input regulated by movements, and of internal core activity of the subcortical Intrinsic Motive Formation (IMF) (Trevarthen & Aitken, 1994).

From 6 weeks, when the neurons of the cerebral cortex are completing the post-natal proliferation of dendritic branches, and the production of synaptic connections is accelerating, a baby will orient strongly to a parent's eyes as they join in turn-taking utterances of *protoconversation*, coordinating coos, prespeech lip and tongue movements, and hand gesture. Protoconversation has a pulse interval approximating to *adagio* (c. 0.9 s), and the infant's vocalizations are organized in *syllables* (0.2-0.5 s, with phrase-final lengthening), and *phrases* (3-5 s) (Lynch et al., 1995; Malloch, 1999; Trevarthen, 1999).

These motor times of dialogue (Jasnow and Felstein, 1986) persist as cross-culturally optimal rhythmic units in poetry and music (Miall & Dissanayake, 2003; Pöppel & Wittmann, 1999). The baby is attracted to the whole constellation of maternal expressions; for example, fixating a mother's eyes more strongly when she vocalizes, or watching her lips to better discriminate her vocal sounds. The neural mechanisms of the hemispheres that mediate this interpersonal awareness appear to be configured approximately as in the adult brain, and they are asymmetric. When the electroencephalic (EEG) activity of an 8-week-old baby was mapped while the infant was looking at a picture of a woman's face, researchers were surprised to discover that the *face recognizing* area in the *right hemisphere*, and both Broca's and Wernicke's areas in the *left hemisphere*, identified in adults respectively with *speaking and hearing speech*, were more active than other parts of the cortex (Tzourio-Mazoyer et al., 2002).

From 3 to 4 months infants amuse themselves with rhythmically moving their bodies, waving limbs, banging objects, clapping hands. Game routines are practiced with companions, confirming favorite action and affect formats or emotional narratives (Ratner & Bruner, 1978). Utterances and expressive movements are repeated with the 3- to 5-second cycle of the prelinguistic phrase, which has a variety of lively and communicative prosodic contours (Malloch, 1999; Stern, 1999). Phrases of baby songs are frequently grouped in stanzas of four lines, lasting 15 to 30 seconds. These are foundations, generated in the brain, for the syntactic rules of language vocal imitation and vocal play when the infant is alone are more obvious and more varied at 3 to 5 months. The baby is experimenting with a growing larynx (Rutkowski & Trollinger, 2004).

In the babbling period, after 6 months, when the infant begins to imitate more elaborate gestures and speech sounds, vowels are inflected in syllables, and frontal consonants are differentiated (Bates & Dick, 2002; Locke, 1993). The infant's vocalizations show the influence of distinctive phonological features of the maternal language from around 6 months. Babbling and hand banging are rhythmic, both repeating up to about 3 pulses per second. These movements may be demonstrated to others as communicative signals and shown off with pride to appreciative others (Trevarthen, 2002). Infants from right-handed families, given a choice at 5 to 6 months, prefer to look at the right of two screens to attend to the matching of the appearance and sound of speech syllables (MacKain et al., 1983). This confirms evoked potential studies that indicate left-hemisphere dominance for hearing speech at this age; the right hemisphere perceives other sounds, including the affective tones of speech, better. Babbling also shows an asymmetry suggesting left hemisphere specialization for expressive vocalization of this articulate kind (Holowka & Petitto, 2002). These hemispheric asymmetries may have significance for the asymmetry that has been reported for acquiring more articulate aspects of musical performance by the left hemisphere, the right hemisphere appearing to be more active in responding to the prosody of emotion in musical sound. There are sex differences in these asymmetries indicating that males, which develop more slowly than females, also have a more marked asymmetry for both verbal and musical articulation and the processing of speech and musical sounds (Falk, 2000).

At 9 months purposeful sharing of interest in what can be done with objects appears (Trevarthen & Hubley, 1978). Infants start to deliberately look and point to indicate a direction of attention and show signs of wanting to participate in the attentions and intentions that familiar partners address to objects by following their gaze or pointing (Bates & Dick, 2002; Bruner, 1983; Carpenter et al., 1998). A baby may know his or her name before 6 months, but at 1 year awareness of others' behavior has transformed and infants now may spontaneously copy arbitrary conventional meanings of expressions, manners, actions, and roles; and they are attracted to objects that others use and identify as special. Vocal or gestural approximations to words in the maternal language emerge within 'acting to mean' after 9 months (Bruner, 1983; Halliday, 1979). At 1 year an infant may comprehend 40 to 100 words and produce up to 20. The early vocabulary reflects both differences in motives of children and the style and speech of older companions, especially, in most cases, the mother. Some children acquire names for objects and actions on objects first, others use interpersonal, social and self-referred phrases more (Barrett, 1981). Again, comparisons may be made to differences in singing or attending to music that depend on how the activities of the infant are responded to and encouraged by adult companions. With exceptional support in learning the drill a one year old may beat a regular tempo. It has been claimed that this ability is uniquely human, and an important foundation for learning the ritual forms of music (Wallin, Merker & Brown, 2000).

A rapid explosion of vocabulary starts about 20 to 30 months after birth, but there are large individual differences in this learning (Bates & Dick, 2002; Locke, 1993). The child has little comprehension of regular syntax or grammar but does make use of elaborate prosody and coordinated sequences of voicing and expression that indicate the possession of regulators of rhythm, serial order, and self-conscious emotions. The language that appears in the first 2 years depends upon the linguistic environment. The brain responds differently for the learned language and another language. A study of 18 very young babies showed no brain activity differences while listening to native language and foreign language sounds. After a few months native and foreign language stimuli produced a difference in left hemisphere EEG activity (Cheour, 1998). In addition, research concerning bilingualism demonstrated that the temporal lobes and temporoparietal cortex of the brain are more activated while listening to stories in one's native language (Perani et al,, 1998). Deaf or hearing children with deaf signing parents learn sign language (Goldin-Meadow & McNeill, 1999; Petitto et al., 2002). These developments have been considered as relevant to the development of musical skills, but it is not likely that musical learning depends on a mechanism adapted for learning speech.

A Brain for Music First, and then Language

The brain mechanisms for music and language appear from recent research to be comparable in general organisation. Why is this so? What are the tasks they share?

Music, like language, develops for communication, and by communication. Both depend upon functions of the brain that engage and regulate expressive movements of the body, that guide these movements by senses that monitor the active Self, and that detect expressions of other persons. Both require the Self to be in command of a coherent conscious intentional agency, and an intersubjective consciousness that can detect intentions, interests and feelings of other persons. Language and music are for the community. They participate in the regulation of interpersonal relationships, in the creation and remembering of habits and rituals by which people cooperate in a society that has many traditions and cultural tools. Of course, both have a vital role in teaching the young, though music in our literate culture is much undervalued as a tool of education. Though individuals may practice what they have learned alone, all the learning of music and language is for social use.

Language is elaborated, with many technical aids derived from writing and from recordings of the voice and gestures of its use, to define, describe, interpret, remember and store shared experiences identified in detail. It is elaborately cognitive and its use involves complex sequences of movement that articulate vocal and gestural messages to distribute attentions and intentional movements in ways that other persons can pick up and reflect upon as if they were their own.

Music does not tell anything specific about the shared world of meaningful actions places, things and events in this way. It expresses the motives and feelings of communication itself with an appeal that bridge the gaps of experience between people alive together in more direct and emotive ways than talking or the distribution of texts. It expresses the motives and feelings of communication itself, with little or no reference to another topic. It is essentially interpersonal, not via extra personal objects, events or occasions.

But music is not just a rule-governed set of discrete audible events corresponding to an arbitrary sequence of motor impulses–it is motivated to tell stories. It carries in its phrases and episodes a purposeful and moving narrative of human intentions and emotions, linking the rhythmic sequences of expressive movement in sound with transitions of emotion that make a kind of dramatic sense, that build expectations, that transform mood and that may leave unforgettable memories of what has happened, what should happen, and how the adventure should end. Thus music, with dance and drama, can make human messages with mimetic force, even though they refer to nothing real in the world outside human action. They give action itself meaning and create the foundations for rituals of practice, including the syntactic rituals that enable language to have persuasive and predictable illocutionary force. In development, musical and poetic messages guide relationships and the sharing of experience before language. A baby engages with other persons without language by movements and perceptions that make a universal foundation for the practiced performances and conventional sound systems of any culture's music. For these functions of human community and the transmission of the collective cultural knowledge, a human brain has two capacities developed beyond any faculties in the brains of other animals.

Why only Human Brains are Musical

The systems for moving the body, for sensing body movements, and for regulating action and awareness by emotions are exceptionally elaborate at birth. Indeed, both the human body and the human brain have new specialisations for communication clearly defined months before birth, in the foetus (Trevarthen, 2001).

Secondly, the human brain has an exceptionally long postnatal development and a great capacity, in a vast system of cerebellar neurons and a greatly enlarged cerebral neocortex, for 'plastic' (but not passive) response to experience–to trace a personal narrative of experience that will serve to guide future creative action (Merker, 2005). This capacity for learning arbitrary skills of cognitive awareness and motor performance depends on the prenatally built systems of sensory reception and selection, of motor expression, and of motivation for seeking new ways of engaging with the world–the Self that regulates complex problems of mobility in a coherent way (Merker, 2003).

The world a newborn infant starts to learn with deliberate intent includes not only its own body, which grows for several decades and changes in size, strength and delicacy of action throughout life, but the body and behaviour of the mother and other human beings. The Self-learning of the body is, from the start, profoundly attracted and influenced by the movements of other persons' bodies, and the detection of their motives and emotions by all sensory means. This is an innate process of detection by sympathy for the dynamic motivations that control human movement, and for the display of movements through the human body in obedience to different states of mind. The direct intersubjective 'attunement' of the child's mind to rhythmic and expressive variables of moving in other humans requires no 'theory of mind', and must be an essential foundation for any such theory, should it develop later as an explanatory system in a more sophisticated and interpretative communication (Trevarthen, 1998; Stern, 2000). The mother's voice detected weeks before birth becomes a bridge linking the emotional self-state-regulating *amphoteronomic* functions and the *synrythmic* communication of conscious states between her and her baby after birth. Thus the song of attachment is created (Panksepp & Bernatzky, 2002).

In early weeks after birth infants listen to the sound patterns made by human bodies moving, both internally (as expressed by vocal sounds) and through action on and displacement of objects with different acoustic affordances (changing pitches, loudness and harmonic composition or timbre). The infant is born into a human world of rhythmic sound full of the qualities of human purposes and feelings, and the usable physics of spaces and things.

Babies express their feelings by rhythmic and sustained vocalisations and by rhythmic gestures of different rapidity and intensity–their expressions obey principles of prospective control with distinctive parameters of power, dynamics and texture or gracefulness that facilitate detection of underlying motives. The expressions of parents have instinctive regulation that is adapted to aid communication with the infant. The infant's learning is self-motivated and supported by the knowing guidance of intuitive parenting. Thus the ground is laid for the learning of an infinite variety of culturally contrived forms of expression and message making, include those we understand to be music.

Musicality and the Two Most Intense Times of Social Learning

The human cerebral cortex is in its most rapid phase of growth and most intense activity in the first 3 years when the child is hopefully seeking new friends to share knowledge and skills, and this growth slows after 5 years. Differentiation of the fine circuitry of the cortex is most intense around the time school starts, between 5 and 7 years. Temporal-frontal connections in the left hemisphere appear to be strengthening between 2 and 5 years, the period of most rapid language learning. Before that, the pyramidal cells of area 44 of the left frontal lobe (Broca's area) develop more basal dendrites between 18 and 36 months, as syllable and word combinations are learned, and as the child stabilizes hand preference for meaningful or conventional acts (Pulvermüller & Schumann, 1994).

The young child is eagerly communicative, perceiving and reacting to the messages in movements of other persons' bodies, learning their languages faster than at any other time of life. The young brain is by no means a passive receptacle for this information; indeed more docile learning begins after this more exuberant period. That is why *mirroring* and *plasticity* are not wholly adequate metaphors for the way knowledge grows. Rather the child's mind system is *elastic*, stretching to accommodate a flood of experiences and working them into concepts and habits with useful shapes. From the first, imitations are a way to test and change other persons' expressions or actions, not simply reproductions of movements. The parts of the brain that grow large and learn most are those that represent most elaborately both the powers of movement and sensation of the distal parts of the body and the emotions of action and discovery. Their development is by luxuriant production of tissues that then become more ordered as they assimilate and imagine, or *think about,* what has been learned (Hobson, 2002).

There is evidence for many age-related periods of rapid change when new kinds of connections are made among action and experience (Trevarthen & Aitken, 2003). EEG coherence studies show that different cortico-cortical tracts linking pairs of locations in the cortex are changing in their efficiency at particular ages, and comparisons between locations in the two hemispheres indicate that cycles of development swing from left to right cortex throughout childhood (Thatcher, 1994). At first, when the infant's body is weak and the most effective activities are the emotion-communicative ones of the face, voice and hands responsive to matching states in other persons, the right hemisphere is most active. At three times–in infancy, the preschool period and adolescence–adjustments are made to new companions, new conventions of behaviour are negotiated and the vitality of interpersonal life is most intense. It seems that at these times the consciousness and emotions of the right hemisphere are more in use than those of the left (Schore, 1994; Siegel, 1999). They are also times of increased musicality, when the voice, gesture and dance of spontaneous expression are seeking rhythmic company and testing the limits of their range of activity with varied intensity, practicing eurhythmia with new resources. New manners, new interests and new languages are acquired.

The correlations between age related changes in ability to acquire fluency and accurate pronunciation in a second language and developments in recognition of individuals from their face or voice suggest that adolescence, like early childhood, is a period of enhanced cerebral adaptation, of motivation for identification of significant social partners and for learning new conventions of communication. Facial recognition, voice recognition, and tonal memory mature at about 7 to 9 years of age, and then there is a dip in this ability at puberty, between 11 and 15 years, and a second period of enhanced ability for a few years after that. The distinctive intonation of a foreign language is learned best by a child between 7 and 8 years, and significantly worse at 11 to 12 (Pulvermüller & Schumann, 1994). The expressive arts, including music and dance, are also more easily learned before puberty. All this can be regarded as a repeat of the toddlers' adventurous play with friends, when Bjørkvold's "children's musical culture" flourishes.

Developmentally Appropriate Music Education Pedagogy

Determining the best possible music pedagogy for children is a complex task. There exist many theories and what may work well for one child may not work well for another child because individual differences in children are the rule rather than the exception. Many writers propose theories and ideas about music development and music education (for example Bamberger, 1991; Gordon, 1997; Hargreaves, 1986; Holgersen & Fink-Jensen, 2002; Moog, 1976; Sloboda, 1985; Swanick & Tilman, 1986; Zimmerman, 1982)[3]. Writers often apply theories such as those of Piaget or Merleau-Ponty to music education pedagogy. Other writers focus on research to determine specific events in development. The major roadblock to determining the best possible music pedagogy is that educators currently do not have an overarching or meta-theory of music learning that is readily accepted. Research advances on learning and the brain in the past 10-15 years provide clues. Solms writes that there is agreement among developmental neurobiologists about how early experiences influence brain connection patterns, especially between mother and infant (Solms, 2006). The influence of early experiences on the brain may fundamentally shape future personality and mental health. However, while brain research has made large advances, it is difficult to find a neurologist willing to suggest that our present level of understanding of the brain and learning is sufficient to provide major changes to music education pedagogy.

A solution to the complex task of determining the best possible pedagogy for young children is offered by the United States association for children K-8 years, the National Association for the Education of Young Children (NAEYC). The NAEYC position statement document comments on general development includes "Because development and learning are so complex, no one theory is sufficient to explain these phenomena" (NAEYC, 1997). There are, however, principles that inform practice. NAEYC uses the term *developmentally appropriate practice* to describe practices based on principles about development, learning, characteristics of individual children, and knowledge of the social and cultural contexts in which the children live. The broad-based review of the literature on early childhood education by the NAEYC generates twelve principles to inform early childhood practice. Several of the NAEYC principles are particularly relevant to this chapter in the way in which the principles align with emotion, music, and a child-centered approach.

1. *Children develop and learn best in the context of a community where they are safe and valued, their physical needs are met, and they feel psychologically secure. In addition, children's development in all areas is influenced by their ability to establish and maintain a limited number of positive, consistent primary relationships with adults and other children.*
2. *Domains of children's development—physical, social, emotional, and cognitive—are closely related. Development in one domain influences and is influenced by development in other domains.*

---

3 For more information the reader is directed to theories, ideas, and other developmental topics reviewed in Flohr (2004) chapter 3 and tables 3.1 and 3.2.

3. *Play is an important vehicle for children's social, emotional, and cognitive development, as well as a reflection of their development.*
4. *Children are active learners, drawing on direct physical and social experience as well as culturally transmitted knowledge to construct their own understandings of the world around them. (NAEYC, 1997)*

Music and movement experiences address social and emotional needs of children. Adult and infant play with singing, rhythms, and clapping provide a primary social bonding that influence development in the physical, social, emotional, and cognitive domains.

Music Aptitude or Ability

We have presented evidence for an innate 'communicative musicality' generated by intrinsic properties of the growing human brain and responsive in its development to the rhythms and expressions in the movements of other humans. There is a history of controversy regarding definitions, measurement, and the implications of music ability (Boyle, 1992). Many related terms such as music aptitude, intelligence, capacity, talent, sensitivity, musicality, and achievement are used in describing music ability. Runfola and Swanwick (2002) note a current use of the term music aptitude. In the 1930s the term in use was musical talent and in subsequent years other terms found popularity. Colwell suggests that it may be best to think about music aptitude, talent, ability, musicality, and related terms as comparable (Colwell, 2002). For the purposes of this chapter music aptitude is defined as the child's potential for learning music. Musical ability is defined as "what a person is able to do musically" (Boyle, 1992, p. 248).

Approaches in Music Education

The link among music, emotion, and movement in early development is supported most by the first three approaches in Table 1–the Manhattan Music Curriculum Project, Montessori, and the Pillsbury Foundation Studies. In these free exploration approaches the emphasis is on child-centered experiences and free musical exploration. They are similar to the active learning approach of Montessori in the way in which experiences are used to give children an opportunity to try out their thought processes by observing, reflecting on their findings, asking questions, and formulating answers.

Table 1. Characteristics of Selected Music Education Preschool Approaches

Group emphasis	Approach	Emphasis	Movement	Exploration
Free exploration	*MMCP*	Creating and exploring.	Moving in groups and in music centers with and without instruments.	Emphasis is on free exploration of instruments.
	Montessori	Children discovering and exploring.	Moving to the piano music (ala Dalcroze).	Creating while playing with and experimenting with instruments. Experimenting with cylinders and monochord. Discovery of sound properties.
	Pillsbury Foundation Studies	Children's world of sound, exploring, discovering, and creating.	Movement in all its natural or spontaneous forms	Emphasis on spontaneous music making of young children.
Guided exploration	*Dalcroze (Music Together)*	The link among movement, music, and musical feeling.	Approached based on movement. Change response to expressive characteristics of music.	Create movements in response to music and stories. Students guided through experiences to emphasize time, space, and musical skills and relationships.
	Gordon	Structured formal guidance to aid in musical development.	Move to rhythm patterns and with games and songs.	Movement through imitation and improvisation using audiation. Experiences designed to increase skills in tonal and rhythmic patterns. Emphasis on Gordon's learning theory.
	Kindermusik	Guided experiences with parents	Moving with simple singing games.	Play activities to encourage creativity. Emphasis on singing and basic experiences in music.
	Kodály	Participation in singing folk songs and musical literacy	Move while participating in singing games.	Free exploration not emphasized. Approach designed to produce music literacy in reading, writing, and production.
	Gruhn (Kindliche Lernwelt)	Guided experiences with parents (informally)	Move to rhythm patterns and with games and songs	Movement creativity through imitation and improvisation using audiation. Based on Gordon approach.
	Music Together	Dalcroze based	Movement emphasis (Dalcroze influence).	Creativity through movement improvisation. Similar to Dalcroze.
	Musikgarten	Guided experiences with parents	Moving with simple singing games.	Play activities to encourage creativity. Emphasis on singing and basic experiences in music.
	Orff	Movement, improvisation	Related to Dalcroze movement ideas.	Improvise with body, voice, and instruments (e.g., xylophones). Most similar to Dalcroze with emphasis on playing instruments.
Performance	*Music for Young Children*	Piano performance	Moving with steady beat.	Free exploration not emphasized. Experiences designed to enhance production of music on the piano.
	Suzuki	Development of the whole child with emphasis on performance skills (violin, piano, or other instrument).	Moving to emphasize performance skills such as bow use.	Free exploration not emphasized. Experiences designed to enhance production of music on the violin or other instrument.

Most of the pedagogies reviewed in Table 1 are categorized as guided exploration approaches. In these approaches children or parents and children are taught in classes with teacher-centered experiences. Although these guided exploration emphasis approaches are mostly delivered with teacher-centered experiences, teachers aligned with any of the approaches may also promote the idea of parent/child interaction. It is unlikely that those teachers would argue against parent/child explorations and playful music making. Gruhn writes that music learning for the children aged 8 months to 25 months of his study is best seen "… as a means to support a human potential and an obvious need for rhythmic structures and expressive sounds. This potential can be developed best by informal guidance that connects listening experience and sound exploration with flow and weight of body movement" (Gruhn, 2002). The final two entries of Table 1 labelled performance are examples of many music education approaches that emphasize the acquisition of musical skills. These approaches are typically teacher-centered and more drill and practice on skills than free exploration of a child's musical possibilities.

Given the opportunity, children enjoy and engage in improvising music with their voices and instruments. Clearly infant musicality must express and serve some other fundamental developmental need of every human being. A developmentally appropriate music curriculum would be sensitive to the ways in which young children interact with caregivers through movement, music, and emotion. Music curriculum based on the ideas presented in this chapter would look far different from the standard curricula. One would find more emphasis on play, especially creative play with music through improvisation and composition. Of all the music experiences for young children, creative experiences are the most often forgotten in classrooms. Two kinds of experiences, free exploration and guided exploration may facilitate improvisation. In free exploration the child is left alone to explore instruments, the voice, pots and pans, or any other sound-making object in the environment. In guided exploration, the teacher or parent serves as a guide to the child's exploration by asking questions or engaging in parallel play with the child. The job of both the parent and teacher is to encourage the child, not get in the child's way, and later guide his/her efforts. The capacity and desire of young children to improvise and create in other ways deserves to be encouraged and nurtured. A developmentally appropriate curriculum will give young children the opportunity to freely explore and play with sound (Flohr, 2004). Music and music education serve a role in the promotion of the individual well-being, confidence, and social adaptation of the young child, and opens the way to a lifelong enjoyment of learning – in making, responding to and sharing music.

REFERENCES

Adamson-Macedo, E. N. (2004). Neo-haptic touch. In, R. L. Gregory (Ed.), *Oxford companion to the mind* (2nd Ed.) (pp. 637-639). Oxford, New York: Oxford University Press.

Adolphs, R. (2003). Investigating the cognitive neuroscience of social behavior. *Neuropsychologia, 41*, 119–126.

Bamberger, J. (1991). *The mind behind the musical ear.* Cambridge: Harvard University Press.

Barrett M (1981). The communicative functions of early child language. *Linguistics, 19,* 273-305.

Bates, E. & Dick, F. (2002). Language, gesture and the developing brain. *Developmental Psychobiology, 40*(3), 293-310.

Bateson, M. C. (1975). Mother-infant exchanges: the epigenesis of conversational interaction. In D. Aaronson and R. W. Rieber (Eds.), *Developmental psycholinguistics and communication disorders; Annals of the New York Academy of Sciences,* Vol. 263 (pp. 101-113). New York: New York Academy of Sciences.

Beebe, B., Jaffe, J., Feldstein, S., Mays, K. & Alson, D. (1985). Inter-personal timing: the application of an adult dialogue model to mother-infant vocal and kinesic interactions. In F.M Field & N. Fox (Eds.), *Social perception in infants (pp.* 217-248) Norwood, N.J.: Ablex.

Bekoff, M. & Byers, J. A. (1998). *Animal play: evolutionary, comparative and ecological approaches.* New York: Cambridge University Press.

Bernstein, N. (1967). *Coordination and regulation of movements.* New York: Pergamon.

Bjørkvold, J.-R. (1992). *The muse within: creativity and communication, song and play from childhood through maturity.* New York: Harper Collins.

Blacking, J. (1976). *How musical is man?* London: Faber and Faber.

Blacking, J. (1988). Dance and music in Venda children's cognitive development. In G. Jahoda & Lewis, I. M. (Eds.), *Acquiring culture: cross cultural studies in child development* (pp. 91-112). Beckenham, Kent: Croom Helm.

Bowlby J. (1988). *A secure base: Parent-Child attachment and healthy human development.* New York: Basic Books.

Boyle, J. D. (1992). Evaluation of music ability. In R. Colwell (Ed.), *Handbook of research on music teaching and learning* (pp. 247-265). New York: Schirmer Books.

Bradley, B. S. (in prep.). Early trios: Patterns of sound and movement in the genesis of meaning among infants in groups. In S. Malloch & C. Trevarthen (Eds.), *Communicative musicality: narratives of expressive gesture and being human.* Oxford: Oxford University Press.

Bruner, J. (1983). *Child's talk: learning to use language.* New York: Norton.

Bruner, J. (1996). The *culture of education.* Cambridge, MA: Harvard University Press.

Carpenter, M., Nagell, K., & Tomasello, M. (1998). Social cognition, joint attention, and communicative competence from 9 to 15 months of age. *Monographs of the Society for Research in Child Development, 63.*

Cheour, M., Ceponiene, R., Lehtokoski, A., Luuk, A., Allik, J., Alho, K., et al. (1998). Development of language-specific phoneme representations in the infant brain. *Nature Neuroscience, 12,* 351-353.

Choksy, L., Abramson, R. M., Gillespie, A. E., Woods, D., & York, F. (2001). *Teaching Music in the Twenty-First Century* (2nd ed.). Englewood Cliffs: Prentice Hall.

Cicchetti, D., & Tucker, D. (1994). Development and self-regulatory structures of the mind. *Development and Psychopathology, 6,* 533-549.

Clark, A. (1997). *Being there: Putting brain, body and world together again.* Cambridge, MA: MIT Press.
Colwell, R. (2002). *Music Education Research.* Paper presented at the Federation of North Texas Universities Music Education Lectures, Denton, TX.
Comenius, J. A. (1592-1670). *The school of infancy.* Translated by D. Benham. London, 1858. Quoted by Quick, 1910.
Craig, C. M. & Lee, D. N. (1999). Neonatal control of nutritive sucking pressure: evidence for an intrinsic tau-guide. *Experimental Brain Research 124*(3), 371-382.
Cross, I., & Moreley, I. (in preparation). The evolution of music: theories, definitions and the nature of the evidence. In S. Malloch & C. Trevarthen (Eds.), *Communicative musicality: Narratives of expressive gesture and being human.* Oxford: Oxford University Press.
Csikszentmihalyi, M. & Csikszentmihalyi, I. S. (Eds.) (1988). *Optimal experience: psychological studies of flow in consciousness.* New York: Cambridge University Press.
Csikszentmihalyi, M. (1990). *Flow. The psychology of optimal experience.* New York: Harper and Row Publishers, Inc.
Csikszentmihalyi, M. (1997). *Finding Flow.* New York: BasicBooks.
Csikszentmihalyi, M. & Schiefele, U. (1992). Arts education, human development, and the quality of experience. In B. Reimer & R. A. Smith (Eds.), *The arts, education, and aesthetic knowing. Ninety-first yearbook of the National Society for the Study of Education,* Vol. 2 (pp. 169-191). Chicago: University of Chicago Press.
Custodero, L. A. (1998). Observing flow in young children's music learning. *General Music Today, 12*(1), 21-27.
Custodero, L. A. (2002). *Connecting with the musical moment: Observations of flow experience in preschool-aged children.* Paper presented at the Children's Musical Connections: International Society for Music Education, Copenhagen, Denmark.
Custodero, L. A., & Johnson-Green, E. A. (2003). Passing the cultural torch: Musical experience and musical parenting of infants. *Journal of Research in Music Education, 51*(2), 102-114.
Damasio, A. R. (1999). *The feeling of what happens: body, emotion and the making of consciousness*. London: Heinemann.
DeCasper, A. J., & Spence, M. J. (1986). Prenatal maternal speech influences newborns' perception of speech sounds. *Infant Behavior and Development, 9*(21), 133-150.
Decety, J. & Chaminade, T. (2003) Neural correlates of feeling sympathy. *Neuropsychologia 41*,127–138.
Delamont, R. S., Julu, P. O. O. & Jamal, G. A. (1999). Periodicity of a noninvasive measure of cardiac vagal tone during non-rapid eye movement sleep in non-sleep-deprived and sleep-deprived normal subjects. *Journal of Clinical Neurophysiology, 16*(2), 146-153.
Dewey, J. (1938/1963). *Experience and education.* New York: Collier.
Dissanayake, E. (1988). *What is art for?* Seattle and London: University of Washington Press.
Dissanayake, E. (2000a). Antecedents of the temporal arts in early mother-infant interaction. In N. Wallin and B. Merker (Eds.), *The origins of music.* Cambridge (pp. 389-410). MA: MIT Press.

Dissanayake, E. (2000b): *Art and intimacy: how the arts began*. University of Washington Press, Seattle and London.

Donald, M. (2001). *A mind so rare: The evolution of human consciousness*. New York, NY and London, England: Norton.

Donaldson, M. (1992). *Human minds: an exploration*. London: Allen Lane/Penguin Books.

Eckerdal, P. & Merker, B. (in preparation). 'Music' and the 'action song' in infant development: an interpretation. In S. Malloch & C. Trevarthen (Eds.), *Communicative musicality: Narratives of expressive gesture and being hum*an. Oxford: Oxford University Press.

Erickson, F. (1996). Going for the zone: Social and cognitive ecology of teacher-student interaction in classroom conversations. In D. Hicks (ed.), *Discourse, learning and schooling (pp.* 29-62). New York: Cambridge University Press.

Falk, D. (2000). Hominid brain evolution and the origin of music. In N. Wallin, B. Merker & S. Brown (Eds.). *The origins of music* (pp. 197-216). Cambridge, MA: MIT Press.

Fernald, A. (1992). Meaningful melodies in mothers' speech to infants. In Papoušek, H., Jürgens, U. & Papoušek, M. (Eds.). *Nonverbal vocal communication: Comparative and developmental aspects (pp.* 262-282). Cambridge: Cambridge University Press/Paris: Editions de la Maison des Sciences de l'Homme.

Fernald, A., Taeschner, T., Dunn, J., Papoušek, M., de Boysson-Bardies, B., & Fukui, I. (1989). A cross-language study of prosodic modifications in mothers' and fathers' speech to preverbal infants. *Journal of Child Language, 16*, 477-501.

Field, T. M. Woodson, R., Greenberg, R., & Cohen, D. (1982). Discrimination and imitation of facial expressions by neonates. *Science, 218*, 179-181.

Fifer, W. P., & Moon, C. M. (1995). The effects of fetal experience with sound. In J.-P. Lecanuet, W. P. Fifer, N. A. Krasnegor & W. P. Smotherman (Eds.), *Fetal development: a psychobiological perspective* (pp. 351-366). Hillsdale NJ: Erlbaum.

Flohr, J. W. (2004). *The musical lives of young children*. Upper Saddle, NJ: Prentice Hall.

Flohr, J. W., Atkins, D., Bower, T. G. R., & Aldridge, M. (2000). Infant music preferences: implications for child development and music. *Music Education Research Reports: Texas Music Educators Association.*

Flohr, J. W., & Hodges, D. (2006). Music and neuroscience. In R. Colwell (Ed.), *Handbook of musical cognition and development*. New York: Oxford University Press.

Freeman, W. (2000). A neurobiological role of music in social bonding. In N. Wallin, B. Merker & S. Brown (Eds.). *The origins of music* (pp. 411-424). Cambridge, MA: MIT Press.

Fröhlich, Ch. (2002). "Herabüben" – Elementare Musik mit Erwachsenen. In Ribke, J. & Dartsch M, (Eds.). *Facetten elementarer Musikpädagogik* (pp. 94 – 110). Con Brio, Regensburg.

Fröhlich, Ch. (in preparation). Vitality in music and dance as basic existential experience: applications in teaching. In S. Malloch & C. Trevarthen (Eds.), *Communicative musicality: narratives of expressive gesture and being human*. Oxford: Oxford University Press.

Gallese, V. (2003). The roots of empathy: The shared manifold hypothesis and the neural basis of intersubjectivity. *Psychopatology*, *36*(4), 171-180.

Gallese, V., Keysers, C., & Rizzolatti, G. (2004). A unifying view of the basis of social cognition. *Trends in Cognitive Sciences, 8*, 396-403.

Gallese, V. & Lakoff, G. (2005). The brain's concepts: the role of the mensory-motor system in reason and language. *Cognitive Neuropsychology, 22,* 455–479.

Gardner, H. (1993). *Frames of mind: a theory of multiple intelligences* (10th ed.). New York: Basic Books.

Gardner, H. (1999). *Intelligence reframed. Multiple intelligences for the twenty-first century.* New York: Basic Books.

Gibson, J. J. (1966). *The senses considered as perceptual systems.* Boston: Houghton Mifflin.

Goldin-Meadow, S. & McNeill, D. (1999). The role of gesture and mimetic representation in making language. In M. C. Corballis & E. G. Lea, (Eds.), *The descent of mind: psychological perspectives on hominid evolution* (pp. 155-172). Oxford: Oxford University Press.

Gordon, E. E. (1997). *A music learning theory for newborn and young children.* Chicago: G.I.A.

Gruhn, W. (2002). Phases and stages in early music learning. A longitudinal study on the development of young children's musical potential. *Music Education Research, 4*(1), 51-72.

Halliday, M. A. K. (1979). One child's protolanguage. In M. Bullowa (Ed.), *Before speech: the beginning of human communication* (pp. 171-190). London: Cambridge University Press.

Hargreaves, D. J. (1986). *The developmental psychology of music.* Cambridge: Cambridge University Press.

Hargraeves, D. (1996). The development of artistic and musical competence. In I. Deliège & J. Sloboda (Eds.), *Musical beginnings: origins and development of musical competence* (pp. 145-170). Oxford: Oxford University Press.

Heimann, M. (1998). Imitation in neonates, in older infants and in children with autism: feedback to theory. In S. Bråten (Ed.), *Intersubjective communication and emotion in early ontogeny* (pp. 89-104). Cambridge: Cambridge University Press.

Hetland, L. (2000). Learning to make music enhances spatial reasoning. *Journal of Aesthetic Education, 34*(3-4), 179-238.

Hobson, P. (2002) *The cradle of thought: exploring the origins of thinking.* London: Macmillan.

Holgersen, S.-E. & Fink-Jensen, K. (2002). *The lived body: Object and subject in research of music activities with preschool children.* Paper presented at the International Society of Music Education Early Childhood Conference, Copenhagen, Denmark.

Holowka, S. & Petitto, L.A. (2002). Left hemisphere cerebral specialization for babies while babbling. *Science, 297*, 15-16.

Imberty, M. (1996). Linguistic and musical development in preschool and school-age children. In I. Deliège & J. Sloboda (Eds.), *Musical beginnings: origins and development of musical competence* (pp. 191-213). Oxford: Oxford University Press.

Imberty, M. (2000). The question of innate competencies in musical communication. In N. Wallin, B. Merker & S. Brown (Eds.). *The origins of music* (pp. 449-462). Cambridge, MA: MIT Press.

Ioannides, A. (2001). Real time human brain function: Observations and inferences from single trial analysis of magnetoencephalographic signals. *Clinical Electroencephalography, 32*, 98-111.

Jaffe, J. Beebe, B., Felstein, S., Crown, C. & Jasnow, M. D. (2001). Rhythms of dialogue in infancy: coordinated timing and social development. *Society of Child Development Monographs, Serial No. 265, Vol. 66*(2). Oxford: Blackwell.

Jaques-Dalcroze, E. (1921). *Rhythm, music and education.* New York: Putnam's Sons.

Jasnow, M. & Feldstein, S. (1986). Adult-like temporal characteristics of mother-infant vocal interactions. *Child Development, 57*, 754-761.

Jeannerod, M. (2006). *Motor cognition: what actions tell the self.* Oxford: Oxford University Press.

Johnson, M. (2005). *Developmental cognitive neuroscience.* (Second Edition) Malden, MA. Blackwell.

Jusczyk, P.W. & Krumhansl, C.L. (1993). Pitch and rhythmic patterns affecting infants' sensitivity to musical phrase structure. *Journal of Experimental Psychology: Human Perception and Performance, 19*(3), 627-640.

Kugiumutzakis, G. (1998). Neonatal imitation in the intersubjective companion space. In S. Bråten, (Ed.), *Intersubjective communication and emotion in early ontogeny* (pp. 63-88). Cambridge: Cambridge University Press.

Kugiumutzakis, G. (1999). Genesis and development of early infant mimesis to facial and vocal models. In J. Nadel & G. Butterworth (Eds), *Imitation in infancy* (pp. 127-185) Cambridge: Cambridge University Press.

Kugiumutzakis, G., Kokkinaki, T., Markodimitraki, M. & Vitalaki, E. (2005). Emotions in early mimesis. In, J. Nadel & D. Muir (Eds.) *Emotional development (*pp. 161-182). Oxford: Oxford University Press.

Kuhl, P. K. (1985). Methods in the study of infant speech perception. In G. Gottlieb & N. A. Krasnegor (Eds.), *Measurement of audition and vision in the first year of postnatal life* (pp. 223-251). New York: Academic Press.

Lakoff, G., & Johnson, M. (1980). *Metaphors we live by.* Chicago: University of Chicago Press.

Lashley, K. S. (1917). The accuracy of movement in the absence of excitation from the moving organ. *American Journal of Physiology, 43*, 169-194.

Lashley, K. S. (1951). The problems of serial order in behavior. In: L. A. Jeffress (Ed.), *Cerebral mechanisms in behaviour* (pp. 112-136). New York: Wiley.

Lecanuet, J.-P. (1996). Prenatal auditory experience. In I. Deliege & J. Sloboda (Eds.), *Musical beginnings: origins and development of musical competence* (pp. 3-34). Oxford, New York, Tokyo: Oxford University Press.

Lee, D. N. (2005). Tau in action in development. In Rieser, J.J., Lockman, J. J., & Nelson, C.A. (Eds.). *Action, perception & cognition in learning and development*, pp. 3-49. Hillsdale, N.J.: Erlbaum.

Lieberman P. (2000). *Human language and our reptilian brain: the subcortical bases of speech, syntax, and thought.* Cambridge, MA: Harvard University Press.

Littleton, D. F. (2002). Music in the time of toddlers. *Zero to Three, 23*(1), 35-40.

Locke, J. L. (1993). *The child's path to spoken language*. Cambridge MA and London: Harvard University Press.

Lynch, M. P., Oller, D. K., Steffens, M. L., and Buder, E. H. (1995). Phrasing in prelinguistic vocalisations. *Developmental Psychobiology, 28,* 3-25.

MacKain, K. S., Studdert-Kennedy, M., Spieker, S. & Stern, D. N. (1983). Infant intermodal speech perception is a left hemisphere functions. *Science, 219*, 1347-1349.

MacLean, P.D. (1990). *The triune brain in evolution, role in paleocerebral functions*. New York, Plenum Press.

Malinowsky, B. (1923). The problem of meaning in primitive languages. In C. K. Ogden & I. A. Richards (Eds.). *The Meaning or Meaning*. London: Routledge and Kegan Paul.

Malloch, S. (1999). Mother and infants and communicative musicality. In Deliège, I., (Ed.). *Rhythms, musical narrative, and the origins of human communication. Musicae Scientiae, Special Issue* (pp. 29-57). Liège, Belgium: European Society for the Cognitive Sciences of Music.

Malloch, S., Sharp, D., Campbell, D. M., Campbell, A. M. & Trevarthen, C. (1997). Measuring the human voice: Analysing pitch, timing, loudness and voice quality in mother/infant communication. *Proceedings of the Institute of Acoustics, 19* (5), 495-500.

Maratos, O. (1982). Trends in development of imitation in early infancy. In T.G. Bever (Ed.), *Regressions in mental development. Basic phenomena and theories* (pp. 81-101). Hillsdale, N.J., Erlbaum.

Mazokopaki, K. & Trevarthen, C. (in prep.). Infants' rhythms: Moving in the absence and presence of music. In S. Malloch & C. Trevarthen (Eds.), *Communicative musicality: narratives of expressive gesture and being human*. Oxford: Oxford University Press.

McPherson, G. E. (Ed.) (2006). The child as musician. A handbook of musical development. Oxford: Oxford University Press.

Meltzoff, A. N. & Moore, M. H. (1977). Imitation of facial and manual gestures by human neonates. *Science, 198*: 75-78.

Meltzoff, A. N. & Moore, M. K. (1998). Infant intersubjectivity. Broadening the dialogue to include imitation, identity and intention. In S. Bråten, (Ed.), *Intersubjective Communication and Emotion in Early Ontogeny* (pp. 47-62). Cambridge: Cambridge University Press.

Ménard, L., Schwartz, J.-L. & Boë, L.-J. (2004). Role of vocal tract morphology in speech development: perceptual targets and sensorimotor maps for synthesized French vowels from birth to adulthood. *Journal of Speech, Language, and Hearing Research, 47*, 1059-1080.

MENC (1991). Beliefs about young children and developmentally and individually appropriate musical experiences. Retrieved June 7, 2002, from *http://www.menc.org/information/prek12/echild.html.*

Merker, B. (2003). Cortex, counter-current context, and dimensional integration of lifetime memory. *Cortex, 40*, 559-576.

Merker, B. (2005). The liabilities of mobility: a selection pressure for the transition to consciousness in animal evolution. *Consciousness & Cognition, 14*(1), 89-114.

Miall, D. S. & Dissanayake, E. (2003). The poetics of babytalk. *Human Nature, 14*(4), 337-364.

Mithen, S. (2005). *The singing Neanderthals. The origins of music, language, mind and body.* London: Weidenfeld and Nicholson.

Murray, L. & Cooper, P. J. (Eds.) (1997). *Postpartum depression and child development.* New York: Guilford Press.

Murray, L. & Trevarthen, C. (1985). Emotional regulation of interactions between two-month-olds and their mothers. In T. M. Field & N. A. Fox (Eds.), *Social Perception in Infants* (pp. 177-197). Norwood, N J: Ablex.

Moog, H. (1976). The development of musical experience in children of preschool age. *Psychology of Music, 4*(2), 38-45.

Moorehead, G. E., & Pond, D. (1978). *Music of young children.* Santa Barbara, CA: Pillsbury Foundation for Advancement of Music Education.

NAEYC (1997). Developmentally appropriate practice in early childhood programs serving children from birth through age 8. Retrieved June 4, 2002, from http://www.naeyc.org/resources/position_statements.

Nadel, J. & Pezé, A. (1993). Immediate imitation as a basis for primary communication in toddlers and autistic children. In J. Nadel & L. Camioni (Eds.), *New perspectives in early communicative development* (pp. 139-156). London: Routledge.

Nadel, J., Carchon, I., Kervella, C., Marcelli, D., & Réserbat-Plantey, D. (1999). Expectancies for social contingency in 2-month-olds. *Developmental Science, 2*(2), 164-173.

Nagy, E. & Molnár, P. (2004). Homo imitans or Homo provocans? Human imprinting model of neonatal imitation. *Infant Behaviour and Development, 27*(1), 54-63.

Oller, D.K., Eilers, R.E., Urbano, R. & Cobo-Lewis, A.B. (1997). Development of precursors to speech in infants exposed to two languages. *Journal of Child Language, 24*, 407-425.

Panksepp, J. (1998). *Affective neuroscience: the foundations of human and animal emotions.* New York: Oxford University Press.

Panksepp, J. & Bernatzky , G. (2002) Emotional sounds and the brain. The neuro-affective foundations of musical appreciation. *Behavioural Processes 60*, 133-155.

Panksepp, J. & Burgdorf, J. (2003). "Laughing" rats and the evolutionary antecedents of human joy? *Physiology and Behavior, 79*, 533-547.

Panksepp, J. & Trevarthen, C. (in preparation). Motive impulse and emotion in acts of musicality and in sympathetic emotional response to music. In S. Malloch & C. Trevarthen (Eds.), *Communicative musicality: narratives of expressive gesture and being human.* Oxford: Oxford University Press.

Papoušek, H. (1996). Musicality in infancy research: biological and cultural origins of early musicality In I. Deliège & J. Sloboda (Eds.), *Musical beginnings: origins and development of musical competence* (pp. 37-55). Oxford, New York, Tokyo: Oxford University Press.

Papoušek, M. (1996). Intuitive parenting: A hidden source of musical stimulation in infancy. In I. Deliège & J. Sloboda (Eds.), *Musical beginnings:origins and development of musical competence* (pp. 88-112). Oxford: Oxford University Press.

Papoušek, M. & Papoušek, H. (1981). Musical elements in the infant's vocalization: Their significance for communication, cognition, and creativity. In L. P. Lipsitt & C. K. Rovee-Collier (eds.), *Advances in infancy research* (pp. 163-224). Norwood, NJ: Ablex.

Papoušek, M., Papoušek, H. & Symmes, D. (1991). The meanings and melodies in motherese in tone and stress languages. *Infant Behavior and Development, 14*, 415-440.

Parsons, L., & Fox, P. (1997). Sensory and cognitive tasks: The cerebellum and cognition. In J. D. Schmahmann (Ed.), *International Review of Neurobiology, Cerebellum and Cognitio* (pp. 255-272). San Diego: Academic Press.

Perani, D., Paulesu, E., Calles, N., Dupoux, E., Dehane, S., Bettinardi, V., Cappa, S., & Fazio, R. M., J. (1998). The bilingual brain: Proficiency and age of acquisition of the second language. *Brain, 121*(10), 1841-1852.

Petitto, L.A., Holowka, S., Sergio, L.E., & Ostry, D. (2002). Language rhythms and baby hand movements. *Nature, 413*:35-36

Pickenhain, L. (1984). Towards a holistic conception of movement control. In H.T.A. Whiting (Ed.). H*uman motor actions: Bernstein reassessed* (pp. 505-530). Amsterdam: Elsevier (North Holland).

Pöppel, E. & Wittmann, M. (1999). Time in the mind. In R. Wilson & F. Keil (Eds.). *The MIT encyclopedia of the cognitive sciences* (pp.836-837). Cambridge, Massachusetts: The MIT Press.

Porges, S. W. (1997). Emotion. An evolutionary by-product of the neural regulation of the autonomic nervous system, In. C. S. Carter, I.I. Lederhendler & B. Kirkpatrick (Eds.). *The integrative neurobiology of affiliation* (pp. 62-78). *Annals of the New York Academy of Sciences, 807*. New York: New York Academy of Sciences.

Pulvermüller F. & Schumann, J. (1994). Neurobiological mechanisms of language acquisition. *Language Learning, 44*, 681-734.

Ratner, N. & Bruner, J. S. (1978). Games, social exchange and the acquisition of language. *Journal of Child Language 5*, 391-400.

Reddy, V. (2003). On being the object of attention: implications for self–other consciousness. *Trends in Cognitive Sciences, 7*(9), 397-402.

Reese, D. (1998, June). Speech Development in Infant and Toddler. Retrieved September 30, 2006, from *http://www.kidsource.com/kidsource/content4/speech.develop.baby.pn.html.*

Robb, L. (1999). Emotional musicality in mother-infant vocal affect, and an acoustic study of postnatal depression. In *Rhythms, musical narrative, and the origins of human communication. Musicae Scientiae, Special Issue, 1999-2000* (pp. 123-151). Liège: European Society for the Cognitive Sciences of Music.

Rousseau, J. J. (1762). *The social contract or principles of political right.* Retrieved 9/30/2006, from http://www.constitution.org/jjr/socon.txt.

Rogoff, B., Paradise, R., Arauz, R. M., Correa-Chávez, M. & Angelillo, C. (2003). Firsthand learning through intent participation. *Annual Review of Psychology, 54*, 175-203.

Runfola, M., & Swanick, K. (2002). Developmental characteristics of music learners. In R. Colwell & C. Richardson (Eds.), *The new handbook of research on music teaching and learning* (pp. 373-397). New York: Oxford University Press.

Rutkowski, J., & Trollinger, V. (2004). Singing. In J. W. Flohr (Ed.), *Musical lives of young children*. Upper Saddle River, NJ: Prentice Hall.

Schilbach, L., Wohlschläger, A.M., Newen, A., Krämer, N, Shah, N. J, Fink, G. R. & Vogeley, K (2006). Being with others: neural correlates of social interaction. *Neuropsychologia, 44*(5), 718-730.

Schögler, B. W. (1999). Studying temporal co-ordination in Jazz duets. In *Rhythms, musical narrative, and the Origins of human communication. Musicae Scientiae, Special Issue, 1999-2000* (pp.75-92). Liège: European Society for the Cognitive Sciences of Music.

Schore, A. N. (1994). *Affect regulation and the origin of the Self: The neurobiology of emotional development.* Hillsdale, NJ: Erlbaum.

Siegel, D. (1999). *The developing mind: Toward a neurobiology of interpersonal experience.* New York/London: Guilford.

Selby, J. M. & Bradley, B. S. (2003). Infants in groups: a paradigm for study of early social experience. *Human Development, 46*, 197-221.

Sloboda, J. A. (1985). *The musical mind: the cognitive psychology of music.* Oxford: Clarendon Press.

Sloboda, J. & Davidson, J. (1996). The young performing musician. In I. Deliège & J. Sloboda (Eds.), *Musical beginnings: origins and development of musical competence (*pp. 171-190). Oxford: Oxford University Press.

Solms, M. (2006). Freud returns. *Scientific American Mind, 17*(2), 28-34.

Sperry, R. W. (1952). Neurology and the mind-brain problem. *American Scientist, 40,* 291-312.

Standley, J. M. (2002). Music therapy in the NICU. Promoting growth and development of premature infants. *Zero to Three, 23*(1), 23-30.

Stern, D. N. (1992). L'enveloppe prénarrative: Vers une unité fondamentale d'expérience permettant d'explorer la réalité psychique du bébé. *Revue Internationale de Psychopathologie, 6*, 13-63.

Stern, D. N. (1999). Vitality contours: The temporal contour of feelings as a basic unit for constructing the infant's social experience. In Rochat, P. (Ed.), *Early social cognition: understanding others in the first months of life* (pp. 67-90). Mahwah, NJ: Erlbaum

Stern, D. N. (2000). *The interpersonal world of the infant: a view from psychoanalysis and development psychology.* (Originally published in 1985. Paperback Second Edition, with new Introduction). Basic Books: New York.

Stern, D. N. (2004). *The present moment:* In *Psychotherapy and everyday life.* New York: Norton.

Swanick, K. & Tilman, J. (1986). A sequence of musical development: a study of children's compositions. *British Journal of Music Education, 3*(3), 305-339.

Takada, A. (2005). Mother-infant interactions among the !Xun: Analysis of gymnastic and breastfeeding behaviors. In B. S. Hewlett & M. E. Lamb (Eds.), *Hunter-gatherer childhoods: evolutionary, developmental, and cultural perspectives* (pp.289-308). New Brunswick, NJ: Transaction Publishers.

Tegnér, A. (1995). *Mors lilla Olle och andra visor.* (Mother´s little Olle and other songs) (12th Edition). Stockholm: Norstedts Förlag.

Thatcher, R.W. (1994). Psychopathology of early frontal lobe damage: dependence on cycles of development. *Development and Psychopathology, 6*, 565-596.

Thompson, E. (2001). Empathy and consciousness. In Thompson, E. (Ed.). *Between ourselves: second-person issues in the study of consciousness* (pp. 1-32). Thorverton, UK: Imprint Academic.

Trainor, L. J. (1996). Infant preferences for infant directed versus noninfant directed playsongs and lullabies. *Infant Behavior and Development, 19*(1), 83-92.

Trainor, L. J. (2002). Lullabies and playsongs. Why we sing to children. *Zero to Three, 23*(1), 31-34.

Trainor, L. J., & Zacharias, C. A. (1997). Infants prefer higher-pitched singing. *Infant Behavior and Development, 21*(4), 799-805.

Trehub, S. E. (1990). The perception of musical patterns by human infants: the provision of similar patterns by their parents. In M. A. Berkley & W. C. Stebbins (Eds.), *Comparative perception: basic mechanisms* (Vol. 1, pp. 429-459). New York: John Wiley.

Trehub, S. E. (2002). Mothers are musical mentors. *Zero to Three, 23*(1), 19-22.

Trehub, S. E. (2004). Music perception in infancy In J. W. Flohr (Ed.), *Musical lives of young children.* Upper Saddle River, NJ Prentice Hall.

Trehub, S. E., Unyk, A. M. & Trainor, L. J. (1993). Maternal singing in cross-cultural perspective. *Infant Behaviour and Development, 16,* 285-295.

Trehub, S. E., Unyk, A. M., Kamenetsky, S. B., Hill, D. S., Trainor, L. J., Henderson, J. L. & Saraza, M. (1997). Mothers' and fathers' singing to infants. *Developmental Psychology, 33*, 500-507.

Trevarthen, C. (1975). Early attempts at speech. In R. Lewin, Ed. *Child alive: new insights into the development of young children* (pp. 62-80) London: Temple Smith.

Trevarthen, C. (1984). How control of movements develops. In H.T.A. Whiting (Ed). *Human motor actions: Bernstein reassessed* (pp. 223-261). Amsterdam: Elsevier (North Holland).

Trevarthen, C. (1986). Development of intersubjective motor control in infants. In M.G. Wade and H.T.A. Whiting (Eds.). *Motor development in children: aspects of coordination and control* (pp. 209-261). Dordrecht, Martinus Nijhof.

Trevarthen, C. (1990). Signs before speech. In: T. A. Sebeok & J. Umiker-Sebeok (Eds.), *The semiotic web* (pp. 689-755). Berlin, New York, Amsterdam: Mouton de Gruyter.

Trevarthen, C. (1998). The concept and foundations of infant intersubjectivity. In S. Bråten (Ed.), *Intersubjective communication and emotion in early ontogeny* (pp. 15-46). Cambridge: Cambridge University Press.

Trevarthen, C. (1999). Musicality and the Intrinsic Motive Pulse: evidence from human psychobiology and infant communication. In I. Deliège (Ed.) *Rhythms, musical narrative, and the origins of human communication. Musicae Scientiae, Special Issue, 1999-2000* (pp. 157-213). Liège: European Society for the Cognitive Sciences of Music.

Trevarthen, C. (2001). The neurobiology of early communication: Intersubjective regulations in human brain development. In A. F. Kalverboer & A. Gramsbergen (Eds.), *Handbook on brain and behavior in human development* (pp. 841-882). Dordrecht, The Netherlands: Kluwer.

Trevarthen, C. (2002) Origins of musical identity: evidence from infancy for musical social awareness. In MacDonald, R., Hargreaves, D. & Miell, D., (Eds.). *Musical identities* (pp. 21-38). Oxford: Oxford University Press.

Trevarthen, C. (2004). Brain development. In R.L. Gregory (Ed.). *Oxford companion to the mind* (Second Edition) (pp. 116-127). Oxford, New York: Oxford University Press.

Trevarthen, C. (2005). Stepping away from the mirror: pride and shame in adventures of companionship reflections on the nature and emotional needs of infant intersubjectivity. In Carter, C., Ahnert, C., Grossman, K., Hrdy, S., Lamb, M., Porges, S., & Sachser, N. (Eds.). *Attachment and bonding: a new synthesis* (pp. 55-84). Dahlem Workshop Report 92. Cambridge, MA: The MIT Press.

Trevarthen, C. (2006). First things first: infants make good use of the sympathetic rhythm of imitation, without reason or language. *Journal of Child Psychotherapy, 31*(1), 91–113.

Trevarthen, C. & Aitken K. J. (1994). Brain development, infant communication, and empathy disorders: Intrinsic factors in child mental health. *Development and Psychopathology, 6*, 599-635.

Trevarthen, C. & Aitken, K. J. (2003). Regulation of brain development and age-related changes in infants' motives: The developmental function of "regressive" periods. In: M. Heimann (Ed.) *Regression periods in human infancy* (pp. 107-184). Mahwah, NJ: Erlbaum.

Trevarthen, C. & Malloch, S. (2002). Musicality and music before three: human vitality and invention shared with pride. *Zero to Three, 23*(1), 10-18.

Trevarthen, C. & Hubley, P. (1978). Secondary intersubjectivity: confidence, confiding and acts of meaning in the first year. In Lock A, (Ed.), *Action, gesture and symbol: the emergence of language* (pp. 183-229). London, New York, San Francisco: Academic Press.

Trevarthen, C. & Reddy, V. (2006). Consciousness in infants. In M. Velman & S. Schneider (Eds.). *A companion to consciousness*. Oxford: Blackwells.

Trevarthen, C., Kokkinaki, T. & Fiamenghi, G. A. Jr. (1999). What infants' imitations communicate: with mothers, with fathers and with peers. In J. Nadel & G. Butterworth (Eds.), *Imitation in infancy* (pp. 127-185). Cambridge: Cambridge University Press.

Trevarthen, C., Aitken, K. J., Vandekerckhove , M., Delafield-Butt, J. & Nagy, E. (2006). Collaborative regulations of vitality in early childhood: stress in intimate relationships and postnatal psychopathology. In D. Cicchetti & D. J. Cohen (Eds.), *Developmental psychopathology, Volume 2, Developmental neuroscience* (pp. 65-126). Second Edition. New York: Wiley.

Trollinger, V. (2004). Preschool children's pitch-matching accuracy in relation to participation in Cantonese-immersion preschools. *Journal of Research in Music Education, 52*(3), 218-233.

Tronick, E.Z., Als H., Adamson L., Wise, S. & Brazelton, T.B. (1978). The infant's response to entrapment between contradictory messages in face-to face interaction. *Journal of the American Academy of Child Psychiatry, 17*, 1-13.

Turner, M. (1996). *The literary mind: the origins of thought and language*. New York/Oxford: Oxford University Press.

Turner, V. (1974). *Dramas, fields and metaphors*. Ithaca, N.Y.: Cornell University Press.

Turner, R.. & Ioannides, A. (in preparation). Brain, music and musicality: inferences from neuroimaging. In S. Malloch & C. Trevarthen (Eds.), *Communicative musicality: narratives of expressive gesture and being human*. Oxford: Oxford University Press.

Tzourio-Mazoyer, N., De Schonen, S., Crivello, F., Reutter, B., et al. (2002). Neural correlates of woman face processing by 2-month-old infants. *Neuroimage, 15*, 454–461.

Varela, F. J., Thompson, E. & Rosch, E. (1991). *The embodied mind.* Cambridge, MA: MIT Press.

Von Hofsten, C. (2004). An action perspective on motor development. *Trends in Cognitive Science, 8*(6, 1), 266-272.

Vygotsky, L.S. & Cole, M. (Eds.) (1978). *Mind in society: the development of higher psychological processes,* Cambridge, Mass: Harvard University Press.

Wallin, N. L., Merker, B. & Brown, S. (Eds.) (2000). *The origins of music* (pp. 389–410). MIT Press, Cambridge MA.

Watanabe, H., Malloch, S. M., Trevarthen, C. Horiuchi, T. & Watanabe, T. (2006). Abstract 414; Workshop 24: *Communicative musicality and premature infants.* 10th World Congress, World Association of Infant Mental Health, Paris, 8-12 July, 2006.

Wittman, M. & Pöppel, E. (1999). Temporal mechanisms of the brain as fundamentals of communication ì with special reference to music perception and performance. In *Rhythms, musical narrative, and the origins of human communication, Musicae Scientiae, Special Issue,* 1999-2000 (pp. 13-28). Liège: European Society for the Cognitive Sciences of Music.

Zimmerman, M. P. (1982). Developmental processes in music learning. In R. Colwell (Ed.), *Symposium in music education. A festschrift for Charles Leonhard.* Urbana, IL: University of Illinois.

In: Neurosciences in Music Pedagogy
Editors: W. Gruhn and F. Rauscher, pp. 101-120
ISBN: 978-1-60021-834-7

Chapter IV

NEURAL CORRELATES OF MUSIC LEARNING AND UNDERSTANDING

Gottfried Schlaug and Marc Bangert

ABSTRACT

Playing an instrument requires the ability to read musical notation and to translate it into sequential and independent bimanual motor actions while receiving and responding to multi-sensory feedback. Several studies have explored the brain basis of these highly specialized sensorimotor, auditory, auditory-spatial, and integrative skills as well as the effects of long-term training of these skills to find evidence for functional as well as structural adaptations in brain regions that are linked to the acquisition and execution of these skills. While primary auditory cortex changes or differences might be related to learning fine spectral/temporal musical discrimination, and motor cortex changes could be related to repeated practice of complicated, independent finger movements, the role of higher-order association and integration of brain regions in perceiving music and making music is less clear. Parietal regions may play a role in the integration of auditory information with other sensory information, while frontal brain regions, particularly the inferior frontal gyrus could play a role in the perceiving single auditory events as part of larger units that follow certain rules, recognizing alterations in sequential auditory-perceptual events, and in mapping auditory actions with motor actions. It is of interest that these parietal and frontal regions that show functional and structural differences between musicians and non-musicians also belong to a network that may constitute the human mirror-neuron system. It is one of our hypotheses that this network of brain regions may constitute the anatomical substrate for some of the cognitive enhancements that have been found in groups of individuals who are undergoing instrumental or non-instrumental music training.

INTRODUCTION

Instrumental music training is an intense, multi-sensory and motor experience that is typically initiated at an early age and requires the practice of multimodal skills throughout a musician's career. Musicians are an ideal human model for studying the behavioral and brain effects of these highly specialized sensorimotor (Amunts et al., 1997; Elbert et al., 1995; Gaser & Schlaug, 2003; Schlaug, 2001 and 2005), auditory (Besson et al., 1994; Gaab & Schlaug, 2003; Keenan et al., 2001; Pantev et al., 1998 and 2001; Schlaug et al., 1995; Schneider et al., 2002; Zatorre et al., 1998), auditory-spatial (Muente et al., 2001), and integrative (Hund-Georgiadis & von Cramon, 1999; Bangert et al., 2006) skills considering the early age at which they commence training and the continuous practice of these skills over the course of a lifetime. Several studies found evidence for structural differences in brain regions that are directly involved in the acquisition of musical skills such as the primary auditory cortex important for auditory perception and discrimination (Schneider et al., 2002; Gaser & Schlaug, 2003), the primary motor cortex, and closely related premotor and supplementary motor regions important for preparation, execution, and control of precise finger movements (Amunts et al., 1997; Gaser & Schlaug, 2003). These structural between-group differences mirror structural changes seen in the primary sensorimotor cortex and cerebellar cortex in animal models that have been subjected to long-term learning of complex motor skills (Anderson et al., 2002). Such microstructural changes of brain plasticity suggest that the between-group differences found in adult musicians compared to non-musicians may be the result of adaptation in response to skill acquisition and long-term practice.

In addition, structural brain differences comparing musicians with matched non-musicians have also been found in brain regions outside primary brain regions (i.e., auditory and motor cortex) such as in the inferior frontal gyrus, the superior parietal lobe as well as the inferior lateral temporal lobe and the cerebellum (Gaser & Schlaug, 2003; Sluming et al., 2002). While primary auditory cortex changes or differences might be related to learning fine spectral and temporal musical discrimination and motor cortex changes to learning complicated and independent finger movements, the functional role of these changes in extra-primary brain regions is less clear. Parietal regions might play a role in sensory encoding as well as in the integration of auditory information with other sensory information, while frontal brain regions, particularly the inferior frontal gyrus could play a role in the integration of auditory events into larger units, learning to make predictions about sequential auditory-perceptual events, recognizing alterations in sequential auditory-perceptual events, and in mapping actions to their associated sounds. It is of interest that these extra-primary cortex brain regions (i.e., inferior frontal gyrus, premotor cortex, superior parietal lobule) belong to a network that has been associated with action observations, action-sound mapping, and cross-modality sensorimotor integration, and may constitute the human homolog of a mirror-neuron system, first described in monkeys. This network is specifically engaged during music making and musical skill learning. The specific and continuous engagement of this network throughout a musicians' lifelong career might be the neural basis for some of the cognitive enhancements seen in musicians or as a response to music training.

STRUCTURAL BRAIN CORRELATES OF LONG-TERM INSTRUMENTAL MUSIC MAKING

Cross-Sectional Comparisons of Adult Musicians and Adult Non-Musicians

As shown in Figure 1, professional keyboard players have significantly more gray matter than amateur musicians and non-musicians in several brain regions including the primary sensorimotor cortex as well as the adjacent superior premotor and anterior superior parietal cortex bilaterally, mesial Heschl's gyrus (primary auditory cortex), the cerebellum, and the left inferior frontal gyrus, and the lateral inferior temporal lobe (Gaser & Schlaug, 2003). The distinction between professional and amateur musicians was based on whether or not a keyboard player's main profession was being a musician (e.g., music teacher, performer). There was also a clear separation between both groups in terms of average practice intensity across their professional life: professional musicians had about double the amount of practice than the amateur musicians (Gaser & Schlaug, 2003).

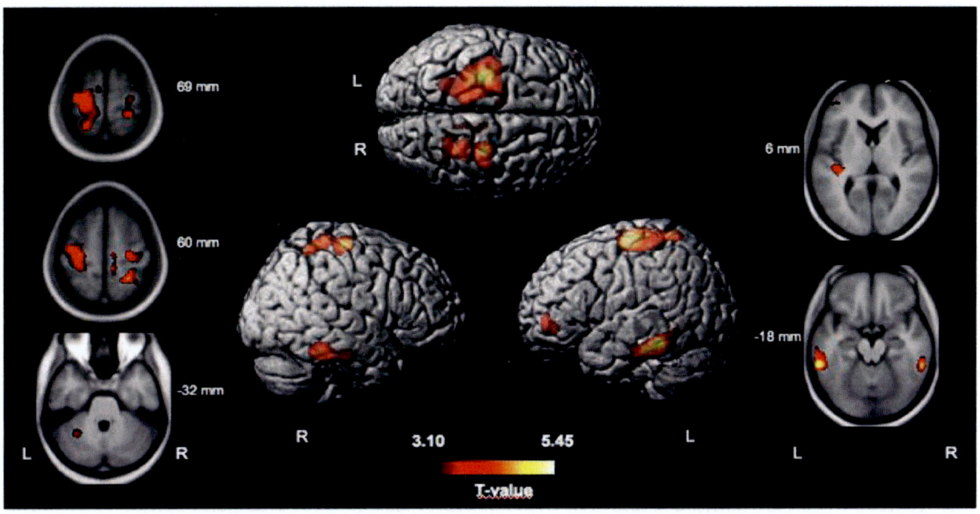

Figure 1. A voxel-based morphometric study comparing gray matter differences between professional musicians, amateur musicians, and non-musicians. The musician status was modeled as a three-level gradation in which professional musicians were ranked highest, amateur musicians were ranked intermediate, and non-musicians were ranked lowest. Only those voxels with a significant positive correlation between musician status and increase in gray matter volume are shown, overlaid on the rendered cortex surface of a single subject.

Most cross-sectional studies comparing musicians with matched non-musicians, have used either age of commencement of musical training or intensity/duration of practice throughout a musician's career as predictors of regional differences suggesting that the longer and the more intense musicians practiced, the more pronounced were the between-group differences (Elbert et al., 1995; Amunts et al., 1997; Schlaug, 2001; Sluming et al., 2002; Schneider et al., 2002; Gaser & Schlaug, 2003, Bangert & Schlaug, 2006). While it may not be surprising to find structural differences in brain regions that subserve skills directly linked to instrumental training and performance (such as the primary sensorimotor cortex that supports the sequential, independent fine motor skill in both hands or the primary auditory cortex that supports auditory discrimination skills), structural differences outside of these primary regions, such as those found in the inferior frontal gyrus, are of particular interest since they may indicate that plasticity can occur in regions that either have control over primary musical functions or serve as multimodal integration regions for musical skills.

There has been much more variability in the functional correlates of these structural brain differences. With regard to the auditory system, studies typically reported a greater functional lateralization and stronger activation of secondary and tertiary auditory brain regions in musicians, whereas non-musicians might have stronger activation of primary auditory regions (Gaab et al., 2003). These effects have also been shown in short-term training studies, both in adult non-musicians and in young children using auditory evoked potentials (Besson et al., 1994; Bosnyak et al., 2004; Shahin et al., 2003, 2004; Trainor et al., 1999). Most studies comparing musicians with non-musicians in simple or complex motor tasks have found that musicians show less activation or more focused activation of the primary motor cortex, but also motor association regions such as premotor and supplementary motor areas (Jaencke et al., 2000; Meister et al., 2005; Haslinger et al., 2004). However, musicians seem to show a co-activation of auditory and motor regions or auditory-motor integration regions when they listen to musical excerpts that they know how to play or when they perform a motor task that resembles playing a musical instrument (Bangert et al., 2006; Lahav et al., 2007).

Further evidence for the brain plasticity hypothesis to explain the musician – non-musician differences, comes from comparisons of musician groups that play different instruments. Pantev and colleagues (2001) found more pronounced cortical responses to trumpet and string tones in the respective players of these instruments, demonstrating that functional brain differences are associated with the particular musical instrument played (Pantev et al. 2001). In the absence of longitudinal data to examine causal relationships, the search for structural brain differences within the heterogeneous group of musicians is a valuable alternative to support the notion of structural brain adaptation. For example, string-players strongly differ from keyboard players in their requirement fine hand/finger motor skills. While string players need to develop fine motor skills of their left hand, keyboard players develop fine hand/finger motor skills in both hands or more in their right hand, since the left hand in keyboard music has more the function of an accompaniment. When comparing string and keyboard players, Bangert and Schlaug (2006) found gross anatomical differences in the precentral gyrus (Figure 2). The majority of the adult keyboard players had an elaborated configuration (called 'Omega Sign') of the precentral gyrus on the left more than right hemisphere, while most of the adult string players had this "atypicality" only on the right.

Bangert and Schlaug (2006) calculated an Asymmetry Index to indicate the differential hemispheric expression of the Omega Sign within and between the different subject groups: In the string players, an AI ≤ -1 (right > left) appeared more frequently than in the controls. The opposite effect was seen in the keyboard players. These results indicate within-musician differences in the Omega Sign expression suggesting an association between acquired sensorimotor skills and features of external brain anatomy.

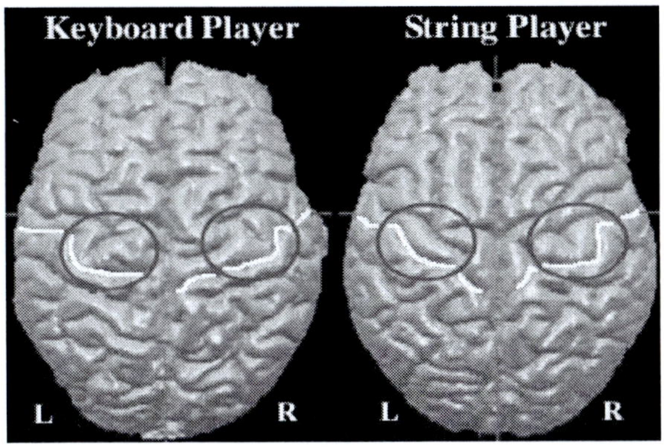

Figure 2. Surface renderings of one typical keyboard and one typical string player. The central sulcus is marked in white. The part of the precentral gyrus that contains the configuration that is similar to the inverted Greek Omega letter is highlighted with red circles. In the two examples above, a prominent omega sign can be seen on the left more than on the right in the keyboard player and on the right only in the string player.

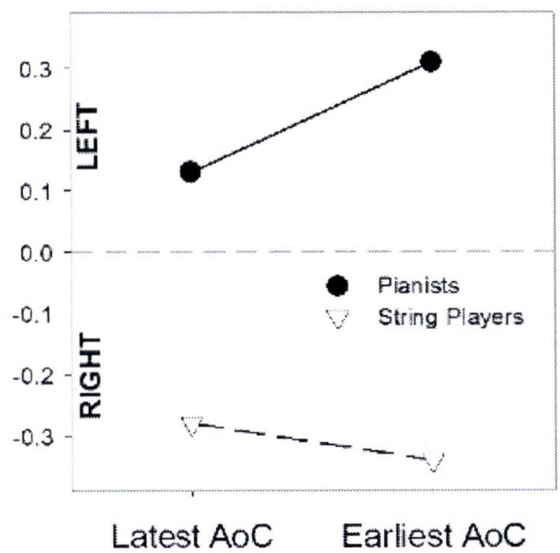

Figure 3. Asymmetry Index from an Omega Sign rating study, comparing musicians with an early Age of Commencement of musical training (AoC = 3.5 yrs, pianists, 4.4 yrs, string players, n=4 each) to musicians with a late Age of Commencement of musical training (AoC = 7.3 years, pianists, 9.1 years, string players, n=4 each).

Although a genetic cause for the Omega Sign expression (i.e., a differential hemispheric expression in this strongly right-handed group of musicians favoring string playing in those individuals that have a prominent Omega Sign in their right hemisphere and favoring keyboard playing in those individuals that have a prominent Omega Sign in their left hemisphere) cannot be completely ruled out, a correlation of the degree of Omega Sign expression with the commencement of musical training (Figure 3), as well as with the cumulative lifetime practice time strongly argues for a structural plasticity mechanism. Considering the wealth of animal (Kleim et al., 1996, Anderson et al., 2002) and human data (Hutchinson et al., 2002; Bengtsson et al., 2005) showing structural changes in response to long-term motor training, these findings indicate differential hemispheric adaptations depending on the instrument played.

There is evidence suggesting that the atypicalities in musicians' brains are not innate but are instead the result of training, since these atypicalities are greater in musicians who began study at a younger age (Elbert et al., 1995; Schlaug et al., 1995) and who practiced with more intensity (Gaser & Schlaug, 2003; Hutchinson et al., 2003; Schneider et al., 2002). Animal studies also support the training-associated brain plasticity view (e.g., Anderson et al., 2002; Dawson et al., 2000; Greenough & Black, 1992). Micro-structural changes (increases in number of synapses, glial cells, and capillary density within the cerebellum and primary motor cortex, as well as new brain cells in the hipppocampus) occur after long-term motor training in rats (Anderson et al., 1994; Black et al., 1990; Isaacs et al., 1992; Kempermann et al., 1997; Kleim et al., 1996; van Praag et al., 1999). The sum of these changes could amount to structural differences detectable on a macro-structural level (see Anderson et al., 2002; Bangert & Schlaug, 2006).

Preliminary Results of a Longitudinal Study in Children Learning to Play a Musical Instrument

To determine whether the structural and functional differences seen in musicians reflect adaptations due to musical training during sensitive periods of brain development, or are instead markers of musical interest and/or aptitude that existed prior to training, it is necessary to examine children and/or adults before the onset of instrumental music training and compare them to a group of control subjects not planning to play and practice regularly on a musical instrument. Over the last years we have been conducting a longitudinal study of the effects of music training on brain development and cognition in young children (Norton et al., 2005; Koelsch et al., 2005; Overy et al., 2004; Schlaug et al., 2003). We have tested a large group of five- to seven-year-old children at baseline prior to music lessons, approximately two-thirds of the children chose to take piano and one-third chose string lessons. We also tested a smaller, untreated control group matched to the instrumental group in age, socioeconomic status (SES), and verbal IQ. Each child underwent a battery of tests including subtests of the Wechsler Intelligence Scale for Children (WISC-III) (for children six years and older) or the Wechsler Preschool and Primary Scale of Intelligence (WPPSI-III) (for children under age six), the Raven's Progressive Matrices tests, auditory discrimination tests, and two motor tests (a finger tapping test and a finger sequencing task described in detail by Walker et al., 2002) as a measure of speed and dexterity in both right and left hands.

Children also underwent a structural and functional magnetic resonance (MR) scan of their brain using a child appropriate magnetic resonance imaging (MRI) protocol. At baseline, we found no pre-existing cognitive, auditory perceptual, motoric, or structural brain differences between children that entered the instrumental and children that entered the control group (Norton et al., 2005), thereby making it unlikely that children who choose to learn an instrument do so because they have an atypical brain, and suggesting that the brain atypicalities seen in adult musicians are most likely the product of intensive music training rather than pre-existing biological markers of musicality. Table 1 shows the mean brain measures. There were no differences in the absolute brain volume, gray matter volume, white matter volume or the midsagittal corpus callosum size. A voxel-based analysis did not show any significant between-group differences in gray or white matter volume or midsagittal corpus callosum size.

Table 1. Morphometric Brain Measures of 5-7 year old children

	Total Brain Volume [cc]	Gray Matter Volume [cc]	White Matter Volume [cc]	Corpus Callosum size [mm^2]
Instrumental Group	1033 (94)	683 (67)	348 (31)	528 (79)
Non-Instrumental Group	1032 (116)	681 (88)	344 (32)	512 (75)

After an average follow-up of 16 months we found significantly greater change scores in the instrumental group than in the control group in skills directly linked to instrumental music training: fine motor skills and auditory discrimination skills including rhythmic and melodic discrimination. No significant between-group differences were seen yet in non-primary domains such as verbal skills, visuospatial skills, and mathematical reasoning. However, we found trends for between-group differences in several of these extra-primary domains (e.g., verbal skills, math skills, visual-pattern skills), suggesting that a longer period of observation might lead to significant between-group differences in these transfer domains. Brain imaging data also support this trend. There was a non-significant trend for a greater increase in corpus callosum size and white matter volume in the instrumental group than in the control group.

In addition to the longitudinal study of children learning to play a musical instrument, we also have preliminary data on a cross-sectional study of nine- to eleven-year-old instrumentalists with an average of four to five years of training and a group of non-instrumental children matched in age, handedness, and SES. Preliminary results from this cross-sectional study showed that the instrumentalists performed significantly better in skills directly linked to instrumental music training: fine motor skills and auditory discrimination skills including rhythmic and melodic discrimination. In addition, we found between-group differences in other domains such as verbal skills and visual pattern matching skills with the instrumentalists performing better in these tests. The nine- to eleven-year-old instrumental group had significantly more gray matter volume that was regionally pronounced in the sensorimotor cortex, but also in the occipital lobe bilaterally (Schlaug et al. 2004). Table 2 shows the average (SD) volume data for total brain, gray matter, and white matter volume. These between-group differences suggest that four to five years of instrumental training might lead to some of the between-group differences that we have seen in comparisons of adult musicians and non-musicians.

Table 2. Morphometric Brain Measures of 9-11 year-old Children

	Total Brain Volume [cc]	Gray Matter Volume [cc]	White Matter Volume [cc]
Instrumental Group	1105 (99)	747 (75)	357 (39)
Non-Instrumental Group	1015 (133)	661 (82)	354 (36)

FUNCTIONAL BRAIN CORRELATES OF MUSIC LISTENING AND MAKING

Musicians and non-musicians differ in their functional brain network that is activated while listening to music. It appears that musicians activate a much more extensive network of brain regions in the parietal and frontal lobe in addition to pure perceptual regions within the temporal lobe even if the required task only involves perceptual processing. It is possible that musicians co-activate frontal regions with any auditory-music input because frontal regions play a role in auditory-motor mapping and in auditory sequencing. The activation of parietal regions might be related to the multimodal nature of music-perceptual information and the integration of auditory information with other sensory information, irrespective of whether or not this information is perceived or imagined. Figure 4 shows a comparison between two groups of adults (musicians and non-musicians) and a group of five- to seven-year-old children, all performing the same rhythm discrimination task while brain activity is being measured with functional MR imaging. Normal maturational changes in the processing of this information can be seen in the comparison between the adult non-musicians and the children, but additional modulations of these maturational changes can be seen in several regions when the group of adult musicians is compared with both the children's group and the adult non-musician group. Those additional regions include the parietal lobe (mainly involving the region of the intra-parietal sulcus), the posterior middle frontal gyrus (Brodmann area 6), and the inferior frontal gyrus (Brodmann areas 44, 45, and 47). In this chapter, we will concentrate on the middle and inferior frontal gyrus, since those regions appear to show the greatest difference between adult musicians and non-musicians. The parietal lobe activation can be seen in both groups of adults. This activation may indicate that with increased experience (either implicit or explicit), regions in the parietal lobe may become active due either to poly-sensory integration of information or the use of global processing strategies in adults as compared to children who may apply more local processing strategies.

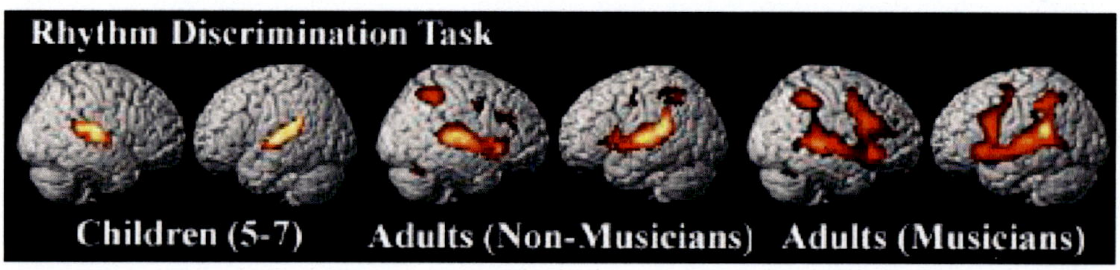

Figure 4. Statistical parametric images superimposed onto a surface rendering of a standardized anatomical brain depicting significant activations during a rhythmic discrimination task in a group of five- to seven-year-old, musically-naïve children, adult non-musicians, and adult musicians.

Time Course of Musical Training – Slow and Rapid Plasticity

Sometimes the only way to clarify which of the musicians' brains' peculiarities are actually induced by practice, is to watch brain plasticity from the very beginning, that is, during learning to play an instrument from scratch. This provides a unique opportunity to tell apart the different parameters that form the network required for playing an instrument. The different contributing parameters are sight-reading, visual-auditory transfer (knowing how a musical score will sound like), anticipatory auditory imagery, motor planning and execution, audiomotor feedback control, visuomotor control, analysis of various somatosensory feedback channels. To train musically naïve subjects to play a musical instrument allows for either selective training (for example, only one of the above feedback modalities), or for cutout training (on the full range of involved modalities, but with selective control of one parameter between groups).

Pascual-Leone et al. (1995) used focal transcranial magnetic stimulation (TMS) to map the motor cortical areas targeting long finger flexor and extensor muscles bilaterally. They showed that as subjects learned a five-finger exercise on the piano over the course of five days under auditory feedback, the cortical representation area targeting the long finger flexor and extensor muscles significantly enlarged (Pascual-Leone et al., 1995; Pascual-Leone, 2001). However, this increase could be demonstrated only when the cortical mapping studies were conducted following a 20- to 30-minute rest after the practice (and test) session. No such modulation in the cortical output maps was noted when maps were obtained before each daily practice session (Pascual-Leone et al., 1999).

Karni et al. (1998) provided evidence that a few minutes of daily practice of a finger opposition sequence induced large, incremental performance gains over a few weeks of training, which was associated with changes in cortical movement representation within M1. The authors suggested that these changes reflected the initial set-up of a task-specific motor program, while the subsequent changes indicated the consolidation of this motor program.

Stewart et al. (2003) focused on the acquisition of sight-reading skills. These authors scanned musically naïve participants before and after a period of 15 weeks during which subjects were taught to read music and play the keyboard. When participants played melodies from musical notation after training, activation was seen in the superior parietal cortex bilaterally. These authors suggest that music reading involves the automatic sensorimotor translation of a spatial code (written music) into a series of motor responses (key presses).

Bangert and Altenmueller (2003) presented an experiment that involved training naïve subjects on a double-dissociated piano task employing only auditory and motor aspects of performing music. They demonstrated that although the learning curve and the respective brain activations kept developing over weeks and months of practice, the fundamental cortical network layout required for the mastering of the instrument emerged after only 15-20 minutes of initial training.

Motor Learning: Memory Consolidation Processes and Practice Strategies

It is known that excellent musicians not only rehearse their musical skills by daily practice, but also use mental imagery strategies. For example, Horowitz used to practice mentally before playing in concerts in order to avoid the feedback of a piano other than his own. Rubinstein did so in order to make daily practice as efficient as possible (Schoenberg, 1987).

The main component of mental rehearsal is motor imagery, i.e. a mental simulation of action, which involves many parts of the sensorimotor system that are active during execution. It therefore provides a possibility for the central nervous system to evaluate the consequences of future actions and to prepare the actual execution properly (Jeannerod, 2001). The relationship between motor imagery and motor action can be deducted from the observation that the timing patterns of both processes are similar (Sirigu et al., 1995), and that the changes in corticospinal excitability involve the same muscles (Fadiga et al., 1999). Functional imaging studies on motor imagery vs. performance in non-musicians have shown that imagery and performance activate essentially the same cortical regions (Porro et al., 1996; Lotze et al., 1999; Gerardin et al., 2000), with the exception, however, of the primary motor cortex. Results regarding the involvement of the primary motor cortex in motor imagery differ across studies, although more studies have shown no activation of primary motor areas.

Meister and collaborators (2004) compared music performance to music imagery in professional musicians with fMRI. During imagery, bilateral activation in the premotor areas, the precuneus and the medial part of BA 40 was found. The only missing brain response (when compared to actual execution) was the contralateral M1 and bilateral posterior parietal cortex (PPC). While mental practice strategies are widely known among musicians and music teachers today, other research that might provide important contributions to the pedagogy of instrumental practice, is not. An example may be the impact of sleep on skill consolidation. After practice, the practiced skill continues to develop off-line during a period of consolidation. This is true for music-perceptual (Gaab et al., 2004) as well as for musical motor tasks (Walker et al., 2005). Storing a motor skill during sleep reorganizes its brain representation toward enhanced efficacy (Fischer et al., 2005). There is evidence for several double dissociations of distinct features of the practiced skill that are consolidated best during specific wake or sleep time windows; for example, different aspects of a procedural memory are processed separately during consolidation.

Representational Maps of the Music Instrument through Links between Functional Areas

Several imaging studies have found activations or activity changes in a dorsal posterior premotor region and in a posterior inferior frontal region with purely perceptual discrimination tasks, combined perceptual-memory tasks, and auditory-motor association tasks (Ohnishi et al., 2001; Zatorre et al., 1998; Gaab et al., 2003, 2006; Bangert & Altenmueller, 2003; Haslinger et al., 2004; Hasegawa et al., 2004; Meister et al., 2004).

It is of particular interest that two, independent voxel-based morphometric studies found more gray matter volume in the inferior frontal gyrus in musicians compared with non-musicians (Sluming et al., 2002; Gaser & Schlaug, 2003; see also Figure 1), although the functional significance of this frontal "auditory" region is still unclear. It has been speculated that the inferior frontal gyrus might play a role in performing tasks that have to do with the sequencing of behaviorally relevant auditory information (Koelsch et al., 2002, 2005; Gaab et al., 2003). Furthermore, they may become active when the integration of information, particularly in the temporal domain, is important for solving a task (Levitin et al., 2003).

Years of practice establish a neuronal correlate of a auditory-sensorimotor connection that has recently been shown by brain imaging studies for the auditory-to-motor (Haueisen & Knoesche, 2001; D'Ausilio et al., 2006), motor-to-auditory (Lotze et al., 2003), and both directions (Bangert & Altenmueller, 2003; Bangert et al., 2006). Similar co-activation phenomena are known from the speech literature, where the classical notion of a functional dissociation of speech perception and speech production has recently been adjusted towards a joint sensorimotor representation (Aboitiz & Garcia, 1997; Watkins et al., 2003; Watkins & Paus, 2004).

To investigate cortical auditory and motor coupling in professional musicians, Bangert and colleagues (2006) compared the functional magnetic resonance imaging (fMRI) activity of seven pianists to seven non-musicians utilizing a passive task paradigm established in a previous learning study (Bangert & Altenmueller, 2003). The tasks involved either passively listening to short piano melodies or pressing keys on a mute MRI-compliant piano keyboard. The professional pianists showed increased activity compared to the non-musicians in a distributed cortical network during both the acoustic and the mute motion-related task (Figure 5). A conjunction analysis (Figure 5, right column) revealed a distinct musicianship-specific network being co-activated during either task type, indicating areas involved in auditory-sensorimotor integration. This network is comprised of dorsolateral and inferior frontal cortex (including Broca's area), the superior temporal gyrus (Wernicke's area), the supramarginal gyrus, and supplementary motor and premotor areas.

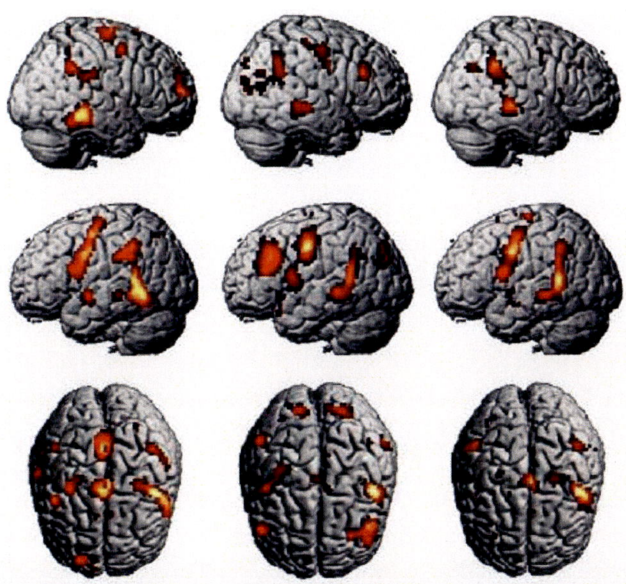

Figure 5. Statistical parametric images superimposed onto a surface rendering of a standardized anatomical brain showing the pianist > non-musicians contrast during three tasks. Left row: auditory task, middle row: motion-related task, right row: conjunction of auditory and motor tasks (for more details see Bangert et al., 2006).

Mirror Systems

A variety of tasks have been shown to activate the inferior frontal gyrus or Broca's area in specific. Broca's area is part of a frontoparietal network that was first described in monkeys and is referred to as mirror neuron system. The so-called "mirror neurons" appear to respond whenever the idea of an action is presented – either visually (observing actions performed by others) or auditorily (listening to action-related sound) or actually executed motorically. Music listening and music making seem to be a model framework for studying functions of the human mirror neuron system. Music involves inherent neural coupling between perception and action — we listen to music and respond to it with motion; play music and watch others performing it; communicate emotions via music and enjoy listening to it. Music making and music listening are unique among possible experimental probes to examine the mirror neuron system because they could engage all possible functions of the system, particularly "seeing-doing" and "hearing-doing" circuits. Music is a strong multimodal stimulus that simultaneously transmits visual, auditory, and motoric information to a specialized mirror neuron network consisting of fronto-parietal brain regions.

Research with monkeys has revealed a neural mechanism that may be responsible for representing associations between actions and sensory stimulation. Rizzolatti and colleagues identified a class of neurons in the rostral ventral premotor cortex (area F5) of monkeys, the presumed homologue of Brodmann Area (BA) 44 and BA 6, that is tuned to the actions of others as well as actions of the self (Buccino et al., 2004; Rizzolatti et al., 2002). These so-called "mirror" neurons respond both when an action is observed and when that same action is performed.

In addition to the sight and performance mirror neurons, a subset of mirror neurons in monkeys also responds to the sound of an action (Keysers et al., 2003; Kohler et al., 2002). Multimodal neurons have been described in several cortical areas and subcortical centers. These neurons, however, responded to specific stimulus locations or directions of movement. The difference with the neurons described by Keysers et al. (2003) is that they do not code space, or some spatial characteristics of stimuli, but actions when they are only heard. These audiovisual mirror neurons therefore represent actions independently of whether these actions are performed, heard or seen. Another remarkable property of audiovisual mirror neurons is that about half of them respond with a similar intensity of discharge whether the action is only heard, only seen or both heard and seen. This finding is important, as it suggests that the neurons code the action in an abstract way, which does not depend on the source of information (auditory or visual) from which the evidence about the presence of the action is taken (Keysers et al., 2003). These "auditory-visual" mirror neurons exemplify high-level abstraction in the representation of action - an identical neural system becomes activated whether a particular action is heard, seen or performed.

Additional support for this physiological finding comes from cytoarchitecture studies (Petrides & Pandya, 1999; 2002; Romanski et al., 1999). Although the areas in question have complex multimodal input, they could be differentiated in terms of some of their inputs. In fact, the 'mirror' functionality of area F5 has led researchers to consider this region the monkey's precursor of Broca's area (Rizzolatti & Arbib, 1998).

Not only visual and auditory-visual mirror neurons have been reported, but also an exclusively auditory responsive domain with high selectivity of these neurons for acoustic stimuli. (Romanski & Goldman-Rakic, 2002). The ability to develop associations between actions and sounds is essential for the production and comprehension of music, or any kind of behaviorally relevant sound information. Several research studies have implicated parts of the frontal cortex as important regions for integrating information from multiple sensory modalities and brain regions (Bremmer et al., 2001; Keysers et al., 2003; Kohler et al., 2002). It is therefore likely that the frontal cortex plays an important role in developing associations across modalities, including action-sound mappings (Bangert & Altenmueller, 2003).

The action representation maintained by mirror neurons has been shown to evolve through learning. The visual stimulus that elicits activity in a particular neuron can change with exposure to new stimuli involving the identical action goal (Rizzolatti & Arbib, 1998). It may be inferred that associations between sound and action are similarly malleable. Schubotz and colleagues hypothesize that Hebbian learning drives a somatotopic organization of premotor brain activity for processing meaningful temporal sequences (Schubotz & von Cramon, 2002; Schubotz et al., 2003). They found that for an identical task involving the anticipation of temporal order, stimuli associated with vocal production (sine tones) generated activation of the inferior ventral premotor cortex whereas stimuli associated with hand manipulation (circles of varying diameter) generated activation of the superior ventral premotor cortex (Schubotz & von Cramon, 2002).

Three studies in musicians point towards a musical mirror system in humans, one using auditory (Ohnishi et al. 2001), the second using visual (Hasegawa et al., 2004), and the third using auditory and visual action-observation (Haslinger et al., 2005). In particular, the inferior portions of the fronto-parieto-temporal activation might, as Haslinger et al. (2005) have speculated, reflect the operation of a visual 'mirror matching' system. In the auditory domain, this is supplemented by the studies mentioned above, and by paradigms that employ only passive listening without accompanying visual observation of piano movements (Bangert et al., 2006; Lahav et al., 2007). The distributions of activations in both auditory and visual 'musical action observation studies' look strikingly similar. The audio-visual or visuo-motor networks may therefore serve as audio-motor networks at the same time.

Conclusions

Making music does not only engage primary auditory and motor regions, but also regions that integrate auditory actions/sounds with motor actions and map actions to sounds. Professional singers learn and continuously practice to associate vocal articulation/pharyngeal movements with meaningful patterns in sound. Similarly, professional keyboard players learn and continuously practice to associate hand/finger movements with meaningful patterns in sound. There is some evidence for the second assumption, since Bangert and Altenmueller (2003) have shown that learning associations between actions and piano tones will lead to activity changes in the frontal/premotor cortex. More research is needed to determine whether parts of the frontal/premotor cortex are a general-purpose locus for processing sequences of sound events that are meaningfully related. This hypothesis would certainly explain a majority of frontal activations that have been seen with pure auditory perceptual tasks, auditory memory tasks, auditory discrimination tasks, and auditory-motor association tasks (Zatorre et al., 1994; Koelsch et al., 2002; Levitin & Menon, 2003; Gaab et al., 2003; Bangert & Altenmueller, 2003). Thus, considering the putative functions of the inferior frontal regions, including its hypothesized role as a supramodal hierarchical processor (Tettamanti & Weniger, 2006) or as a hub of the mirror neuron system (Lahav et al., 2007), it might not be surprising to see that some of the most pronounced differences between musicians and non-musicians, both structurally and functionally, are in the inferior frontal region.

Acknowledgements

The research discussed in this chapter was supported by grants from the National Science Foundation (NSF), the International Foundation for Music Research (IFMR), the German Research Foundation (DFG) SPP1001, the BMBF-grant 01GO0202 (CAI), and the Grammy Foundation. Christian Gaser, Andrea Norton, Katie Overy, Ellen Winner, Martin Kotynek-Friedl, DJ Lee, Monika Ellenberg, Hauke Egermann, Ilona Wiedenhöft, Kerstin Möhlenbrink, Martin Kanowski, Eckart Altenmueller, Thomas Peschel and Stefan Koelsch have contributed to the research presented in this chapter.

References

Aboitiz, F. & Garcia, R. (1997). The anatomy of language revisited. *Biological Research, 30(4)*, 171-183.

Amunts, K., Schlaug, G., Jaencke, L., Dabringhaus, A., Steinmetz, H., Schleicher, A., Zilles, K. (1997). Motor cortex and hand motor skills: structural compliance in the human brain. *Human Brain Mapping, 5*, 206-215.

Anderson, B.J., Eckburg, P.B., & Relucio, K.I. (2002). Alterations in the thickness of motor cortical subregions after motor skill learning and exercise. *Learning and Memory, 9*, 1-9.

Anderson, B.J., Li, S., Alcantara, A., Isaacs, K.R., Black, J.E., & Greenough, W.T. (1994). Glial hypertrophy is associated with synaptogenesis following motor-skill learning but not with angiogenesis, following exercise. *Glia, 11*, 73-80.

Backhaus, J. & Junghanns, K. (2006). Daytime naps improve procedural motor memory. *Sleep Medicine, 7(6),* 508-512.

Bangert, M. & Schlaug, G. (2006) Specialization of the Specialized in Features of External Human Brain Morphology. *European Journal of Neuroscience, 24*, 1832–1834.

Bangert, M., Peschel, T., Rotte, M., Drescher, D., Hinrichs, H., Schlaug, G., Heinze, H.J., & Altenmueller, E. (2006). Shared networks for auditory and motor processing in professional pianists: evidence from fMRI conjunction. *NeuroImage, 30(3),* 917-926.

Bangert, M., & Altenmueller, E.O. (2003). Mapping perception to action in piano practice: a longitudinal DC-EEG study. *BMC Neuroscience, 4*, 26.

Bengtsson, S.L., Nagy, Z., Skare, S., Forsman, L., Forssberg, H., & Ullen, F. (2005). Extensive piano practicing has regionally specific effects on white matter development. *Nature Neuroscience 8(9)*, 1148-1150.

Besson, M., Gaita, F., & Requin, J. (1994). Brain waves associated with musical incongruities differ for musicians and non-musicians. *Neuroscience Letters, 168*, 101-105.

Black, J.E., Isaacs, K.R., Anderson, B.J., Alcantra, A.A., Greenough, W.T. (1990). Learning causes synaptogenesis whereas motor activity causes angiogenesis in cerebellar cortex of adult rats. *Proceedings of the National Academy of Sciences United States of America, 87*, 5568-5572.

Bosnyak, D.J., Eaton, R.A., & Roberts, L.E. (2004). Distributed auditory cortical representations are modified when non-musicians are trained at pitch discrimination with 40 Hz amplitude modulated tones. *Cerebral Cortex, 14*, 1088-1099.

Bremmer, F., Schlack, A., Shah, J., Zafiris, O., Kubischik, M., Hoffman, K.P. et al. (2001). Polymodal motion processing in posterior parietal and premotor cortex: A human fMRI study strongly implies equivalencies between humans and monkeys. Neuron, 29, 287-296.

Buccino, G., Binkofski, F. & Riggio, L. (2004). The mirror neuron system and action recognition. *Brain and Language, 89*, 370-376.

Charness, M. & Schlaug, G. (2000). Cortical activation during finger movements in concert pianists, dystonic pianists, and non-musicians. *Neurology 54*, A221.

Cohen, D.A., Pascual-Leone, A., Press, D.Z., & Robertson, E.M. (2005). Off-line learning of motor skill memory: A double dissociation of goal and movement. *Proceedings of the National Academy of Sciences, 102 (50)*, 18237–18241.

D'Ausilio, A., Altenmueller, E., Olivetti Belardinelli, M., & Lotze, M. (2006). Cross-modal plasticity of the motor cortex while listening to a rehearsed musical piece. *European Journal of Neuroscience 24*, 955–958.

Dawson, G., Ashman, S.B. & Carver, L.J. (2000). The role of early experience in shaping behavioral and brain development and its implications for social policy. *Development and Psychopathology, 12*, 695-712.

Elbert, T., Pantev, C., Wienbruch, C., Rockstroh, B. & Taub, E. (1995). Increased cortical representation of the fingers of the left hand in string players. *Science*, 270, 305-306.

Fadiga, L., Buccino, G., Craighero, L., Fogassi, L., Gallese, V., & Pavesi, G. (1999). Corticospinal excitability is specifically modulated by motor imagery: a magnetic stimulation study. *Neuropsychologia 37*, 147–158.

Fischer, S., Nitschke, M.F., Melchert, U.H., Erdmann, C., Born, J. (2005). Motor Memory Consolidation in Sleep Shapes More Effective Neuronal Representations. *Journal of Neuroscience 25(49)*, 11248 –11255.

Gaab, N., Paetzold, M., Becker, M., Walker, M.P., & Schlaug, G. (2004). The influence of sleep on auditory learning: a behavioral study. *Neuroreport 15*, 731-734.

Gaab, N. & Schlaug, G. (2003). The effect of musicianship on pitch memory in performance matched groups. *Neuroreport, 14*, 2291-2295.

Gaab, N., Gaser, C., Zaehle, T., Jaencke, L. & Schlaug, G. (2003). Functional anatomy of pitch memory – an fMRI study with sparse temporal sampling. *Neuroimage, 19*, 1417-1426.

Gaser, C., & Schlaug, G. (2003). Brain structures differ between musicians and non-musicians. *Journal of Neuroscience, 23*, 9240-9245.

Gerardin, E., Sirigu, A., Lehericy, S., Poline, J.B., Gaymard, B., Marsault, C., Agid, Y., Le Bihan, D. (2000). Partially overlapping neural net-works for real and imagined hand movements. *Cerebral Cortex 10,* 1093– 1104.

Greenough, W.T. & Black, J.E. (1992). Induction of brain structure by experience: Substrates for cognitive development. In M.R.Gunnar & C.A.Nelson (Eds.). *The Minnesota Symposia on Child Psychology: Vol 24. Developmental behavioral neuroscience* (pp. 155-200). Hillsdale, NJ: Erlbaum.

Hasegawa, T., Matsuki, K., Ueno, T., Maeda, Y., Matsue, Y., Konishi, Y., & Sadato, N. (2004). Learned audio-visual cross-modal associations in observed piano playing activate the left planum temporale. An fMRI study. *Cognitive Brain Research, 20*, 510-518.

Haslinger, B., Erhard, P., Altenmueller, E., Hennenlotter, A., Schwaiger, M., Einsiedel, H., Rummeny, E., Conrad, B., Ceballos-Baumann, A.O. (2004). Reduced recruitment of motor association areas during bimanual coordination in concert pianists. *Human Brain Mapping, 22*, 206-215.

Haueisen, J. & Knoesche, T.R.(2001): Involuntary motor activity in pianists evoked by music perception. *Journal of Cognitive Neuroscience, 13(6),* 786-792.

Hund-Georgiadis, M. & von Cramon, Y. (1999). Motor-learning-related changes in piano players and non-musicians revealed by functional magnetic-resonance signals. *Experimental Brain Research, 125*, 417-425.

Hutchinson, S., Lee, L.H.L., Gaab, N. & Schlaug, G. (2003). Cerebellar volume of musicians. *Cerebral Cortex, 13*, 943-949.

Isaacs, K.R., Anderson, B.J., Alcantara, A.A., Black, J.E., & Greenough, W.T. (1992). Exercise and the brain: angiogenesis in the adult rat cerebellum after vigorous physical activity and motor skill learning. *Journal of Cerebral Blood Flow and Metabolism 12*,110-119.

Jacobs, K.M. & Donoghue, J.P. (1991). Reshaping the cortical motor map by unmasking latent intracortical connections. *Science 251*, 944–947.

Jeannerod, M. (2001). Neural simulation of action: a unifying mechanism for motor cognition. *NeuroImage 14*, S103–S109.

Karni, A., Meyer, G., Rey-Hipolito, C. et al. (1998). The acquisition of skilled motor performance: fast and slow experience-driven changes in primary motor cortex. *Proceedings of the National Academy of Sciences of the USA 95*, 861–868.

Karni, A., Meyer, G., Jezzard, P. et al. (1995). Functional MRI evidence for adult motor cortex plasticity during motor skill learning. *Nature 377*, 155–158.

Keenan, J.P., Thangaraj, V., Halpern, A.R., & Schlaug, G. (2001). Absolute pitch and planum temporale. *Neuroimage, 14*, 1402-1408.

Kempermann, G., Kuhn, H.G., & Gage, F.H. (1997). More hippocampal neurons in adult mice living in an enriched environment. *Nature 386*, 493-495.

Keysers, C., Kohler, E., Umiltà, M.A., Nanetti, L., Fogassi, L., & Gallese, V. (2003). Audiovisual mirror neurons and action recognition. *Experimental Brain Research, 153*, 628–636.

Kleim, J.A., Lussnig, E., Schwarz, E.R., Comery, T.A., & Greenough, W.T. (1996). Synaptogenesis and fos expression in the motor cortex of the adult rat after complex motor skill acquisition. *Journal of Neuroscience, 16*, 4529-4535.

Kohler, E., Keysers, C., Umiltà, M.A., Fogassi, L., Gallese, V., & Rizzolatti, G. (2002). Hearing sounds, understanding actions: Action representation in mirror neurons. *Science, 297*, 846-848.

Koelsch, S., Gunter, T.C., v Cramon, D.Y., Zysset, S., Lohmann, G., & Friederici, A.D. (2002). Bach speaks: a cortical "language-network" serves the processing of music. *Neuroimage 17(2*), 956-966.

Koelsch, S., Fritz, T., Schulze, K., Alsop, D., & Schlaug, G. (2005). Adults and children processing music: an fMRI study. *Neuroimage, 25*, 1068-1076.

Lahav, A., Saltzman, E., Schlaug, G. (2007). Action representation of sound: audiomotor recognition network while listening to newly-acquired actions. *Journal of Neuroscience, 27,* 308-314.

Levitin, D.J. & Menon, V. (2003). Musical structure is processed in language areas of the brain: a possible role for Brodmann Area 47 in temporal coherence. *Neuroimage, 20*, 2142-2152.

Liberman, A. (1995). *Speech: A special code* (learning, development, and conceptual change series). CIT Press.

Lotze, M., Montoya, P., Erb, M., Hulsmann, E., Flor, H., Klose, U., Birbaumer, N., & Grodd, W. (1999). Activation of cortical and cerebellar motor areas during executed and imagined hand movements: an fMRI study. *Journal of Cognitive Neuroscience, 11*, 491–501.

Lotze, M., Scheler, G., Tan, H.R., Braun, C., & Birbaumer, N. (2003). The musician's brain: functional imaging of amateurs and professionals during performance and imagery. *Neuroimage 20(3),* 1817-29.

Meister, I.G., Krings, T., Foltys, H., Boroojerdi, B., Muller, M., Topper, R., & Thron, A. (2004). Playing piano in the mind – an fMRI study on music imagery and performance in pianists. *Brain Research. Cognitive Brain Research, 19*, 219-228.

Merzenich, M.M., Recanzone, G., Jenkins, W.M. et al. (1988). Cortical representational plasticity. In P. Rakic & W. Singer (Eds.) *Neurobiology of Neocortex* (pp. 41–47). John Wiley & Sons Limited. New York.

Muente, T.F., Kohlmetz, C., Nager, W., & Altenmueller, E. (2001). Neuroperception: Superior auditory spatial tuning in conductors. *Nature, 409*, 580.

Norton, A., Winner. E., Cronin, K., Overy, K., Lee, D. & Schlaug, G. (2005). Are there pre-existing neural, cognitive, or motoric markers for musical ability? *Brain and Cognition, 59,* 124-134.

Ohnishi, T., Matsuda, H., Asada, T., Aruga, M., Hirakata, M., Nishikawa, M., Katoh, A., & Imabayashi, E. (2001). Functional anatomy of musical perception in musicians, *Cerebral Cortex, 11*, 754-760.

Overy, K., Norton, A., Cronin, K., Gaab, N., Alsop, D., Winner, E., & Schlaug, G. (2004). Imaging melody and rhythm processing in young children. *NeuroReport, 15*, 1723-1726.

Pantev, C., Roberts, L.E., Schulz, M., Engelien, A., & Ross, B. (2001). Timbre-specific enhancement of auditory cortical representations in musicians. *NeuroReport 12(1),* 169-174.

Pantev, C., Oostenveld, R., Engelien, A., Ross, B., Roberts, L.E., & Hoke, M. (1998). Increased auditory cortical representation in musicians. *Nature, 392*, 811-814.

Pascual-Leone, A. (2001). The brain that plays music and is changed by it. *Annals of the New York Academy of Sciences, 930*, 315-329.

Pascual-Leone, A., Nguyet, D., Cohen, L.G., Brasil-Neto, J.P., Cammarota, A., & Hallett, M. (1995). Modulation of muscle responses evoked by transcranial magnetic stimulation during the acquisition of new fine motor skills. *Journal of Neurophysiology, 74(3),* 1037-1045.

Pascual-Leone, A., Tarazona, F., & Catala, M.D. (1999). Applications of transcranial magnetic stimulation in studies on motor learning. *Electroencephalography and Clinical Neurophysiology, Suppl. 51*, 157–161.

Pascual-Leone, A., Grafman, J. & Hallett, M. (1994). Modulation of cortical motor output maps during development of implicit and explicit knowledge. *Science 263*, 1287–1289.

Petrides, M. & Pandya, D.N. (1999). Dorsolateral prefrontal cortex: comparative cytoarchitectonic analysis in the human and the macaque brain and corticocortical connection patterns. *European Journal of Neuroscience, 11*, 1011-1036.

Petrides, M. & Pandya, D.N. (2002). Comparative cytoarchitectonic analysis of the human and the macaque ventrolateral prefrontal cortex and corticocortical connection patterns in the monkey. *European Journal of Neuroscience, 16*, 291-310.

Porro, C.A., Francescato, M.P., Cettolo, V., Diamond, M.E., Baraldi, P., Zuiani, C., Bazzocchi, M., & di Prampero, P.E. (1996). Primary motor and sensory cortex activation during motor performance and motor imagery: a functional magnetic resonance imaging study. *Journal of Neuroscience, 16*, 7688– 7698.

Price, C.J., Wise, R.J., Warburton, E.A., Moore, C.J., Howard, D., Patterson, K., Frackowiak, R.S., & Friston, K.J. (1996). Hearing and saying. The functional neuro-anatomy of auditory word processing. *Brain, 119*, 919-931.

Rizzolatti, G., & Arbib, M.A. (1998). Language within our grasp. *TINS, 21(5)*, 188-194.

Rizzolatti, G., Fogassi, L., & Gallese, V. (2002). Motor and cognitive functions of the ventral premotor cortex. *Current Opinion in Neurobiology, 12*, 149-154.

Romanski, L.M., Bates, J.F., & Goldman-Rakic, P.S. (1999). Auditory belt and parabelt projections to the prefrontal cortex in the rhesus monkey. *Journal of Comparative Neurology, 403*, 141-157.

Romanski, L.M., Goldman-Rakic, P.S. (2002). An auditory domain in primate prefrontal cortex. *Nature Neuroscience, 5*, 15-16.

Schlaug, G. (2001). The brain of musicians: A model for functional and structural adaptation. In R. Zatorre & I. Peretz (Eds.), The biological foundations of music. *Annals of the New York Academy of Sciences, 930*, 281-299.

Schlaug, G., Jaencke, L., Huang, Y., & Steinmetz, H. (1995). In vivo evidence of structural brain asymmetry in musicians. *Science, 267*, 699-701.

Schlaug, G., Lee, D.J., Overy, K., Cronin, K., Norton, A., & Winner, E. (2004). Does brain anatomy predict musicianship? Poster presented at Human Brain Mapping, Budapest, 2004 (published in *Neuroimage, Vol 22S*)

Schlaug, G., Winner, E., Norton, A., Cronin, K., & Overy, K.(2003). Effects of music training on children's brain and cognitive development. *Poster presented at Society for Neurosciences*, New Orleans, November, 2003.

Schneider, P., Scherg, M., Dosch, H.G., Specht, H.J., Gutschalk, A., & Rupp, A. (2002). Morphology of Heschl's gyrus reflects enhanced activation in the auditory cortex of musicians. *Nature Neuroscience, 5*, 688-694.

Schoenberg, H. (1987). *Great pianists*, Fireside Books, St. Louis, USA, 1987.

Schubotz, R.I., & von Cramon, D.Y. (2002). Predicting perceptual events activates corresponding motor schemes in lateral premotor cortex: An fMRI study. *NeuroImage, 15*, 787–796.

Schubotz, R.I., von Cramon, D.Y., & Lohmann, G. (2003). Auditory what, where, and when: a sensory somatotopy in lateral premotor cortex.. *NeuroImage, 20*, 173-185.

Seitz RJ, Roland PE, Bohm C et al. (1990). Motor learning in man: a positron emission tomographic study. *Neuroreport, 1*, 57–60.

Shahin, A., Bosnyak, D.J., Trainor, L.J., & Roberts, L.E. (2003). Enhancement of neuroplastic P2 and N1c auditory evoked potentials in skilled musicians. *Journal of Neuroscience, 23*, 5545-5552.

Shahin, A., Roberts, L.E., Trainor, L.J. (2004). Enhancement of auditory cortical development by musical experience in children. Neuroreport, 15, 1917-1921.

Sirigu, A., Cohen, L., Duhamel, J.R., Pillon, B., Dubois, B., Agid, Y., & Pierrot-Deseilligny, C. (1995). Congruent unilateral impairments for real and imagined hand movements. *NeuroReport 6*, 997– 1001.

Sluming, V., Barrick, T., Howard, M., Cezayirli, E., Mayes, A., & Roberts, N. (2002). Voxel-based morphometry reveals increased gray matter density in Broca's area in male symphony orchestra musicians. *NeuroImage, 17*, 1613-1622.

Stewart, L., Henson, R., Kampe, K., Walsh, V., Turner, R., & Frith, U. (2003). Brain changes after learning to read and play music. *NeuroImage. 20(1),* 71-83.

Tettamanti, M., Weniger, D. (2006). Broca's area: a supramodal hierarchical processor? *Cortex, 42(4),* 491-494.

Trainor, L.J., Desjardins, R.N., & Rockel, C. (1999). A comparison of contour and interval processing in musicians and nonmusicians using event-related potentials. *Australian Journal of Psychology, 51,* 147-153.

Van Praag, H., Christie, B.R., Sejnowksi, T.J., & Gage, F.H. (1999). Running enhances neurogenesis, learning, and long-term potentiation in mice. *Proceedings of the National Academy of Sciences of the USA, 96,* 13427-13431.

Walker, M.P., Stickgold, R., Alsop, D., Gaab, N., & Schlaug, G. (2005). Sleep-dependent motor memory plasticity in the human brain. *Neuroscience 133,* 911–917.

Walker, M.P., Brakefield, T., Morgan, A., Hobson, J.A., & Stickgold, R. (2002). Practice with sleep makes perfect: sleep-dependent motor skill learning. *Neuron, 35*, 205-211.

Wang, X., Merzenich, M.M., Sameshima, K., et al. (1995). Remodeling of hand representation in adult cortex determined by timing of tactile stimulation. *Nature, 378*, 71–75.

Watkins, K. & Paus, T. (2004). Modulation of motor excitability during speech perception: the role of Broca's area. *Journal of Cognitive Neuroscience, 16(6)*, 978-987.

Watkins, K.E., Strafella, A.P., & Paus, T. (2003). Seeing and hearing speech excites the motor system involved in speech production. *Neuropsychologia 41(8)*, 989-994.

Zatorre, R.J., Evans, A.C., & Meyer, E. (1994). Neural mechanism underlying melodic perception and memory for pitch. *Journal of Neuroscience, 14*, 1908-1919.

Zatorre, R.J., Perry, D.W., Beckett, C.A., Westbury, C.F., & Evans, A.C. (1998). Functional anatomy of musical processing in listeners with absolute pitch and relative pitch. *Proceedings of the National Academy of Sciences, 95*, 3172-3177.

In: Neurosciences in Music Pedagogy
Editors: W. Gruhn and F. Rauscher, pp. 121-143
ISBN: 978-1-60021-834-7

Chapter V

Motor Learning and Instrumental Training

Eckart Altenmueller and Gary E. McPherson

Abstract

This chapter presents accessible and relevant information from research literature concerned with motor development, motor control, sensorimotor interaction, and motor coordination involved in musical performance and practice. It describes the neuroanatomy and neurophysiology of motor systems in general and the neural basis of sensorimotor learning in the different stages of motor development. The brain mechanisms of skill acquisition and the effects of music making on brain networks as well as brain structure are explained. Furthermore, insights into the neurobiological correlates of training strategies such as mental practice demonstrate the relationship between mental (auditory) imagery and acoustic perception. Our chapter also discusses the expertise model of musical achievement in terms of the neurosciences and the neuro-hormonal reward mechanisms linked to the neurotransmitter dopamine. Based on these principles, a final section provides implications for teaching and learning musical instruments.

Introduction

Our aim in this chapter is to review current knowledge concerning the brain mechanisms involved in motor development, motor control, sensorimotor interaction, and motor coordination in the context of musical performance and practice. We will describe the neural foundations of training strategies such as mental practice, demonstrate the relation between mental (auditory) imagery and acoustic perception, and discuss the expertise model of musical achievement with respect to its neurobiological background. The principles outlined in the chapter are provided, based on our belief that a basic understanding of the complex neurobiological processes underlying children's training through to expert musical performance will stimulate new insights into the practice and theory of music pedagogy.

There can be no doubt that making music is one of the most demanding tasks for the human central nervous system. It involves the precise execution of very fast and, in many instances, extremely complex physical movements that must be coordinated with continuous auditory feedback.

Practice is required to develop new skills and carry out these complex tasks. During childhood and adolescence, the sensorimotor systems that children utilize to represent and respond to objects and experiences are still maturating and are therefore not fully developed with respect to myelination, structure of the nervous tissue and nerve conduction velocities. During childhood fine motor skills improve alongside other aspects of physical and intellectual development with the result that there are limits to the maximum speed, force and coordination skills of children compared to adults. As an example, maximum finger tapping speed increases with age from around 3.5 taps per second in the dominant hand at age 5 to a maximum of 5.2 taps per second by the age of 20 (Heinen et. al., 1998). Examples such as this demonstrate how bodily maturation, which varies considerably among different individuals, can significantly impact on the range of sensorimotor skills required in music performance.

Motor skills are best automated by countless repetitions whereas aural skills are typically refined through a broad variety of listening experiences. Both types of skills are not represented in isolated brain areas, but rather depend on the multiple connections and interactions established during training within and between the different regions of the brain. The general ability of our central nervous system to adapt to changing environmental conditions and newly imposed tasks during its entire life span is referred to as *plasticity*. In music, learning through experience and training is accompanied by development and changes which not only take place in the brain's neuronal networks as a result of a strengthening of neuronal connections but also in its overall gross structure. Unfortunately, it is still not completely understood how practice habits and sensorimotor maturation influence each other. With respect to brain plasticity it is known that music practice enhances myelination, grey matter growth and fiber formation of brain structures involved in the specific musical task (Elbert et al., 1995, Amunts et al., 1997, Gaser & Schlaug, 2003, Schlaug et al., 2005, see chapter 4).

There are two main reasons why researchers believe these effects on brain plasticity are more pronounced in instrumental music performers than in other skilled activities. First, musical training usually starts very early, sometimes before age six when the adaptability of the central nervous system is highest, and second musical activities are strongly linked to positive emotions, which are known to enhance plastic adaptations. We would be wise to keeping mind however, that the methodologies used in contemporary brain research produce a bias. For example, since effects are usually demonstrated within group investigations of classical musicians, it could be argued that these brains have a similar acculturation due to the canonical nature of the training. Classical pianists for example, tend to study études of Hanon, Czerny and Chopin that are likely to produce uniform brain adaptations which in turn, then, dominate the individual changes. In other pursuits such as the visual arts, creative writing, architecture and composing music, individualized training may produce more diverse effects which may be masked within group statistics.

NEUROANATOMY AND NEUROPHYSIOLOGY OF MOTOR SYSTEMS

Playing a musical instrument requires highly refined motor skills that are acquired over many years of extensive training, and that have to be stored and maintained as a result of further regular practice. Auditory feedback is needed to improve and perfect performance. Performance based music making, therefore, relies primarily on a highly developed auditory-motor integration capacity, which can be compared to the oral-aural loop in speech production. In addition, somato-sensory feedback constitutes another basis of high level performance. Here, the kinesthetic sense, which allows for control and feedback of muscle and tendon-tension as well as joint positions which enable continuous monitoring of finger-, hand- or lip-position in the frames of body and instrument coordinates (e.g., the keyboard, the mouthpiece), is especially important. In a more general context, the motor system of music performance can be understood as a sub-specialty of the motor systems for planned and skilled voluntary limb movements.

Voluntary skilled limb movements involve four cortical regions in both hemispheres: the *primary motor area* (M1) located in the *precentral gyrus* directly in front of the central sulcus; the *supplementary motor area* (SMA) located anterior to the M1 of the frontal lobe and the inner (medial) side of the cortex; the *cingulate motor area* (CMA) below the SMA and above the corpus callosum on the inner (medial) side of the hemisphere; and the *premotor area* (PMA), which is located adjacent to the lateral aspect of the primary motor area (see Figure 1).

SMA, CMA and PMA can be described as *secondary motor areas*, because they are used to process movement patterns rather than simple movements. In addition to the cortical regions, the motor system includes the subcortical structures of the basal ganglia, and the cerebellum. The sensory areas are necessary in order to maintain the control of movements. Their steady kinesthetic feedback information is required for any guided motor action. The sensory areas are located in the *primary somato-sensory area* (S1) behind the central sulcus in the parietal lobe. This lobe is involved in many aspects of movement processing and is an area where information from multiple sensory regions converge. In the posterior parietal area, the body-coordinates in space are monitored and calculated and visual information is transferred into body-coordinates. For musicians, this area is typically activated during tasks involving multi-sensory integration, for example during sight-reading and the playing of complex pieces of music (Sergent, 1993; Haslinger et al., 2005).

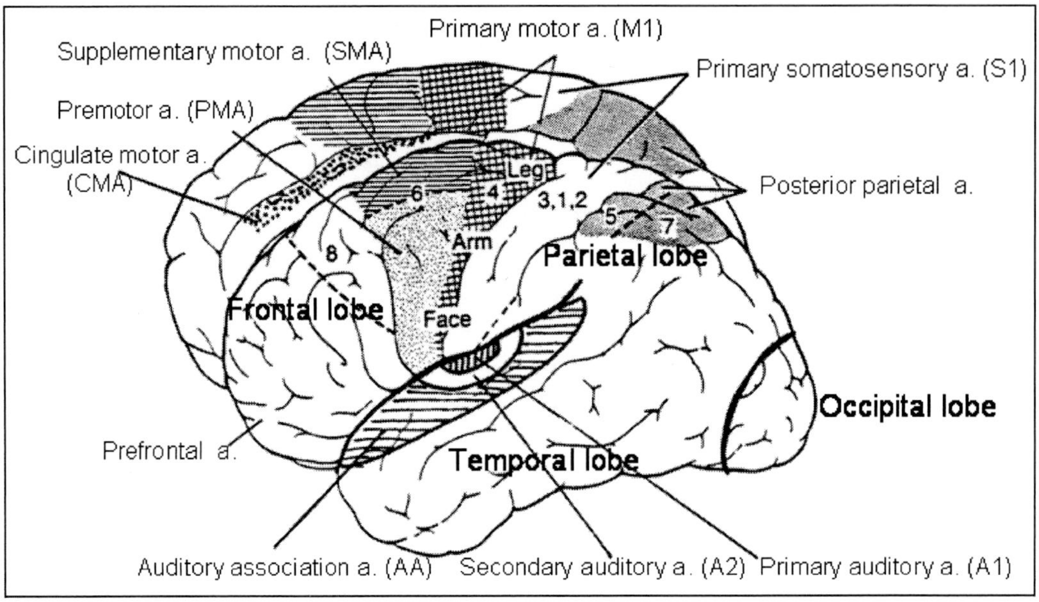

Figure 1. Brain regions involved in sensory and motor music processing. (The abbreviation "a" stands for "area") Left hemisphere is shown in the foreground (lower right) and right hemisphere in the background (upper left). The numbers relate to the respective Brodmann's areas, a labelling of the cortical areas according to the fine structure of the nervous tissue.

The *primary motor area* (M1) represents the movements of body parts in a separate, but systematic order. The representation of the leg is located on the top and the inner side of the hemisphere, the arm in the upper portion, and the hand and mouth in the lower portion of M1. This representation of distinct body parts in corresponding brain regions is called *somatotopic* or *homuncular order*. Just as the *motor homunculus* is represented upside down, so too is the *sensory homunculus* on the other side of the central sulcus (see Figure 2a and 2b). The proportion of both the motor and the sensory homunculus are markedly distorted since they are since they are determined by the density of motor and sensory innervation of the respective body parts. For example, control of fine movements of the tongue requires many more nerve fibers transmitting the information to this muscle as compared to the muscles in the back. Therefore, the hand, the lips and the tongue require almost two-thirds of the neurons in this area. However, as further explained below, the representation of the body parts may be modified by usage. Moreover, the primary motor area does not simply represent individual muscles: multiple muscular representations are arranged in a complex way so as to allow the execution of simple types of movements rather than the activation of a specific muscle. This is a consequence of the fact that a two-dimensional array of neurons in M1 has to code for three dimensional movements in space (Gentner & Classen, 2006). Put more simply, our brain does not represent muscles but rather movements.

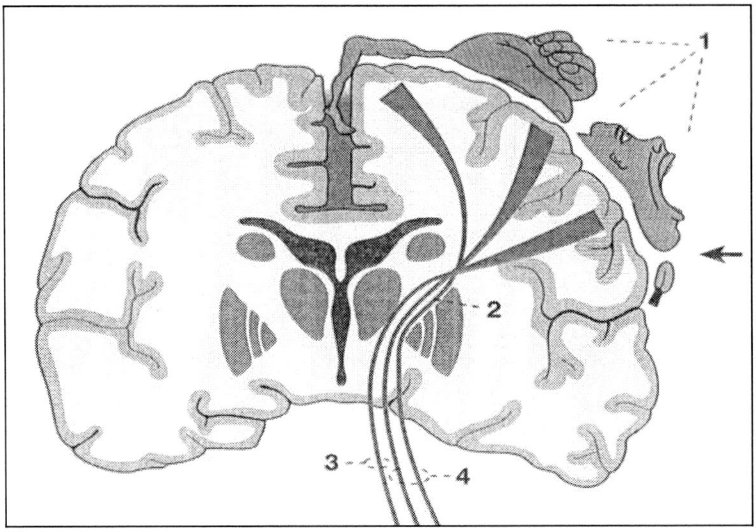

Figure 2a. The motor representation in the primary motor area of the precentral gyrus. A cross-section of the brain is shown. The uneven distribution of nerve cells, subserving motor functions of the oral tract and the face and the hands compared to the other body parts is clearly shown.

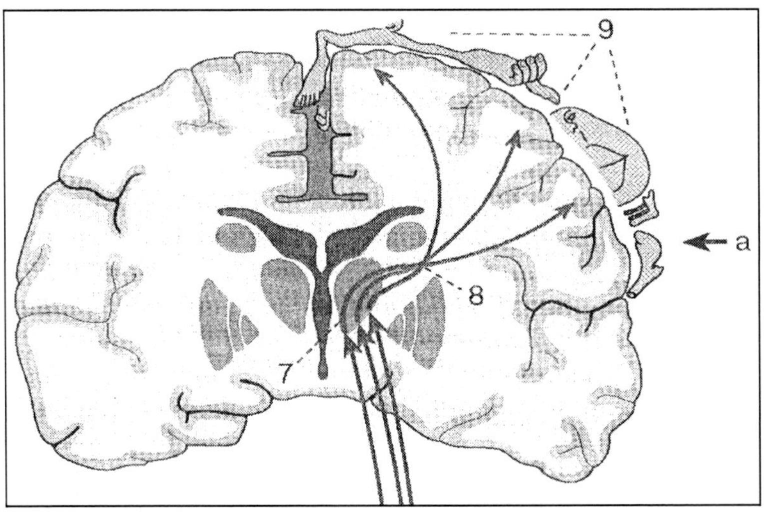

Figure 2b. The sensory representation of the body surface in the primary somatosensory area of the postcentral gyrus. Similarly to the motor representation of the body, the lips, face and hands are predominant.

The *supplementary motor area* (SMA) is mainly involved in the coordination of the two hands, in the sequencing of complex movements and in the triggering of movements based on internal cues. It is particularly engaged when the execution of a sequential movement depends on internally stored and memorized information. The SMA can be subdivided into two distinct functional areas. In the anterior SMA, it would seem that the planning of complex movement patterns is processed. The posterior SMA seems to be predominantly engaged in two-handed movements and, in particular, in the synchronization of both hands during complex movement patterns.

The function of the *cingulate motor area* (CMA) is still under debate. Electrical stimulation and brain imaging studies demonstrate its involvement in movement selection in situations when movements are critical to obtain reward or punishment. This points towards a close links between the cingulate gyrus and the emotion processing limbic system. In summary, it would seem that the CMA plays an important role in mediating cortical cognitive functions and limbic-emotional functions. The *premotor area* (PMA) is primarily engaged when externally stimulated behavior is being planned and prepared. It is therefore involved in the learning, execution and recognition of limb movements and seems to be particularly concerned with the processing of visual information necessary for movement planning.

The *basal ganglia*, located deep inside the cerebral hemispheres, are inter-connected reciprocally via the thalamus to the motor and sensory cortices, thus constituting a loop of information flow between the cortex and the basal ganglia. They are indispensable for any kind of voluntary actions that are not highly automated. Their special role consists in the control of voluntary action by selecting appropriate motor actions and by comparing the goal and course of those actions with previous experience. In the basal ganglia, the flow of information between the cortex and the limbic emotion system, in particular the amygdala, converges. It is therefore assumed that the basal ganglia processes and controls the emotional evaluation of motor behavior in terms of expected rewards or punishment. Finally, the *cerebellum* contributes essentially to the timing and accuracy of fine-tuned movements.

Development of Motor Control in Childhood and Adolescence

Motor learning depends on the development and maturation of all functions involved in motor control. These include the movement apparatus with its bones, muscles, joints and tendons as well as the peripheral and central nervous system and the sensory organs. Humans are unique in their "premature" birth, in that brain volume amounts only to about 25% of the adult volume, whereas in primates brain volume at birth amounts to about 40% of the adults. At age two, this discrepancy and slow maturation is even more evident. Whereas two year old primates have reached 90% of the adults' brain volume, in humans it is still only 35 %. The maturation of the brain is not accomplished until early adulthood. In a landmark study, Sowell and colleagues (2001) demonstrated that the myelination of the frontal lobe, which is involved in complex social interactions, is the last to develop and may not be completed until age 20! Interestingly, these brain regions are involved in complex social interactions. This corresponds to everyday experience where social interactions become more refined only after puberty.

With respect to the maturation of motor abilities most infants develop in the same order and at approximately the same age. Researchers generally agree that some forms of motor ability are genetically pre-programmed within all infants. The environment does play a role in the development, with an enriched environment often reducing the learning time and an impoverished one doing the opposite. Table 1 shows the development of infants in a sequential order. Ages in this table are approximate and it is normal for these to vary by a month or two in either direction.

Typical childhood development is defined as a predictable and orderly process; that is, individual children develop at about the same rate over time as their peers. Although there are individual differences in children's personalities, activity levels, and timing of developmental milestones, this basic principle of development appears to follow a similar universal path.

Table 1. Typical stages of infants' motor development

2 months – able to lift head independently
3 months – can roll over
4 months – can sit propped up without falling over
6 months – is able to sit up without support
7 months – begins to stand while holding on to things for support
9 months – can begin to walk, still using support
10 months – is able to momentarily stand independently without support
11 months – can stand alone with more confidence
12 months – begins to walk alone without support
14 months – can walk backward without support
17 months – can walk up steps with little or no support
18 months – able to manipulate objects with feet while walking, such as kicking a ball.

Development Depends on Maturation and Learning

Maturation refers to the sequential characteristic of biological growth and development. Biological changes occur in a sequential order and enable children to acquire new abilities. Changes in the brain and nervous system account largely for maturation. These changes help children to improve their cognition (thinking) and motor (physical) skills. Children must therefore mature to a certain point of readiness before they can progress to new skills. For example, a four-month-old cannot use language because the infant's brain has not matured enough to allow it to talk. By two years of age, the brain has developed further and with help from others, the child will have the capacity to say and understand words. Also, children cannot write or draw until they have developed the motor control to hold a pencil or crayon. Although maturational patterns are innate – that is, genetically programmed – environmental stimulation and learning as a result of children's experiences largely determine whether they will reach an optimal level of development. Therefore, a stimulating environment and varied experiences are therefore essential if children are to develop to their full potential

Growth and Development is a Continuous Process

As children develop, they add new skills to the skills already acquired which become the basis for further achievement and mastery of more sophisticated skills. Most children follow a similar pattern and one stage of development lays the foundation for the next. For example, in motor development, there is a predictable sequence of development that occurs before walking such that infants will lift and turn their head before being able to turn over. Infants must first learn to move their limbs (arms and legs) before grasping an object. Mastery of climbing stairs involves increasing skills from holding on to walking alone. By the age of four, most children can walk up and down stairs with alternating feet. Between ages four and six, children have developed the manual control to hold a violin and a bow and to co-ordinate complicated movements of the right and the left hand. Prior to this age, bilateral synchronous mirror movements in both arms and hands are predominant, based on a strong information exchange between the left and the right motor cortex via the callosal body. As a prerequisite to move both upper extremities at the same time independently inhibitory neuronal networks have to be established, suppressing the strong information exchange between both cerebral hemispheres.

Growth and Development Proceed from the General to the Specific

In motor development, the infant will be able to grasp an object with the whole hand before using only the thumb and forefinger. The infant's first motor movements are very generalized, mostly bilateral, undirected, and reflexive, waving arms or kicking before being able to reach or creep toward an object. Motor development occurs from large muscle movements to more refined (smaller) muscle movements. As a general principle the development of fine motor skills is always based on the inhibition of gross motor skills. For example, the isolated depression of one finger on a keyboard requires inhibitory impulses to the adjacent fingers, not to depress a key. Inhibition is a general principle of maturation in the nervous system. In the adult brain about 90% of the synaptic connections are inhibitory.

There are individual rates of growth and development. Each child is different, and the rates at which individual children grow are therefore also different. Although the patterns and sequences for growth and development are usually the same for all children, the rates at which individual children reach varying developmental stages differ. Understanding these individual differences in rates of development should cause us to be careful about solely using and relying on age and stage characteristics to describe or label children, even though there is a range of ages for any developmental task to take place. Some children will walk at ten months while others walk a few months older at eighteen months of age. Some children are more active while others are more passive. This does not mean that the passive child will be less intelligent as an adult. There is no validity to comparing one child's progress with or against another child. Rates of development are also not uniform within an individual child. For example, a child's intellectual development may progress faster than his or her motor, emotional or social development.

An understanding of the principles of development helps us to plan appropriate activities and stimulating and enriching experiences for children, in addition to providing the basis for understanding how to encourage and support young children's learning of musical instruments.

Observing the Brain's Functions

Until the 1970s, the only method available to study brain functions was to study patients with temporary or permanent brain lesions; that is, patients with damaged or diseased brains. The behavior of patients with localized brain lesions was observed and from the pattern of behavioral deficits it was concluded that the impaired brain regions was a necessary, though not sufficient prerequisite for the respective brain function. This way, the term *amusia* (see chapter 9) was coined, which denotes an isolated loss of musical functions mostly after right temporal brain damage. However, detailed information concerning active musical performance was not collected. During the last two decades, knowledge of brain regions involved in complex tasks such as playing a musical instrument has increased enormously. This is mainly due to the development of novel technologies that allow non-invasive assessment of the intact brains' function. Additionally, new imaging techniques, derived from the Magnetic Resonance Imaging (MRI)-Technology precisely demonstrate minute changes in brain structure. Voxel-Based Morphometry (VBM), for example, provides detailed information of the thickness of the grey matter in the layers of neurons in the cerebral cortex. Using this technique longitudinal follow up-studies have demonstrated changes in grey matter volume in the range of cubic millimeters as a result of musical training (Gaser & Schlaug, 2003). Diffusion Tensor Imaging (DTI) on the other hand is a way to assess direction and volume of fiber tracts in the white matter of the brain. In pianists this method has shown changes in myelination of the callosal body, connecting the two brain hemispheres (Bengtsson et. al., 2005).

Other imaging tools such as positron emission tomography (PET), functional MRI (fMRI), electroencephalography (EEG), event related potentials (ERP), magnetoencephalography (MEG), and transcranial magnetic stimulation (TMS) allow the functional assessment of the brain rather than the structures. As Don Hodges stresses (see chapter 1), each of these approaches has strengths and weaknesses so it is always important to pool findings from different approaches for a more comprehensive understanding. Therefore, combining methods that have excellent time resolution (e.g., EEG, MEG, TMS) in combination with methods that have superior spatial resolution (e.g., PET, fMRI) are presently the state of the art in brain imaging. However, even though the technology of brain imaging has advanced enormously during the past twenty year, one should keep in mind, that colorful "brain scans" must be regarded with some caution. First, in humans, the link between the mechanisms on the microscopical neuronal level and on the macroscopical level demonstrating activated or inhibited brain regions are not well understood. In other words, brain scans do not explain the prerequisites of human perception or motor production in detail. Rather, they demonstrate relationships between human behavior and changes in brain metabolisms or electrical properties. To give an example: methods like fMRI and PET which are based on indirect measures such as an increase in blood flow and oxygen or glucose consumption do not distinguish whether the neuronal activity that causes the augmented blood-flow etc., is excitatory or inhibitory which are obviously contrasting functional properties.

Second, there is increasing evidence that all human cognitive processes are based on complex and extremely rapid interactions of neuronal populations in distant brain regions.

This highly dynamic and rapidly adapting signaling of neuronal populations is still not directly accessible for any neuroscience research. Methods like calculations of neuronal oscillations or inter-electrode correlations from EEG-data collected at the surface of the scull mirror not precisely enough the real mechanisms at the neuronal level in the cerebral cortex.

All-together, the remaining gap in understanding of neuronal interactions and its behavioral correlates and consequences constitutes a challenge in brain science which has to be addressed in the future.

Learning of Fine Motor Skills

Our knowledge concerning the regions and mechanisms of the brain involved in sensorimotor learning is still incomplete. Overall, musicians appear to process new incoming stimuli more effectively compared to non-musicians. As will be outlined below, musical sensory motor learning affects neuronal development in several ways. However, to better understand the influences of musical sensorimotor learning on brain development and brain organization, more long-term longitudinal studies of musically trained children are needed.

According to newly emerging evidence (for a review see Halsband & Lange, 2006) all structures involved in motor control participate in the acquisition of new sensorimotor skills. It has been assumed since the nineteenth century that the cerebellum plays an important role in the acquisition of new motor skills. Such information was based on studies in which patients suffering from lesions of the cerebellum were found to be unable to increase the speed of a sequence of complex finger movements after practice. More recent evidence demonstrates that the cerebellum is involved in the selection, the sequence and the timing of movements.

Another functional system, which is equally important for the development and learning of fine finger movements, is the basal ganglia. Patients suffering from Parkinson's disease have deficits in learning new motor tasks, and, although they can improve the speed of complex movement patterns during practice sessions, they do not learn as quickly, and do not reach the level of performance of normal controls (for a review see Seger, 2006). In professional musicians, subtle degradations of the functions of the cerebellum, for example as a consequence of long term alcohol consumption or of the basal ganglia in the beginning stages of Parkinson's syndrome, frequently selectively impair highly virtuoso performance, thus demonstrating the importance of possessing a 100% intact central nervous system to obtain peak performance quality.

It has been known for some time that the activity in the SMA and in the premotor area of the brain are enhanced as a result of increasing complexity of finger movement sequences (Roland et. al., 1980). Using fMRI, Karni and colleagues (1995) investigated adult subjects learning of complex finger sequences which are similar to those necessary for piano playing. After 30 minutes of practice, the representation of the fingers in the primary motor cortex increased. However, without further training, this effect diminished after one week with the hand representation returning to its previous size. In contrast, continuous practice resulted in a stable enlargement of the hand area in the primary motor cortex. This effect was specific for the daily trained sequence of complex finger-movements, and did not occur when the subjects improvised complex finger movements which were not subsequently repeated. Parallel to the enlargement of the hand area in the primary motor cortex, the size of the cerebellar hand representation diminished, suggesting that the cerebellum plays an important role only in the initial phase of motor learning.

More recent investigations have compared brain activation in professional pianists and non-musicians. Hundt-Georgiadis and v. Cramon (1999), for example, studied brain activation during 35 minutes of short-term motor learning of a complex finger-tapping task using the dominant right hand. They found that activation of the primary motor area during motor learning increased only among the pianists. Non-musicians developed a temporary increase in activity during the first 7 to 14 minutes of training, but, subsequently, a rapid attenuation of this activation. Furthermore, compared with non-musicians, the motor learning of these pianists was accompanied by only small contributions from the supplementary motor areas and the cerebellar cortices. This finding is compatible with the idea that the anterior SMA is essential for setting up and executing complex motor programs prior to automatic performance. The pianists were able to rapidly and thoroughly learn the complex finger sequence, reaching a high degree of automaticity during the first minutes of motor learning. This finding is supplemented by findings involving professional pianists who showed less activation in the primary and secondary motor cortex during an over learned motor task than controls, suggesting greater efficiency (Jaencke et al., 2000). This is plausible since fewer neural resources are required once a task such as a scale or arpeggio is learned and automated.

An interesting finding common to all of the studies cited above is that the motor cortex of the *untrained* hand is, at the same time, contributing to motor learning. This explains why we see improved performance on motor tasks in an untrained left hand in both pianists and non-musicians. One of the most fascinating features of the human sensorimotor system therefore relates to this phenomenon. Despite clear *somatotopic* organization of the motor cortex, a movement can be learned with one extremity and performed with the other. For example, Rijntjes and colleagues (1999) investigated subjects writing their signature with their right index finger and with their right big toe. The results of their fMRI study show that the movement parameters for these highly trained movements are stored in the premotor and supplementary motor cortex adjacent to the right hand area in the primary motor cortex, but are also accessible for the foot area. Thus, somatotopy in secondary motor areas (SMA, PM) seems to be defined functionally - as abstract movement information independently from the executing limb (movement-ideas or *Bewegungs-Ideen*) - and not on the basis of anatomical representations. Whereas these studies refer to explicit motor learning by trial and error and sensory feedback, implicit motor learning seems to be processed in a different way. When subjects were unaware of a motor learning task – because their attention was drawn to a different problem – activity of the basal ganglia correlated with motor learning.

All of the studies mentioned above do not take into account one special quality of musicianship, namely the strong coupling of sensorimotor and auditory processing required for performing music. Practicing an instrument involves assembling, storing, and constantly improving complex sensorimotor programs through prolonged and repeated execution of motor patterns under the controlled monitoring of the auditory system. Many professional pianists for example report that their fingers move more or less automatically when they are listening to piano music played by a colleague. In a cross-sectional experiment, we demonstrated a strong linkage between auditory and sensorimotor cortical regions develops as a result of many years of practice (Bangert et. al., 2006, see also chapter 4). Using fMRI, professional pianists were asked to listen to simple piano tunes without moving their fingers or any other body part. Figure 3a demonstrates the increase in activation of professional pianists in comparison to non-musicians. There is an impressive activation of the motor cortex demonstrating the sub-conscious or automated auditory-motor co-activation.

Furthermore, in a longitudinal study, it was possible to follow up the formation of such neuronal multisensory connections along with piano training in beginning pianists. Non-musicians, who had never played an instrument before, were trained on a computer piano twice a week over a period of five weeks. They listened to short piano melodies of a three-second duration played in a five tone range, and after a brief pause, were required to replay the melodies with their right hand as accurately as possible. After 10 minutes of training, listening to piano tunes produced additional activity in the central and left sensori-motor regions. In turn, playing on a keyboard produced additional activity in the auditory regions of both temporal lobes. These early signs of cortical plasticity during the first training session were not stable, but stabilized within the subsequent five weeks of training. In the movement task, the most remarkable effect after five weeks was the development of additional activation of the right anterior temporal and frontal lobe. Since it has been demonstrated that this area is involved in the perception of pitch sequences, such activation might reflect the auditory imagery of sounds while moving the fingers on a soundless keyboard. In this context, it should be mentioned that the results of this experiment support the idea of the direct effectiveness of mental training on subtle sensory motor activation patterns represented in the central nervous system (Bangert & Altenmueller, 2003). Furthermore it has to be emphasized that these experiments in novice pianists impressively demonstrate how dynamically brain adaptations accompany these multi-sensorimotor learning processes.

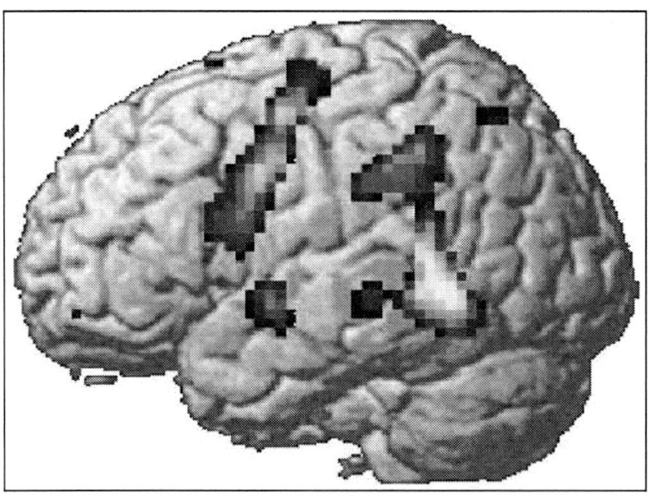

Figure 3a. Additional brain activity (grey zones) of skilled pianists compared to non-pianists when listening to piano tunes without moving their fingers. The primary motor cortex of the precentral area and auditory association areas are lighting up demonstrating an unconscious co-representation of heard tunes as movement patterns (modified from Bangert et al. 2006).

Activation of motor co-representations can occur in trained pianists not only by listening to piano tunes, but also by observing a pianist's finger movements while watching a video. The brain mechanisms of such learning through observation have been clarified in recent years. When monkeys observed the actions of con-species, for example grasping peanuts, exactly the same brain areas were active as if the observing monkeys were performing the observed action themselves. Additionally, in the observing monkeys, a region in the parietal lobe was activated, which is believed to represent the knowledge that "it is not me, who is performing the action". Quite appropriately, this neuronal network was termed a *"mirror neuron network"*. It appears that in humans mirror neurons are not only important for any kind of learning but also for empathy and other social feelings, for example when we observe the suffering of someone we love.

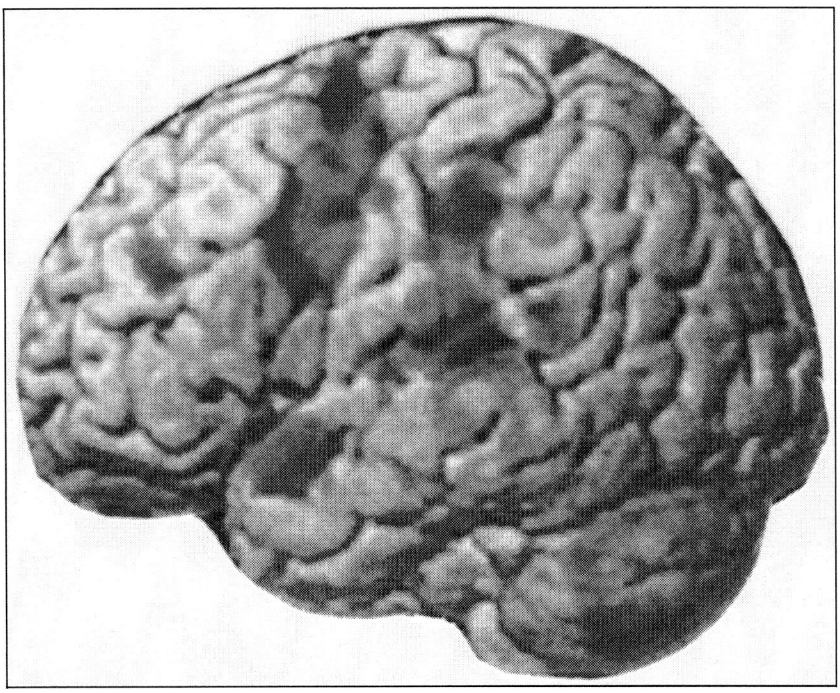

Figure 3b. Additional brain activity (grey zones) of skilled pianists compared to non-pianists when observing pianist movements in a soundless video. Again, the precentral area and auditory association areas are lighting up demonstrating the mirror-system: the observed movements are unconsciously activated, such that the subjects did not move their fingers. Furthermore the auditory areas are activated demonstrating a visual-auditory co-representation of seen movements. This effect demonstrates impressively the powerful humans imitation system and may be utilized by teachers, when demonstrating at the instrument (modified from Haslinger et al., 2005).

In Figure 3b the increases in brain activation of trained pianists are shown whilst they are observing video sequences of a moving hand at the piano as compared to the activation of musically naïve subjects (Haslinger et al., 2005). Besides the motor hand area in the primary motor cortex, secondary auditory cortices in the temporal lobe and the cerebellum are activated, thus impressively demonstrating such a mirror-system in humans. As a consequence for musical practice it follows that careful demonstration at the instrument may enhance learning. Such a teaching method based on demonstration and imitation is widely used at all levels of musical training, and would appear to be particularly effective in cases where teachers demonstrate an action or series of actions that are carefully and methodically observed by the student. Teachers would probably generally agree that modeling of this type is typically more effective than verbal explanations, because not only does it offer learners an opportunity to compare their own performance with that of a more experienced musician, but also to gauge their level of achievement directly as they see and hear a more sophisticated performance (Bandura, 1997; Kohut, 1985). In this sense, if used effectively, modeling can be an efficient and very direct technique for scaffolding a learner to a higher level of functioning.

Practicing through listening and observation can be considered as special cases of *mental training*. Narrowly defined, mental training is understood as the vivid imagination of movement sequences without physically performing them. As with observation of actions, principally the same brain regions are active as if the imagined action is performed; that is, the primary motor cortex, the supplementary motor cortex and the cerebellum (Kuhtz-Buschbeck et al., 2003). In a study investigating mental training of finger movement sequences of different complexities brain activation increased along with the degree of difficulty of the imagined motor task. Furthermore, when continuing mental practice over a period of several days, the involved brain regions showed plastic adaptations. Although these adaptations are less dramatic than if the motor tasks were practiced physically, mental training produced a clear improvement in task performance as assessed in finger tapping tests.

Many issues concerning the brain mechanisms of sensorimotor learning and processing during music performance remain to be clarified. It is still unclear, for example, how and where the rapid adjustment of the sensorimotor system to different spatial coordinates is processed when musicians switch between instruments of different size (e.g., from alto flute to piccolo or from violin to viola). This phenomenon is referred to as response size or movement schema. Besides the above mentioned limb-independent storage of movement information in the secondary motor areas, the cerebellum may contribute to this ability to adjust movements to altered spatial coordinates, which differ in absolute size, but preserve relative spatial relationships and temporal sequence of movements. Another unsolved problem is the neuronal basis of the transition from guided slow movements, which are performed under steady sensory control, to fast, ballistic movements, which have to be performed without on-line sensory feedback. It is assumed that different brain regions produce these two types of movements and that the transition from one type to the other may be incomplete. This might explain why practicing guided movements whilst slowly and systematically increasing the tempo may finally hamper the execution of this movement at a very fast tempo. Although recognized as a problem by many instrumental teachers, no research data are available on this topic. From an empirical standpoint, we would recommend, even at an early stage, that musicians should rehearse small segments of the movement pattern in a fast tempo. However, at the same time, the precise automatization of difficult movements has to be practiced in a precisely guided slow tempo. Some eminent pedagogues have intuitively designed exercises addressing precisely this problem. Alfred

Cortot's edition of the preparing studies for Chopin's piano etudes Opus 10, No. 1 is an excellent example (see Figure 4).

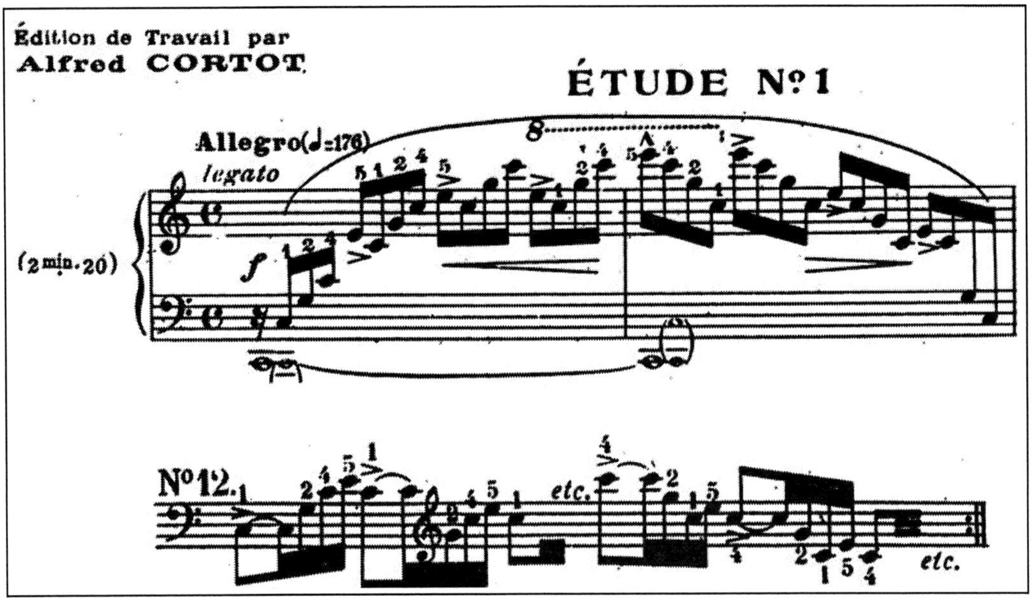

Figure 4. Examples of Cortots preparatory exercises of split movement patterns in order to facilitate motor learning of rapid passages (From Editions Salabert, Paris, No 5.004, page 8, 1915 with permission).

Plasticity of Sensory Motor Systems: Musicians' Brains are Different

During the past decade, brain imaging has provided important insights into the enormous capacity of the human brain to adapt to complex demands. These adaptations are referred to as *brain plasticity* and do not only include the connections or firing rates of neurons, - the "software" of our brain, - but also the "hardware", namely the fine structure of nervous tissue and even the visible gross structure of brain anatomy (see also chapter 4). Brain plasticity is best observed in complex tasks with high behavioral relevance for the individual such that they cause strong emotional and motivational activation. Plastic changes are more pronounced in situations where the task or activity is intense and the earlier in life it has been developed. Obviously, the continued activities of accomplished musicians provide in an ideal manner the prerequisites of brain plasticity. It is therefore not astonishing that the most dramatic brain plasticity effects have been demonstrated in professional musicians (for a review see Muente et al., 2002).

Our understanding of the molecular and cellular mechanisms underlying these adaptations is far from complete. Brain plasticity may occur on different time axes. For example, the efficiency and size of synapses may be modified in a time window of seconds to minutes, while the growth of new synapses and dendrites may require hours to days. An increase in gray matter density, which mainly reflects an enlargement of neurons, needs at least several weeks. White matter density also increases as a consequence of musical training. This effect is primarily due to an enlargement of myelin cells: The myelin cells, wrapped around the nerve fibers (axons) are contributing essentially to the velocity of the electrical impulses traveling along the nerve fiber tracts. Under conditions requiring rapid information transfer and high temporal precision these myelin cells are growing and as a consequence nerve conduction velocity increases. Finally, brain regions involved in specific tasks may also be enlarged after long-term training due to the growth of structures supporting the nervous function, for example blood vessels that are necessary for the oxygen and glucose transportation required to sustain nervous function.

Comparison of the brain anatomy of skilled musicians with that of non-musicians shows that prolonged instrumental practice leads to an enlargement of the hand area in the motor cortex (Amunts et. al., 1997) and to an increase in grey matter density corresponding to more and/or larger neurons in the respective area (Gaser & Schlaug, 2003). These adaptations appear to be particularly prominent in all instrumentalists who have started to play prior to the age of ten and correlate positively with cumulative practice time. Furthermore, in professional musicians, the normal anatomical difference between the larger, dominant (mostly right) hand area and the smaller, non-dominant (left) hand area is less pronounced when compared to non-musicians. These results suggest that functional adaptation of the gross structure of the brain occurs during training at an early age.

Similar effects of specialization have been found with respect to the size of the corpus callosum. Professional pianists and violinists tend to have a larger anterior (front) portion of this structure, especially those who have started prior to the age of seven (Schlaug et al., 1995a). Since this part of the corpus callosum contains fibers from the motor and supplementary motor areas, it seems plausible to assume that the high demands on coordination between the two hands, and the rapid exchange of information may either stimulate the nerve fiber growth – the *myelination* of nerve fibers that determines the velocity of nerve conduction – or prevent the physiological loss of nerve tissue during aging.

It is not only motor areas however, that are subject to anatomical adaptation. By means of magnetoencephalography (MEG), the number of nerve cells involved in the processing of sensory stimulation in individual fingers can be monitored. Using this technique, professional violinists have been shown to posses enlarged sensory areas corresponding to the index through to the small (second to fifth) fingers of the left hand (Elbert et. al., 1995) even though their left thumb representation is no different from that of non-musicians. Again, these effects were most pronounced in violinists who started their instrumental training prior to the age of ten.

A further example of functional specialization reflected by changes in gross cortical anatomy can be found in musicians possessing absolute pitch. In musicians who possess this ability, the upper back portion of the left temporal lobe (*Wernicke region*) is larger in comparison to those musicians without absolute pitch (Schlaug et. al., 1995b). Using MEG, the functional specialization of the auditory cortex has been demonstrated by Pantev (Pantev et. al., 1998). When compared to subjects who had never played an instrument, the number of auditory nerve cells involved in the processing of piano tones, but not of pure sinusoidal tones, was about 25% greater in pianists. Recently, Sarah Bengtsson and her colleagues

(2005) have found an increase in the size of the posterior part of the corpus callosum in professional pianists. This finding seems to correspond to a more effective connectivity between the left and right auditory regions in the respective temporal lobes.

It is not only cortical structures that are enlarged by early and prolonged instrumental training. Subcortical structures also seem to be highly affected. In professional musicians, the cerebellum, which contributes significantly to the precise timing and accuracy of motor commands, is also enlarged.

In summary, when training starts at an early age (before about seven years), these plastic adaptations of the nervous system affect brain anatomy by enlarging the brain structures that are involved in different types of musical skills. When training starts later, it modifies brain organization by re-wiring neuronal webs and involving adjacent nerve cells to contribute to the required tasks. These changes result in enlarged cortical representations of, for example, specific fingers or sounds within existing brain structures. In the following section, the behavioral correlates of the neuronal adaptations with respect to practice will be briefly outlined.

Expertise Theory – Does Practice Make Perfect?

Perhaps the most important study on practice to emerge during the past couple of decades was undertaken with students at the Berlin Academy of Music by Ericsson and his colleagues in 1993. Ushering in a new way of thinking about practice, Ericsson et. al. proposed the concept of "deliberate practice" as a means of studying goal-oriented, structured and effortful facets of practice in which motivation, resources and attention determine the amount and quality of practice undertaken. They argued that a major distinction between professional and amateur musicians (and perhaps successful versus unsuccessful learners) is the amount of deliberate practice undertaken during the many years required to develop instrumental skills to a high level (Ericsson, 1996; Ericsson & Lehmann, 1996). They also proposed that highly skilled musicians exert a great deal more effort and concentration during their practice than less skilled musicians, and are more likely to image, monitor and control their playing by focusing their attention on what they are practising and how it can be improved (Ericsson, 1996; Ericsson & Lehmann, 1996).

Using this approach, researchers have uncovered evidence showing a close association between practice and expertise. In fact, many researchers would argue that the amount of practice a musician has accumulated is predictive of that musician's level of ability; that is, there is a close association between the number of hours of deliberate practice musicians have accumulated and their level of expertise (Lehmann & Gruber, 2006). This line of research therefore refutes the common assumption that highly talented musicians have not practiced as much as their less talented peers. Rather, expert musicians appear to have consistently invested similar amounts of time – about 10,000 hours of dedicated practice over 10 or more years – to have any chance of attaining eminence on their instrument. It should be kept in mind that this rule is a necessary rather than a sufficient condition for success. For many music educators this finding may appear trivial, mirroring daily life experience in a music school.

Despite these findings, many details of expertise theory remain to be clarified. For example, what causes children to dedicate so many hours of their young lives to deliberate practice of their instrument instead having fun with their schoolmates? Motivational factors are important, as are the ability to think in sound, to improvise, and to transpose pieces all of which lead to a rewarding playful atmosphere during skill acquisition. It is therefore is crucial that children learn to represent music in multiple ways, so that, – on the one hand – they are able to develop their sensitivity to the expressive properties of music and, on the other hand find continuous interest in their activity (McPherson, 2005). In addition, the quality of deliberate practice is critical and may depend largely on strategies which allow the child to monitor, control and reflect on his or her own progress in a self-rewarding manner. These ingredients are essential for a quality, well rounded music education (McPherson, 2005; 2006; McPherson & Zimmerman, 2002).

The neurobiological basis of this behaviour, which is characterized by curiosity, stamina and the ability to strive for rewarding experiences and which results in incentive goal directed behaviour such as deliberate practice is predominantly the transmitter substance Dopamine (Depue & Collins 1999). Most nerve cells sensitive to this neurotransmitter are found in a small part of the brain which is localized behind the basis of the frontal cortex, the so-called meso-limbic system, an important part of the “emotional” brain. Dopamine is widely recognized to be critical to the neurobiology of reward, learning and addiction. Virtually all drugs of abuse, including heroin, alcohol, cocaine, and nicotine activate dopaminergic systems. So called "natural" rewards such as musical experiences and other positive social interactions likewise activate dopaminergic neurons and are powerful aids to attention and learning (Keitz et al., 2003). There is ample evidence that the sensitivity to dopamine in the meso-limbic brain regions is largely genetically determined resulting in the enormous variability in reward-dependent behaviour. The genetic “polymorphism” of dopaminergic response explains the different motivational drives we observe in children with similar social and educational backgrounds.

It is intriguing that there is a strong link of dopaminergic activity to learning and memory, which in turn promotes plastic adaptations of the brain in areas involved in the learned task. Subtle signs of plastic adaptations can be demonstrated as early as in five to seven year old children after one year of music training (Schlaug et al., 2005). Children with instrumental music training showed a tendency towards an increase in grey matter volume in the cerebral cortex as compared to a non-musician group. However, more data needs to be accumulated before firm conclusions can be drawn. In adults, music expertise clearly is related to plastic adaptations of the brain as outlined above.

Conclusion: Implications for Practice

In the preceding paragraphs we have demonstrated the neurobiological foundations of motor learning and practice. Here the data shall be summarized in order to delineate some practical rules which might be useful for the daily work of instrument music teachers and of music educators more generally. One has to bear in mind however, that the results of scientific experiments and the knowledge of brain processes can never replace the profound practical knowledge that teachers have acquired over generations and hundreds of years in teaching millions of gifted and not so able children. For most pedagogues therefore, the "rules" are not new, and can therefore be understood more as affirmation of the traditions of good practice. We believe however, that it will be useful for teachers to understand some of the underlying brain mechanisms which support their practices.

1. How to Practice is Best Learned by Practicing

As all skilled human motor activities, practicing is largely based on procedural knowledge. How to practice and when to stop practicing is best learned by experience. Practicing can be considered as a self-organizing process, frequently starting with uneconomical activation of large neuronal pools in the sensorimotor brain regions. Optimizing the movement patterns occurs under continuous sensory feedback from the ears, the eyes, the muscles, tendons, joints and from the skin. The integration of this information into movement patterns is the most important step in procedural learning. It is mainly based on the formation of neuronal networks, for example the connections between auditory and motor areas and in a stepwise reduction of cortical activity and augmentation of subcortical activity in the basal ganglia and the cerebellum.

2. Start Early But Not Too Early With Playing An Instrument

At the same age, there can be large individual differences in the development of children's motor and cognitive abilities. If the motor system is not sufficient developed, a child may experiences great difficulties in bimanual coordination or execution of rapid movements. Starting too early can lead to needless frustration and anxiety. On the other hand, during the early years the brain is more adaptive and plasticity effects may contribute crucially to the development of manual, auditory and sensory skills necessary for music making.

3. Breaks and Sleep Are Essential for Motor Learning

When playing a musical instrument, the central nervous system is mainly involved in processing a huge amount of incoming information from the ears and eyes, and from the sensory organs in muscles, tendons, joints and skin. The consolidation of the networks necessary for programming movement sequences occurs mainly in the breaks after playing and during sleep. As a consequence, the more complex a task is, the shorter the practice time should be scheduled in one session and the longer the breaks should be planned. Sleep is another factor supporting procedural memory formation. Therefore, sufficient sleep should be encouraged, especially when an instrumentalists of any ability levels is working hard to master new repertoire.

Generally, a practice session should be terminated, when signs of fatigue appear. It is important to consider that over-practice (practice into bodily or mental fatigue) not only leads to no improvement, but to an active worsening of motor programs. This is due to a blurring of central nervous sensorimotor representations, when muscular fatigue appears. Furthermore, a lack of attention causes a higher probability of uneconomical movements or production of false notes which, as a consequence, are stored in the procedural memory.

4. Use the Human Mirror System

The human mirror system is a powerful tool to facilitate skill learning. Auditory and visual cues presented to students activate their sensorimotor representations and can directly lead to the formation of motor programs. This is the basis of imitation learning. On the other hand, sloppy and careless demonstrations may produce a negative effect for students, decreasing their sensorimotor programs as they adopt bad habits modelled by their teacher. Teachers therefore need to be able to demonstrate skills in a variety of ways in order to ensure that their students are able to comprehend the difference between effective and ineffective performance techniques.

We would like to conclude our chapter with a general remark. Processes involved with instrumental musical training are probably one of the most complex of all human activities. Essentially, these are not restricted to the sensorimotor brain circuits but also involve emotion, memory and imagination. The best trained musicians with the best working sensorimotor networks will not move their listeners if imagination, colour, fantasy and emotion are not a part of their artistic expression. These qualities are often not trained solely within a practice studio, but depend on and are possible linked to experiences from daily life, to human relationships, to a rich artistic environment and to empathy and emotional depth. These factors, which profoundly influence the aesthetic quality of music performance, are at present far from being accessible to any neuroscientific research.

References

Amunts, K., Schlaug, G., Jaencke, L., Steinmetz, H., Schleicher, A., Dabringhaus, A., & Zilles, K. (1997). Motor cortex and hand motor skills: structural compliance in the human brain. *Human Brain Mapping 5*, 206–215.

Bandura, A. (1997). *Self-efficacy: The exercise of control.* New York: W. H. Freeman & Company.

Bangert, M. & Altenmueller, E. (2003). Mapping perception to action in piano practice: a longitudinal dc-eeg-study. *BMC Neuroscience 4*, 26–36.

Bangert, M., Peschel, T., Rotte, M., Drescher, D., Hinrichs, H., Schlaug, G., Heinze, H. J., & Altenmueller, E. (2006). Shared networks for auditory and motor processing in professional pianists: Evidence from fMRI conjunction. *NeuroImage, 15*, 917-26. (2006).

Bengtsson, S. L., Nagy, Z., Skare, S., Forsman, L., Forssberg, H., & Ullen, F. (2005). Extensive piano practicing has regionally specific effects on white matter development. *Nature Neuroscience, 8*(9), 1148-1150.

D'Ausilio A, Altenmueller E, Olivetti Belardinelli M, & Lotze M. (2006). Cross modal plasticity of the motor cortex while listening to a rehearsed musical piece. *European Journal of Neuroscience 24(3)*, 955-958.

Depue, R.A. & Collins, P.F. (1999). Neurobiology of the structure of personality: Dopamine, facilitation of incentive motivation, and extraversion. *Behavioral Brain Science, 22*, 491-569.

Elbert, T., Pantev, C., Wienbruch, C., Rockstroh, B., & Taub, E. (1995). Increased cortical representation of the fingers of the left hand in string players. *Science, 270*, 305-307.

Ericsson, K. A. (Ed.). (1996). *The road to excellence: The acquisition of expert performance in the arts and sciences, sports, and games*. Mahwah, NJ: Erlbaum.

Ericsson, K. A., Krampe, R. T., & Tesch-Römer, C. (1993). The role of deliberate practice in the acquisition of expert performance. *Psychological Review, 100*, 363-406.

Ericsson, K. A. & Lehmann, A. C. (1996). Expert and exceptional performance: Evidence of maximal adaptation to task constraints. *Annual Review of Psychology, 47*, 273-305.

Gaser, C. & Schlaug, G. (2003). Brain structures differ between musicians and non-musicians. *Journal of Neuroscience 23*, 9240–9245.

Gentner, R. & Classen, J. (2006). Modular organization of finger movements by the human central nervous system. *Neuron 52*, 731-742.

Halsband, U., & Lange, R.K. (2006). Motor learning in man: a review of functional and clinical studies. *Journal of Physiology Paris 99*, 414-424.

Haslinger, B., Erhard, P., Altenmueller, E., Schroeder, U., Boecker, H., & Ceballos-Baumann, A. O. (2005). Transmodal sensorimotor networks during action observation in professional pianists. *Journal of Cognitive Neuroscience 17,* 282–293.

Heinen. F., Fietzek, U. M., Berweck, S., Hufschmidt, A., Deuschl, G., & Korinthenberg, R. (1998). Fast corticospinal system and motor performance in children: conduction proceeds skill. *Pediatric Neurology 19*, 217-221.

Hundt-Georgiadis, M., & von Cramon, D. Y. (1999). Motor-learning- related changes in piano players and non- musicians revealed by functional magnetic-resonance signals. *Experimental Brain Research 125*, 417– 425.

Jaencke, L., Shah, N.J., & Peters, M. (2000). Cortical activations in primary and secondary motor areas for complex bimanual movements in professional pianists. *Cognitive Brain Research 10*, 177-183.

Karni, A., Meyer, G., Jezzard, P., Adams, M. M., Turner, R., & Ungerleider, L. G. (1995). Functional MRI evidence for adult motor cortex plasticity during motor skill learning. *Nature 377*, 155–158.

Keitz, M., Martin-Soelch, C., & Leenders, K.L. (2003). Reward processing in the brain: a prerequisite for movement preparation? *Neural Plasticity 10,*121-128.

Kohut, D. L. (1985). *Musical performance: Learning theory and pedagogy.* Englewood Cliffs, New Jersey: Prentice-Hall.

Kuhtz-Buschbeck, J. P., Mahnkopf, C., Holzknecht, C., Siebner, H., Ulmer, S., & Jansen, O. (2003). Effector-independent representations of simple and complex imagined finger movements: a combined fMRI and TMS study. *European Journal of Neuroscience 18*, 3375–3387.

Lehmann, A. C., & Gruber, H. (2006). Music (pp. 457-488). In K. Anders Ericsson, N. Charness, P. J. Feltovich & R. R. Hoffman (Eds.), *The Cambridge handbook of expertise and expert performance.* Cambridge: Cambridge University Press.

Logothetis, N.K., Pauls, J., Augath, M., Trinath, T., & Oeltermann, A. (2001). Neurophysiological investigation of the basis of the fMRI signal. *Nature, 395 (6843)*, 150-157.

McPherson, G.E. (2005). From child to musician: skill development during the beginning stages of learning an instrument. *Psychology of Music 33*, 5-35.

McPherson, G. E., & Davidson, J. W. (2006). Playing an instrument (pp. 331-351). In G. E. McPherson (Ed.), *The child as musician: A handbook of musical development.* Oxford: Oxford University Press.

McPherson, G. E., & Zimmerman, B. J. (2002). Self-regulation of musical learning: A social cognitive perspective (pp. 327-347). In R. Colwell & C. Richardson (Eds.), *The New handbook of research on music teaching and learning*. Oxford: Oxford University Press: New York.

Muente, T. F., Altenmueller, E. , & Jaencke, L. (2002). The musician's brain as a model of neuroplasticity. *Nature Neuroscience 3, 473–478.*

Pantev. C., Oostenveld. R., Engelien. A., Ross, B., Roberts, L.E., & Hoke, M. (1998). Increased auditory cortical representation in musicians. *Nature. 392(6678),* 811-814.

Pascual-Leone, A., Dang, N., Cohen, L. G., Brasil-Neto, J. P., Cammarota, A., & Hallett, M. (1995). Modulation of muscle responses evoked by transcranial magnetic stimulation during the acquisition of new fine motor skills. *Journal of Neurophysiology* 74, 1037–1045.

Ridding M.C., Brouwer B., & Nordstrom M. A. (2000). Reduced interhemispheric inhibition in musicians. *Experimental Brain Research 133*, 249-253.

Rijntjes, M, Dettmers, C., Buchel, C., Kiebel, S., Frackowiak, R. S., & Weiller, C. (1999). A blueprint for movement: Functional and anatomical representations in the human motor system. *Journal of Neuroscience 19*, 8043–8048.

Roland, P. E., Larsen, B., Lassen, N. A., & Skinhoj, E. (1980). Supplementary motor area and other cortical areas in the organization of voluntary movements in man. *Journal of Neurophysiology 43*, 118–136.

Schlaug, G., Jaencke, L., Huang, Y., & Steinmetz, H. (1995a). Increased corpus callosum size in musicians. *Neuropsychologia 33*, 1047–1055.

Schlaug, G., Jaencke, L., Huang, Y., Steinmetz, H. (1995b). In vivo-evidence of structural brain asymmetry in musicians. *Science 267,* 699-701.

Schlaug, G., Norton, A., Overy, K., & Winner, E. (2005). Effects of music training on the child's brain and cognitive development. *Annals of the New York Academy of Sciences vol. 1060*, 219-30.

Seger, C.A. (2006). The basal ganglia in human learning. *Neuroscientist 12,* 285-290.

Sergent, J. (1993). Mapping the musician brain. *Human Brain Mapping 1*, 20-38.

Sowell, E.R., Thompson, P.M., Tessner, K.D., & Toga, A.W. (2001). Mapping continued brain growth and gray matter density reduction in dorsal frontal cortex: Inverse relationships during postadolescent brain maturation. *Journal of Neuroscience 21*, 8819-8829.

In: Neurosciences in Music Pedagogy
Editors: W. Gruhn and F. Rauscher, pp. 145-169
ISBN: 978-1-60021-834-7

Chapter VI

NEUROSCIENCE OF MUSIC AND EMOTION

Gunter Kreutz and Martin Lotze

ABSTRACT

The chapter discusses recent studies of neural representations of musical emotions. Despite an increasing number of empirical studies on the emotional effects of music, there is paucity of brain research that has identified the underlying neural networks. It is argued that the representation of musical emotions might be based on similar structures as compared to emotions in other domains. Consequently, it is hypothesized that musical emotions recruit networks of emotion processing that are known to be involved in both visual and auditory (speech) perception and that are responsible for psychological as well as physiological responses to emotional stimuli. Beside the provision of theoretic and empirical accounts, the review addresses methodological issues of current imaging techniques that may be particularly detrimental for the study of neural correlates of musical emotions. However, research findings so far support the assumption of largely overlapping networks required for the processing of both general and music-specific emotions. They also corroborate the notion of closely interconnected networks for cognitive and emotional processing even within the same neural structures. Specifically, emotional processing of music most likely involves limbic and paralimbic structures that include amygdala, hippocampus, parahippocampal gyrus, insula, temporal poles, ventral striatum, orbitofrontal cortex, and the cingulate cortex. It is further assumed that the behavioral distinction between perceptual and experiential (or feeling) aspects of musical emotions might be established within these neural structures. More research is needed, for example, to address the generation of musical emotions in the performer as well as developmental aspects and individual differences in neural representations of musical emotions.

MUSIC, EMOTION, AND THE HUMAN BRAIN

Emotions are critical to our understanding of music. They have long been the subject of debate among philosophers, artists and scientists in Western culture and other musical traditions around the world. According to the Greek philosophers, for example, music listening could influence the physiology and character of an individual and even affect subsequent behavior (e.g. Freeman, 2000). These authors not only stressed the therapeutic and educational power of music, but also warned against the negative consequences of musical malpractice for individuals and society.

Since the end of the 19th century, research methods in psychology and physiology have been able to elucidate philosophical speculations. For once, the relationship between musical structure and emotional expression has proved a viable approach. Music is thought to convey emotional information via an interactive, communicative process involving composer, performer and listener. It has been shown that the structural cues in musical scores that reflect the intentions of the composer interact with the cues that are under the control of the performer to shape the emotional expression that is communicated to the listener. For example, the expression of 'sadness' in a piece of music conveyed by the slow tempi and soft timbres indicated by the composer can be enhanced by the deliberate choices of articulation and nuances of timing made by the performer (e.g. Juslin, 2001).

Until recently, the plethora of research on musical emotions has been directed towards the effects of listening to music. More active musical behaviors such as singing, playing musical instruments, or dancing may produce different kinds of emotional response. For example, the extent to which and how emotional experiences might interfere with performance *per se*, has long been debated, since performers are also listening to themselves, in the first place.

This chapter focuses on neuroscientific approaches to music and emotion, which has become an important domain of interdisciplinary research (Juslin & Sloboda, 2001; Kreutz, *in press*). This relatively new field draws from the philosophical and behavioral research traditions rooted in the social sciences on one hand, and the advances of modern neuropsychological and imaging techniques emerging from the natural sciences on the other. The question arises as to the structure and function of neural correlates in the perception and experience of musical emotions. Some might belong to the auditory modality in the sensory-motor system, while others are neither specific to auditory processing nor the processing of music.

Whether the neural basis of emotional appraisal of music is innate or rather acquired through acculturation is subject to ongoing discussion (Peretz & Sloboda, 2005). Beside the nature-nurture problem, there are more puzzling phenomena in relation to musical emotions. For example, on the one hand, the reliable recognition of emotions, unlike the perceptual processing of music *per se*, appears independent of musical training and occurs in time windows so short that such experiences qualify as *reflexes* (Peretz, Gagnon & Bouchard, 1998; Bigand, Filipic & Lalitte, 2005). On the other hand, emotions are often seen as dynamically unfolding processes, in which physiological (Krumhansl, 1997) as well as brain activation changes occur over time (Koelsch, Fritz, v.Cramon, Müller & Friederici, 2006). These dynamic changes appear to be associated with the experienced intensity of emotions, sometimes culminating in pleasant sensations such as "chills" (Grewe, Nagel, Kopiez & Altenmueller, 2005; Panksepp, 1995) that may indicate the release of endorphins (McKinney, Tims & Kumar, 1997).

In another vein, repeated exposure to complete pieces of music often modulates appreciation, giving rise to so-called *exposure effects* (Samson & Peretz, 2005). Note that recognizing and experiencing musical emotion are different: not only do they involve different temporal aspects of processing, but they also might reflect different modes of processing that interact with time, situation, context, and the musical biographies of individual listeners.

Many researchers argue that emotional responses to music should be at least partially similar to emotional responses in other domains such as, for example, vision and speech (Peretz, 2001). To this end, introspective and behavioral approaches as well as peripheral physiological measurements of hemodynamic, respiratory, galvanic skin responses, and other parameters of bodily function have been used (Bartlett, 1996). The precise relationships between subjective experience and physiological concomitants during the processing of emotions are largely unknown. However, it is likely that brain research will provide important new information to reduce the amount of unexplained variances in psychophysiological studies of emotion in general and musical emotion in particular. Moreover, studying musical emotions from a neuroscientific perspective seems particularly promising with respect to the time course and dynamics of human emotion processing (Koelsch, 2006; Grewe et al., 2005; for a review of continuous measurement techniques used in behavioral studies see Schubert, 2001).

A recurrent observation is that the same brain structures may serve different roles in functional architectures within and across domains. Features of the brain such as plasticity as the result of learning and its ability to reorganize neural functions in the case of physical damage or disease suggest that any modeling of brain function needs to account for this flexibility. Thus the neuroscientific perspective on music and emotion necessitates interdisciplinary approaches, in which general models and theories of emotion, music theory, and brain research combine (Peretz, 2001). Although the implications of this approach to learning and education will be dealt with in the last chapter at the end of this book, it can be stated from the outset, that emotions – as a means rather than an end – appear to play a major role in modulating human learning processes. Investigating the underlying neural structures and the functions of musical emotion might thus be an important step towards understanding human learning processes within and beyond the domain of music.

In the remainder of this chapter, we will first deal with the neural structures that have been implicated in emotion processing. Then we will address empirical evidence relating to the neural correlates of musical emotion. The final goal of this chapter is to summarize our current knowledge of brain structures and functions related to the emotional processing of music, as well as to suggest some possible topics for future research.

Neural Correlates of Emotion Processing

Historically, emotion in the human brain has been considered with respect, first, to hemispheric differences, and, second, to the distinction between high-level cortical structures and deeper-level limbic structures. In fact, as early as 1912, Mills suggested that cognition and emotion were functionally specified and separated, postulating an analytic-cognitive dominance for the left and an emotional dominance for the right hemisphere (Mills, 1912). Later in the first half of the 20th century, McLean (1949) proposed the idea of a structure-function relationship between the limbic system and the generation of emotions (for an overview of the historical development of this hypothesis, see LeDoux, 1996).

Whereas the traditional view on brain processing differentiates the cognitive and emotional domains (Drevets & Raichle, 1998), there are other approaches that integrate the two. Recent studies provide evidence of a substantial overlap between the neural structures involved in both cognition and emotion. Indeed, cognitive and emotive systems might be far more interconnected and indistinct at neuronal levels (Wager & Feldmann Barrett, 2004). Some researchers contend that emotion even precedes cognition as the former is critical to motivating some cognitive and behavioral processes such as decision-making or executing movements. Specifically, the origins of attention control are difficult to conceive of in the absence of affect. According to this view, emotions arise from cognitive appraisals of situations, some of which are relevant to survival (Lazarus, 1991). However, emotion and cognition may interact in more complex ways. For example, Schachter and Singer (1962) observed that after participants received injections of epinephrine, their specific interpretation of a 'felt' emotion depended on contextual information rather than the physiological state of arousal induced by the drug. From this view, it would be less plausible to assume that experiences of emotions are more or less hard-wired cognitive interpretations of physiological states as implicated, for example, by the classic theory by William James (1890).

The idea of emotion as preceding cognition has also emerged from theories of the evolution of the human brain. The basic affective experience that arises when a self-relevant event occurs has been termed *core affect* (Russell & Barrett, 1999). Wager and Feldman Barrett (2004) suggest that with this core affect "... arise the physiological and motivated response tendencies that have been shaped over the course of our evolution to promote adaptive cognitions and behaviors. Thus, in this view, emotion and cognition are not opponents in a zero-sum tug of war. Rather, they are synergistic partners in the game of adaptive self-regulation, each shaping the direction of the other" (Wager & Feldman Barrett, 2004).

To conclude, the idea of a clear separation between cognition and emotion on the basis of structure-functional differentiation appears inadequate. Therefore, broader concepts of one or possibly several emotion-related neural systems at deeper levels in the brain interacting with the high-level cortical structures that are engaged in processes such as, for example, perception, decision-making, or motor-control, are favored by most researchers (Damasio, 1998; LeDoux, 1996).

Valence and Arousal

One important proposal for distinguishing emotions has been the concept of so-called *valence*. The valence of a given stimulus indicates whether it is likely to induce an approach towards or a withdrawal from that stimulus. In neural terms, the left frontolimbic areas should represent positive, the right negative emotional content (Davidson, 1984). However, appetitive/aversive stimuli might not always induce approach/withdrawal behaviors. The precise relationship between the antecedents and consequences of emotional experiences are often modulated by a variety of interfering cognitive processes. The present view of emotional processing, therefore, is moving away from "thinking in terms of an integrated neural system that codes all emotional processes towards thinking in terms of individual neural systems coding distinct dimensions of emotion or different affect programs" (Murphy, Nimmo-Smith & Lawrence, 2003, p. 208).

Emotions differ not only with respect to their valence, varying from negative to positive, but also in their *arousal potential*, varying from low to high. The idea is that these dimensions are considered independent of each other, which suggests that they can be conceptualized as orthogonal in a two-dimensional space (Russell, 1980). While some emotions, such as sadness and peacefulness, are low in associated levels of activity, other emotions, such as anger and happiness, are considered highly activating. On the basis of these two dimensions, valence and arousal, the psychological literature most commonly differentiates between a small number of basic emotions: fear, anger, disgust, happiness, surprise and sadness (Ekman & Friesen, 1982). Research on facial expressions has shown that these basic emotions are communicated across, as well as within, various ethnicities and have been found to be associated with psychophysiological changes (Ekman, Levenson, & Friesen, 1983).

Multivariate analysis of verbal responses to emotional stimuli has shown that most of the variance in descriptions of emotions can be explained by valence and arousal (Mehrabian & Russel, 1974). Moreover, verbal reports and peripheral physiological responses are significantly and differentially correlated along the two dimensions of valence and arousal (Greenwald, Cook & Lang, 1989). In particular, startle reflex amplitudes, for example, increase with negative valence and decrease with positive valence (for a comprehensive review see Bradley and Vrana, 1993), whereas skin conductance response (SCR) increases with subjective arousal, independently of emotional valence. Note that the latter observation has been confirmed recently in studies using music stimuli (Khalfa, Peretz, Blondin & Manon, 2002). Nyklicek, Thayer, and van Doornen (1997) observed in their study using impedance cardiography (ICG) that a respiratory component was significantly related to subjective arousal, while other cardiovascular measures (inter-beat-interval and left ventricular ejection time) decreased for all emotions except for happiness, which the authors interpreted as a possible indication of physiological response related to valence. In another ICG-study, Kreutz, Bongard, and von Jussis (2002) found that the 'happy' or 'sad' tone of music excerpts was differentiated in both expert musicians and non-musicians by a weaker sympathetic influence on the hemodynamic response in the latter group.

Neuroscientific approaches to identifying networks associated with valence and arousal have partially corroborated the view of a dimensional representation of emotion (e.g., Lang,

Bradley, Fitzsimmons et al., 1998). For example, Anders et al. (2004) conducted a correlation analysis between activity in different brain areas and individual emotional valence and arousal responses. They found that responses along the valence dimension were positively correlated with activity in the insular cortex, and arousal was correlated with thalamic activity. The second important finding of this study was the functional segregation of brain structures underlying the peripheral physiological responses and verbal reports: startle reflex augmentation was associated with amygdala activity and skin conductance responses with frontomedial activity (Critchley, Elliott, Mathias, & Dolan, 2000).

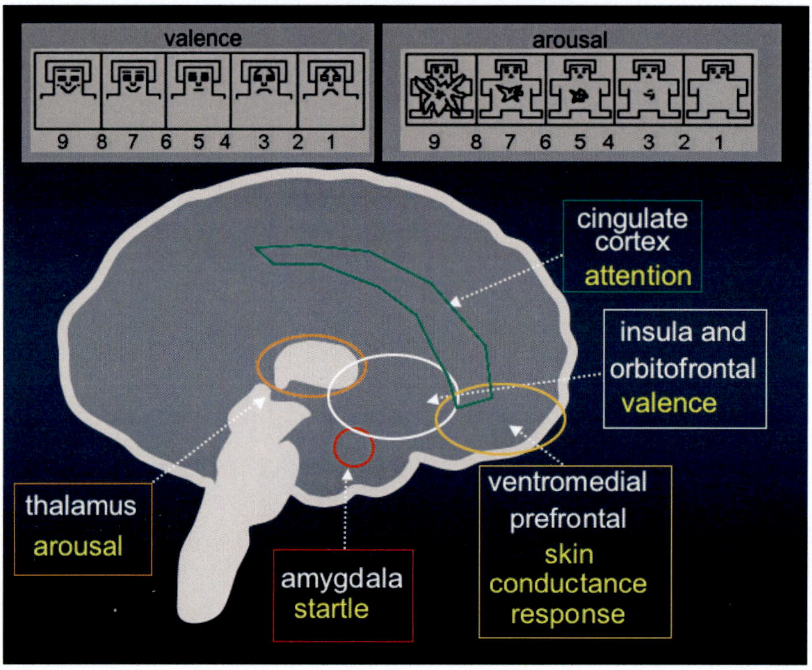

Figure 1. By using correlation analyses interactions between fMRI-activation and other measures used for quantifying the emotional intensity perceived can be detected. These are peripheral physiological measures such as skin conductance response (sweat) or startle amplitude (eye blink increase) but also ratings along the dimensions of valence and arousal. Anders et al. (2004) showed that fMRI-activation in the thalamus correlated positively with arousal, in the amygdala with the startle response, in the ventromedial prefrontal cortex with the skin conductance response and in the insula (and the adjacent orbitofrontal cortex: Wildgruber, Hertrich, Riecker et al., 2004; Lotze, Heymans, Birbaumer et al., 2006b) with the valence perceived. Additionally, activation in the anterior cingulate cortex is associated with attention to the stimulus that is more cognitive in the dorsal and more emotional in the ventral part (Bush, Luu & Posner, 2000).

Lesion and Imaging Studies

Recent studies of impaired emotional processing in patients with circumscribed lesions of the bilateral amygdala suggest impairments of the recognition and experience of facial expressions of fear in these patients (Adolphs, Gosselin, Buchanan, Tranel, Schyns & Damasio, 2005). By contrast, lesions of the insula and the surrounding basal ganglia lead to problems in the recognition of expressions representing disgust (for a review see Calder, Lawrence and Young, 2001). Interpretations of lesion studies in the context of studies of emotion (or otherwise) are restricted, because their sites often extend beyond the specific areas of interest. To bypass this problem, repetitive Transcranial Magnetic Stimulation (rTMS) has been recently developed as a method to artificially induce temporary lesions in healthy participants to allow investigations of the functional relevance of confined cortical areas for a given task (Cohen, Celnik, Pascual-Leone et al., 1997; Lotze, Markert, Sauseng et al., 2006a). Therefore, the focalized stimulation capacity of this method overcomes some of the limitations that arise from lesion studies. However, additional tools are needed to explore emotion processing so as to account for potential interactions between networks distributed across the brain.

Methodological Issues in Brain Imaging

Functional imaging techniques have been introduced to observe patterns of activation across the entire brain. These methods usually involve measuring changes in electric, magnetic, or metabolic activity over time and across cerebral regions that, depending on the location and size of area investigated, involve thousands to millions of individual nerve cells. These spatiotemporal changes of activity are then often correlated with behavioral as well as peripheral physiological measures. Note that the observed patterns of activation only indirectly represent underlying neural processing and, more importantly, are not necessarily *caused* by (correlated) behavioral information, or vice versa. Instead, relationships between brain activation on the one hand, and subjective or peripheral measures on the other, could well be mediated by intervening processes.

One useful method to identify structure-function associations is characterized as the subtractive approach. Activation patterns represented in so-called brain maps are obtained in different experimental conditions. Then, activation values taken during each condition are subtracted using either brain map as reference. The remaining activation might be interpreted as related to the specific experimental condition. This strategy has been applied in studies of the perceptual and emotional processing of visual materials. It has been shown, however, that disentangling the different functions of neural networks in different brain maps is particularly difficult when studying emotional processing (Murphy et al., 2003). For example, certain patterns of activation might well reflect aspects of visual processing that are induced by the attentional or emotional content of the perceived stimulus rather than being attributable to the emotion circuit (Lang et al., 1998). In other words, interpretations from the subtraction approach are clearly limited, because it is often difficult to associate the observed patterns of activation in the brain with specific functions in stimulus processing.

The methodological problems just mentioned extend to a more conceptual level. It must be noted that the theoretical separation of brain functions appears somewhat artificial given the extreme overlap of different elements of processing in real life situations. For instance, performing and listening to music involve multiple affective processes that are difficult to differentiate. These include, for example, perceiving and feeling the expression of a piece, interacting with others, memorizing and performing music from memory, analyzing music, being motivated to learn an instrument and much more.

Moreover, functional imaging techniques such as Positron Emission Tomography (PET) and functional Magnetic Resonance Imaging (fMRI) are problematic for reasons of experimental economy. Due to disadvantageous signal-to-noise ratios, effect sizes are usually small, while scanning protocols and data evaluation are subject to highly elaborate – and for the participant, uncomfortable – procedures. Furthermore, scanning noise produced by fMRI may have a deleterious effect on auditory processing, if the focus of the study includes music.

Another common problem, related to the low signal-to-noise ratio just mentioned, may arise when there are so few participants in a study that researchers are forced to use low activation thresholds (i.e. when observed patterns of brain activation are deemed significant). In such circumstances, increased blood oxygenation in one hemisphere may be wrongly interpreted as a specialization of this hemisphere for the given task. Obviously, cross-validation by replication is impaired if inadequate activation threshold levels are used.

A final methodological problem in the context of emotion studies arises from the neuroanatomical construction of the limbic system. Measurements of this system are subject to substantial artifacts based on the different densities of adjacent structures: there are air cavities next to the basal forebrain. Differences in lateralization of the amygdala, for example, might therefore result from certain scanning parameters, which can vary from study to study (Robinson, Windischberger, Rauscher & Moser, 2004).

Emotion-Related Neural Structures as Revealed by Imaging Studies

The most recent review articles on studies focusing on the functional specificity of areas related to the emotional processing system suggest no hemispheric lateralization of activation during the processing of basic emotions (Phan, Wager, Taylor, & Liberzon, 2002; 2004a; Murphy et al., 2003). Instead, after entering results from 106 imaging studies into meta-analytical procedures, almost none of the hypotheses on localization of emotional processing were upheld. Neither was emotion processing found to be restricted to classical limbic areas, nor did hemispheric lateralization of emotional valence emerge (Murphy et al., 2003). However, a further review of 55 neuroimaging studies on emotional processing converged in identifying a number of emotion-related neural circuits that were located in the bilateral amygdala, the bilateral insula (especially its anterior part: AI), the anterior cingulate cortex (ACC), and the medial and lateral ventral prefrontal cortex (mPFC/vPFC) (Phan et al., 2002).

Converging evidence from lesion and imaging studies suggests that fear and disgust are processed in different areas. Whereas fear activated the bilateral amygdala, disgust activated most strongly the insula and the basal ganglia (pallidum) (Murphy et al., 2003). By contrast, the same meta-analysis suggests that happiness and sadness did not reliably activate distinct representation sites. Finally, the processing of anger appeared to be associated with activation in the lateral orbitofrontal cortex. In sum, the available imaging (and lesion) data point towards a neural system of emotion processing that is far more complex than that which had been envisaged by some earlier proposals.

The meta-analyses mentioned above (Phan et al., 2002, Murphy et al., 2003) also suggest the amygdala to be associated particularly with the processing of fear, while specific patterns of activation were also observed in response to other emotions with both positive and negative valence. These findings converge with results from lesion studies of animals (Hitchcock & Davis, 1986, 1991), in that fear-related activations in the amygdala in humans are highly correlated with the startle response (Anders et al., 2004). Note that bilateral simultaneous activation of this structure and the parahippocampus is also associated with the perception of non-pleasant musical material (Blood, Zatorre, Bermudez & Evans, 1999) and is negatively correlated with the perception of intensely pleasurable episodes, so-called *chills*, while listening to music (Blood & Zatorre, 2001).

In the search for neural correlates during highly violent computer gaming the amygdala as well as the anterior cingulate cortex (ACC) were found to be deactivated (Mathiak et al., 2006). Aggression may decrease activation in areas associated with fear, but it increases activation in areas associated with the cognitive and attentional load during affective tasks, such as the more dorsal part of the cingulate gyrus (Bush et al., 2000). The ACC is closely interconnected with the amygdala and the mPFC (Devinsky, Morrell, & Vogt, 1995). Both the ACC and mPFC are highly correlated with changes in SCR and arousal. Whereas Murphy et al. (2003) found no indication that ACC plays a specialized role in emotion processing, Phan, Wager, Taylor and Liberzon (2004a) reported that in most studies ACC-activation was associated with a feeling of sadness. This report is corroborated by the findings from a study involving depressed patients who had recovered from their condition following pharmacological therapy (Mayberg, Brannan, Mahurin et al., 1997) and demonstrated altered ACC-activation from before to after the treatment.

A functionally relevant connection between ACC-activation and the intensity of subjectively experienced pain has been shown using fMRI-feedback methods. In particular, decreasing activation in the ACC correlated with a decrease in perceived pain (deCharms, Maeda, Glover et al., 2005). As the possibility that increases in ACC activity increased pain can be ruled out, rather than *vice versa,* such studies suggest ways of investigating potential causal links between sensory stimulation and activation in specific regions of interest. This method is already used for studying the modulation of one particular emotion: fear (Veit et al., 2006).

The insula and surrounding operculum contain primary cortical representations of smell and taste (Rolls, Critchley & Treves, 1996), somatosensory and viscero-sensation (Lotze, Wietek, Birbaumer et al., 2001), and pain perception (Apkarian, Bushnell, Treede, & Zbieta, 2005). For this reason, it has been termed the "limbic sensory cortex", and it is associated with the subjective feeling of emotional states. The anterior insula seems especially to reflect the self-conscious aspect of the rated intensity of a given emotion as observed for the visual modality in a parametric activation study. This technique involves regression analyses of brain maps using, for example, the subjective valence values induced by emotional pictures as predictors (Anders et al., 2004). In one particular study, the valence attributed to observed expressive gestures correlated with the activation of the anterior insula and the adjacent lateral orbitofrontal lobe (Lotze et al., 2006b). Whereas the anterior insula is associated with emotional valence, the dorsal part of this structure is more active during cognitive tasks (for a meta-analysis, see Wager & Feldmann Barrett, 2004). It seems that cognitive and emotional aspects of higher-level stimulus processing are represented in tightly adjacent regions in the insula.

Interactions between cognitive and emotional processing have also been observed in the mPFC. Activation of the *dorsal* mPFC has been observed during the cognitive regulation of emotional behavior (Ochsner, Ray, Cooper et al., 2004a), and when participants made judgments about other people's emotional states (Ochsner, Knierim, Ludlow et al., 2004b). In contrast, activity in the *ventral* mPFC has been associated with monitoring one's own feelings (Phan, Taylor, Welsh et al., 2004b) and the physiological changes that accompany particular emotional responses (Damasio, 1996). With respect to music, the modulation of emotions in the *dorsal* mPFC might be especially important since suppression of any emotional feeling during musical perception is most likely to be related to activation in this area. Activity in the *ventral* mPFC, however, is assumed to be associated with bodily responses such as shivers, sweating, and increases of heart rate during intensely pleasurable music (Blood & Zatorre, 2001).

The nucleus accumbens, located in the inferior part of the striatum, is active when individuals are rewarded with a pleasant odour (Gottfried, O'Doherty, & Dolan, 2002) but also when they experience revenge in a social interactive procedure after being cheated (Singer, Seymour, O'Doherty et al., 2006). The nucleus accumbens is often associated with self-rewarding processes (Menon & Levitin, 2005) and might to play some role in the motivation for learning to play musical instruments. Recently, activations of the nucleus accumbens were observed in emotional responses during listening to pleasant music (Blood & Zatorre, 2001; Kreutz, Ott & Wehrum, 2006).

In sum, investigations of emotion processing across specific domains suggest higher specialization of some areas than the meta-analyses are able to demonstrate. These studies also corroborate the view that cognitive and emotional networks are closely interconnected even within the same anatomical regions.

Empirical Research on Neural Correlates of Musical Emotion

Binaural Listening Tasks

Peretz et al. (1998) investigated musical judgments in affective and non-affective tasks, while participants listened monaurally to tonal and atonal melodies. The authors observed that tonal melodies were found more pleasant when processed in the left hemisphere, whereas atonal melodies were found more unpleasant when processed in the right hemisphere. These asymmetries were not present when listeners rated the correctness of melodies (Peretz et al., 1998). By contrast, Leichner and Bröscher (1999) concluded from their study using a similar research method, that valence was primarily evaluated by the right hemisphere. The authors used pre-selected examples of classical music to represent four categories, namely positive activating, positive calming, negative activating, and negative calming. Valence ratings were similar for both positive and negative music when stimuli were presented to the left hemisphere (right ear), and dissimilar, when the music was presented to the right or both hemispheres. Moreover, ratings on the 'activating-calming' dimension were markedly different for positive and negative music depending on mode of presentation. When presented to the right hemisphere or both hemispheres, negative music was found less activating (i.e. more calming) than positive music. However, when the same excerpts were presented to the left hemisphere, judgments were reversed in the sense that, although not significantly different, positive music was found less activating and negative music was found more activating (i.e. less calming). The authors speculate that there is a hemispheric dominance for some surface features of music that are relevant to emotional processing (Leichner & Bröscher, 1999).

Aphasia and Amusia

Broca's area is an extremely important region of the brain, in the vicinity of the auditory cortices; it has traditionally been implicated in speech production. Consequently, patients with lesions in this area, particularly in the left hemisphere, often suffer from impaired speech production. However, it is note-worthy that some patients with *aphasia* are less affected in their musical abilities. Moreover, singing the lyrics of a song is easier for them than speaking the same words (Kaplan & Gardner, 1990). By contrast, lesions in the right hemisphere are more likely to impair the representation of melody contours (Peretz, 1990) and pitch processing (Murayama, Kashiwagi, Kashiwagi & Mimura, 2004), a condition that has been termed *amusia* (see also chapter 9).

Thus there seems to be some dissociation between the representation of speech and music that emerges from the laterality of cortical processing. Lesion studies are informative in this dissociation. Amusia may involve problems with perceptual processing as well as other skills such as memory for music. Importantly, however, amusia cannot – in most cases – be attributed to general sensory, attentional, or working memory impairments. Only rarely are characteristics of amusia such as *tone-deafness* observed in healthy people (Foxton, Dean, Gee, Peretz & Griffiths, 2004; Sloboda, Wise & Peretz, 2005)

Amusia has been found specifically in patients with bilateral anterior temporal lesions (Satoh, Takeda, Murakami et al., 2005). Other patients show deficits in the retrieval of musical material after right hemisphere posterior medial stroke including the inferior right frontal gyrus, the posterior temporal lobe and the inferior parietal lobe (Schön, Lorber, Spacal & Semenza, 2003). Peretz (1990) summarized a sequence of studies addressing music and emotion processing in a female "patient IR, who suffers from a longstanding bilateral brain damage to the auditory cortex" (Peretz, 1990, p. 116). Interestingly, IR shows normal speech, intelligence and memory capacities, while being unable to recognize familiar tunes or learn novel melodies. One of the most remarkable aspects of her musical deficits is that the recognition of the emotion conveyed by a given melody or musical excerpt is spared. For example, in one task, IR was asked to rate the 'happy' or 'sad' emotional tone in one half of a set of melodies while judging the degree of 'familiarity' in the other half, using the same response format for each scale. The results indicated that IR recognizes musical emotions just like healthy control participants, while being unable to recognize, discriminate, or learn to identify any of the tunes. Samson and Peretz (2005) studied patients suffering from temporal lobe lesions; their findings suggest that right temporal lobe structures play a critical role in priming and recognition memory for melodies as reflected by the absence of an exposure effect. In sum, these studies suggest cortical contributions to musical preference and liking judgments (Samson & Peretz, 2005).

Griffiths, Warren, Dean, and Howard (2004) observed a double dissociation between musical cognition and emotional processing in a 52-year-old patient – a radio announcer – who suffered from a focal lesion in the left insula, hippocampus and amygdala after a stroke. He retained most of his speech and motor functions, and his hearing had returned to normal 12 months later. However, he did not recover the particular pleasure he had previously experienced while listening to Rachmaninov preludes. Yet when he was tested for amusia by means of a standardized test battery, his perceptual abilities were found to be normal. The authors concluded that unilateral lesions in three structures of the limbic system resulted in a specific impairment for the feeling of pleasure elicited by music in this patient (Griffiths et al., 2004).

Brain Imaging Studies on Music Listening

Brain imaging studies on emotional responses to music have addressed a variety of research questions related to both the effects of music as well as the influences of individual differences on emotional responses. Some studies looked at brain activations while listening to music differing in emotional valence (Blood et al. 1999; Blood & Zatorre, 2001; Koelsch & Fritz, 2003), emotional tone, e.g. 'happy' or 'sad' (Kreutz, Russ, Bongard & Lanfermann, 2003), and in structural characteristics giving rise to emotional processing (Koelsch, Fritz, Schulze, Alsop & Schlaug, 2005; Pallesen, Brattico, Bailey et al. 2005). In recent studies, participants listened to music while viewing emotionally standardized pictures (Baumgartner, Esslen, & Jaencke, 2005; Spreckelmeyer, Kutas, Urbach et al., 2006). Few studies so far have investigated the influence of long-term individual differences such as musical training on brain responses to musical emotions (Koelsch et al., 2005; Pallesen et al., 2005).

Blood and Zatorre (2001) had their participants listen to music of their own choice, while peripheral cardiovascular and electrodermal activity as well as regional Cerebral Blood Flow (rCBF) changes were recorded using Positron Emission Tomography (PET). The purpose of using participant-selected music was to induce (pleasurable) *chills* reliably while listening. Subjective chill responses were then correlated with physiological measures. Participants' responses while listening to a self-selected piece were compared with responses while listening to the favorite piece of another participant. Musical chill-related brain responses were identified in a number of regions including ventral striatum, midbrain, amygdala, orbitofrontal cortex, and ventro medial prefrontal cortex (VMPF). These structures are implicated in reward and emotion by previous research based partially on animal models (see Figure 2).

In their PET study Brown, Martinez and Parsons (2004) investigated the effects of unfamiliar music on a group of listeners who appeared to enjoy the experimenters' choice of music. Again, increased rCBF was found in limbic and paralimbic structures, specifically the anterior cingulate cortex (ACC). The anterior insula, posterior hippocampus, superior temporal poles and nucleus accumbens, part of the ventral striatum, were activated when listening to music but not in a rest condition.

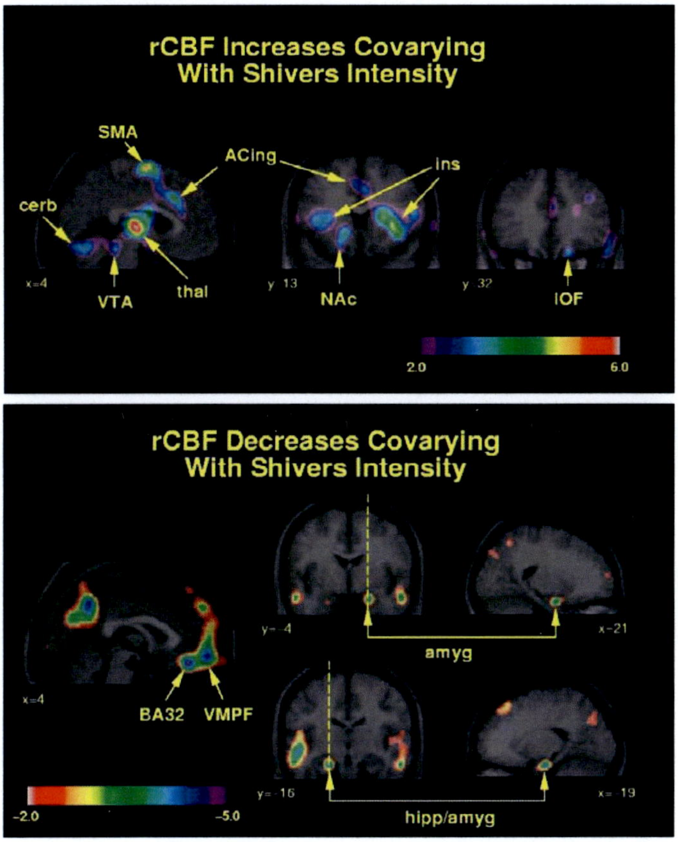

Figure 2. Neuroanatomical regions demonstrating significant rCBF correlations with chills intensity ratings. Regression analyses were used to correlate rCBF from averaged PET data for combined subject-selected and control music scans with ratings of chills intensity (0 to 10). Correlations are shown as t-statistic images superimposed on corresponding average MRI scans. The t-statistic ranges for each set of images are coded by color scales below each column, corresponding to *a–c* (positive correlations with increasing chills intensity), and *d–f* (negative correlations). *a* (sagittal section, x = 4 mm) shows positive rCBF correlations in left dorsomedial midbrain (Mb), right thalamus (Th), AC, SMA, and bilateral cerebellum (Cb). *b* (coronal section, $y = 13$ mm) shows left ventral striatum (VStr) and bilateral insula (In; also AC). *c* (coronal section, $y = 32$ mm) shows right orbitofrontal cortex (Of). *d* (sagittal section, $x = 4$ mm) shows negative rCBF correlations in VMPF and visual cortex (VC). e (sagittal section, $x = 21$ mm) shows right amygdala (Am). *f* (sagittal section, $x = -19$ mm) shows left hippocampus/amygdala (H/Am) (Figure 2 from PNAS, 98(20), p.11821, copyright (2001) National Academy of Sciences, U.S.A.).

Blood et al. (1999) addressed the valence dimension (pleasantness/unpleasantness) of emotion in a PET-study in which music stimuli with systematically varied degrees of dissonance were used. This variation resulted from the tonal relationship between a melody and its chord accompaniment. If the melody and accompaniment were in the same tonality, they were perceived as more consonant and more pleasant than when melody and accompaniment were derived from different tonalities, or when pitches with no harmonic relationship to the melody were included in the accompaniment. Parametric analyses of brain activation modulated by perceived pleasantness suggested that increasing unpleasantness activated the right parahippocampal gyrus and the precuneus, whereas increasing pleasantness correlated with activation in frontopolar, orbitofrontal regions and the ACC.

Notably, increases of rCBF in the orbitofrontal cortex, the temporal pole, and the superior frontal gyrus were observed in perceptual responses to positively valenced visual, auditory, and olfactory stimuli across all three modalities (Royet, Zald, Versace et. al., 2000). Although non-musical environmental (including speech) sounds were used to test the auditory modality, these results suggest links between music and the emotion-related patterns of activation in specific cortical and limbic structures. This particular link between auditory, music, and emotion processes appears most evident in the activation of the temporal poles (Brown et al., 2004; Kreutz et al., 2003; Royet et al., 2000), which may well be an important structural gateway between limbic and cortical areas.

Brain responses to manipulations of the harmonic structure of chord sequences (Koelsch et al., 2005), rather than of the degree of consonance as in the study by Blood et al. (1999), are believed to include areas of emotional processing, specifically the orbital frontolateral cortex (OFLC), which has been implicated in the evaluation of the emotional significance of sensory information (see Koelsch, 2005, p. 414). The study appears to lend some support to Leonard Meyer's theory of musical emotion as arising from violations of expectancy in the progress of musical sequences (Meyer, 1956). Note that brain activations including amygdala, retrospinal cortex, brain stem, and cerebellum were observed even when responses to isolated dissonant and minor chords were compared with responses to major chords (Pallesen et al., 2005). It seems possible that single musical events followed by silence suffice to induce expectations that might be felt as disrupted, leading to emotional responses (Bigand et al., 2005; Peretz et al., 1998).

Instead of computerized sounds, as in the study by Blood et al. (1999), Koelsch et al. (2006) used commercial recordings as basic materials in their fMRI-study. Excerpts were taken from instrumental dance music, e.g. *Tango Argentino*. From these excerpts, dissonant versions were constructed by digitally manipulating the sound files to introduce massive distortions while maintaining the overall physical time and amplitude characteristics of the music. Contrasts of brain activation in response to the original versus dissonant music showed that the original music activated extensive structures in the inferior frontal gyrus, anterior superior insula, ventral striatum, Heschl's gyrus and Rolandic operculum. Dissonant music, however, activated the amygdala, hippocampus, parahippocampal gyri and temporal poles. All of these sites, except for the hippocampus, were found to be more strongly activated after comparing BOLD signals from the first and the last 30 seconds of the music stimulation, which suggests dynamically changing activations over time. The authors conclude that activation of the amygdala during distorted music is corroborated by a previous report of impaired responses to music representing 'fear' in patients with amygdala resections (Gosselin, Peretz, Noulhiane et al., 2005). They also point out that, according to their own review of the literature, the same neural structures activated during dissonant music are also involved in the processing of pleasant stimuli, including the study by Blood and Zatorre (2001) mentioned above.

Kreutz et al. (2006) presented 25 classical music excerpts representing 'happiness', 'sadness', 'fear', 'anger', and 'peace', to listeners who rated each excerpt for emotion, valence and arousal. Ratings were entered into a parametric modulation analysis of activations in the entire brain. Results showed that valence as well as positive emotions were associated with activations in cortical and limbic areas including the anterior cingulum, basal ganglia, insula, and nucleus accumbens. Subsequent analyses of activation in other regions of interest largely supported these findings. Negative emotions, however, did not yield significant activations at group level (Kreutz et al., 2006).

Several EEG studies using various sets of musical stimuli (Altenmueller, Schuermann, Lim, & Parlitz, 2002; Schmidt and Trainor, 2001; Tsang et al., 2001) provide support for the hemispheric specialization hypothesis for emotional valence (Heilman, 1997), while. That is, music stimuli, which are judged positive or negative in valence, elicited asymmetrical frontal EEG activity. For example, Schmidt and Trainor (2001) found that music expressing joy and happiness elicited relatively greater EEG activity in the left fronto-temporal lobe, whereas music expressing fear and sadness elicited more activity in the right hemisphere. Moreover, faster tempi and the major mode produced greater responses in the left hemisphere, whereas slower tempi and minor mode were associated with greater responses in the right hemisphere (Tsang et al., 2001). Using a related technique (direct-current EEG), Altenmueller et al. (2002) played 160 music excerpts from four different genres to 16 right-handed students, who provided valence judgments for each stimulus. Positively valenced stimuli elicited bilateral fronto-temporal activations with preponderance of the left hemisphere, whereas negatively valenced stimuli elicited bilateral activations predominantly in the fronto-temporal right hemisphere. Females showed greater valence-related differences than did males. The authors conclude that the emotional responses observed in the EEG-patterns are independent of fine-grained musical structure. In consequence, the frontal temporal lobes seem to be involved in emotional evaluation and judgment rather than the perceptual analysis of emotional information (Heilman, 1997). Significant increases of frontal cortical activity as reflected in the 4-8 Hz power band of the EEG were observed over the first year of life in a developmental study on infants' responses to affective music (Schmidt, Trainor & Santesso, 2003). Music excerpts, varying in valence and arousal (happy, sad, fear), were also found to induce significant decreases in overall activity between groups of 3-month-old versus 12-month-old children. The authors suggest that taken together their findings, which showed an emerging asymmetry of activation in the presence of an overall decrease of EEG-power, indicate maturation of cortical music processing as well as a "calming" influence of music by the end of the first year of life (Schmidt, Trainor & Santesso, 2003).

Goydke, Altenmueller, Moeller, and Muente (2004) conducted event-related potential (ERP) experiments to investigate the preattentive discrimination of emotions in instrumental timbres. ERPs derive from EEG measures usually in time windows shorter than one second in duration. More specifically, these authors used a particular component of the ERP signal, the so-called mismatch negativity (MMN) as an index of the brain's capacity for identifying more or less subtle changes of sound. Non-musician listeners were presented with 'happy' and 'sad' versions of violin and flute sounds of 600 milliseconds duration, which also varied in pitch. The key finding was that emotional tone, like changes in pitch and timbre, triggered MMN. However, it is not clear, why the authors characterize the difference between 'happy' and 'sad' tones as more subtle than differences in pitch or timbre. It may be that MMN indexed not the perception of emotion *per se*, but rather – as it usually does – mere changes in the structural features of sound. These in turn feed into emotional processing at deeper levels of the brain. However, this does not preclude the possibility that cortical processing is essential to emotional experience (Damasio, 1998).

Brain Imaging Studies on Music Performance

Performing music – whether by singing, playing musical instruments, or dancing – is one of the most complex areas of human engagement in general. One might imagine that emotion in music performance must be based on similar neural processes to those involved in listening, on the grounds that performers listen to themselves. However, there are at least two hypotheses as to how the emotional effects of music performance might differ from those experienced in relation to less active musical behaviors such as listening. On the one hand, it may be assumed that cognitive processes interfere with emotional processes at both preattentive and attentive levels. Thus monitoring and integrating motor-sensory, tactile, kinesthetic, visual and auditory information, attentional and memory processes, etc., might *reduce* the intensity of emotional experiences. On the other hand, it could be argued, conversely, that similar perceptual and cognitive processes *enhance* emotional experiences during performance, because performance gestures appear to be partially conveying emotional information as one of their functions. However, to date it appears that investigations of the neural correlates of emotion in music performance are not nearly as advanced as, for example, research on cognitive processes in this domain.

Notably, psychological research on professional performance has emphasized the role of negative emotions in music performance such as performance anxiety (for a recent review see Steptoe, 2001), whereas a surprisingly small amount of research has addressed (or confirmed) potentially more positive emotional influences on musical activities (e.g., Kreutz, Bongard, Grebe, Rohrmann & Hodapp, 2004). It should be assumed that making music is associated with high levels of motivation and self-reward in order to be sustained as a lifetime commitment for many professional as well as amateur musicians. Indeed, social psychological research suggests that musicians often experience particular emotional states as characterized in the concept of *flow* (Csikszentmihalyi, 1990). In all accounts, however, emotional brain responses to music performance appear to be mediated by a range of factors including individual differences and situational context.

Singing is an obvious activity in which speech and music combine. There are musical aspects of speech, as reflected in so-called *prosody* (Bostanov & Kotchoubey, 2004; Mitchell, Elliott, Barry et al., 2003). Briefly, speech prosody is defined by the (intentional and unintentional) variation of fundamental pitch, intensity, and spectral information during speech production. Prosody is thought to influence the emotional tone of any given utterance. Imaging studies have shown that recognition and perception of prosody appears to be lateralized to the right hemisphere and is associated with activity in the lateral orbitofrontal lobe (Wildgruber et al., 2004). This area seems also important bilaterally for the recognition of valence in visual stimuli (Lotze et al., 2006b).

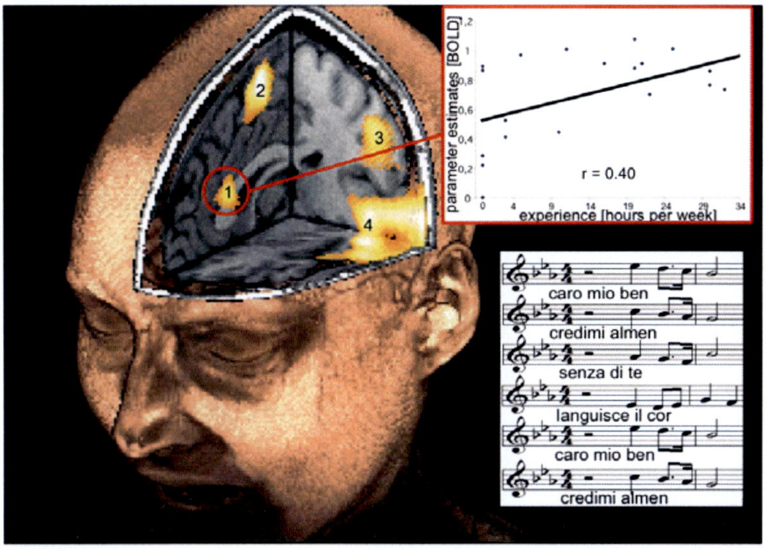

Figure 3. Functional map of a group of 25 participants with different experience in singing during singing an Italian 'aria antiche' ("caro mio ben"). Group statistical map was coregistered and then projected on an individual head. During singing the anterior cingulate cortex (1), the SMA (2), the bilateral sensorimotor cortex (3, only seen on the left side) and the primary and secondary auditory cortex (4) showed activation. Interestingly, activity in the anterior cingulate correlated positively ($r = 0.40$; $p < 0.05$) with the amount of practice per week. Some subjects were naïve but some (n=15) were professional singers (8 singing students and 7 opera singers) who performed singing up to 30 hours a week.

Figure 3 depicts patterns of brain activation in 25 participants while singing an Italian bel canto aria. Note that activation in the ACC depends on levels of singing expertise. Significant increase in ACC activation was shown in experienced singers, while naïve participants who had not received singing training (n = 5) showed no such increased activation. When interpreting these findings, it is worth noting that the ACC is also involved in the recall of emotions (Phan et al., 2004a), a capacity that might be crucial to the conveying of emotional expression in professional singing.

Conclusions

In music psychology, cognitive approaches to music listening and performance dominated during the last third of the 20th century, while research on musical emotion was scarce (cf. Dowling & Harwood, 1986; Lerdahl & Jackendoff, 1983). Nevertheless, over the course of the last decade emotion has now become an established topic. Some authors speculate, with regard to unsolved issues concerning the cognitive processing of music that "there is an even deeper mystery within brain organization to which all these cognitive issues are subservient" (Panksepp & Bernatzky, 2002, p. 134).

Experimental research on emotion is often designed to include measurements of subjective, brain imaging and peripheral physiological responses within single protocols (for a theoretical discussion, see Peper & Lüken, 2002). Investigations of the psychophysiological correlates of musical emotion follow this general trend. However, behavioral approaches in themselves (without or in combination with simultaneous physiological recordings) are of continued importance for several reasons. First, introspection is a key motivation for addressing the biological underpinnings of musical experiences and to ascertain the deep emotional impact of music on potentially every human (e.g. Gabrielsson, 2001). Second, behavioral information, which can be verbal or non-verbal, appears crucial in the generation and construction of research hypotheses and the development of experimental designs that include imaging methods. Third, subjective responses are indispensable when interpreting the physiological concomitants of human emotions (Panksepp & Bernatzky, 2002).

Hypothesis-driven approaches clearly have advantages over exploratory studies, as the former often lead to more precise interpretations. Inevitably, a much wider empirical basis is needed for using brain-imaging methods in the context of musical behavior and emotion. It appears that our current knowledge of neural correlates of musical emotions is based on very few studies each with very few participants. Therefore, in addition to those methodological problems discussed in section 2 of this paper, the interpretation of findings in the field of music appears even more constrained than that of findings from other domains such as vision or speech/language, for which a much larger body of research exists.

It appears that the richness and subtlety of emotional experiences of music stand in sharp contrast to current research on emotion, which has often only focused on, for example, broad categories of so-called basic emotions, or music of drastically different degrees of valence and arousal potential. Similarly, there is a dearth of empirical evidence relating to biographically or culturally mediated influences on musical emotions. To be sure, music might not always induce positive responses in our daily lives. Instead, loud music, for instance, is a prominent stressor with potentially hazardous health implications to the nervous system. However, it is worth noting, as Koelsch (2005) argues, that music is among the very few stimuli that can reliably elicit (strong) *positive* emotions in the laboratory, even across a wide range of listeners, including amusics (Peretz et al., 1998). Moreover, the findings of studies seem to converge in suggesting the involvement of "limbic and paralimbic structures (such as amygdala, hippocampus, parahippocampal gyrus, insula, temporal poles, ventral striatum, orbitofrontal cortex, and the cingulate cortex)" (Koelsch, 2005, p. 412) in musical emotion processing. Note that none of these structures appears to be specialized exclusively for music. However, the existence of music-specific modules for emotion processing remains a plausible hypothesis for future research (Griffiths et al., 2004).

To summarize, the view of musical affect as subordinate to cognition is challenged by the results of investigations into the neural correlates of musical experience that explicitly address emotion. Whether emotional responses precede cognitions, or vice versa, or if music (including aspects of emotion processing) represents an independent modality, are matters of ongoing debate (Peretz & Zatorre, 2005). In any case, music seems an "excellent paradigm to explore the interactions between neocortically mediated cognitive processes and subcortically mediated affective responses" (Peretz & Sloboda, 2005, p. 410).

REFERENCES

Adolphs, R., Gosselin, F., Buchanan, T.W., Tranel, D., Schyns, P. & Damasio, A.R. (2005). A mechanism for impaired fear recognition after amygdala damage. *Nature, 433*, 68-72.

Altenmueller, E. (1986). Brain electrical correlates of cerebral music processing in the human. *European Archives of Psychiatry and Neurological Science, 235*, 342-354.

Altenmueller, E., Schuermann, K., Lim, V.K. & Parlitz, D. (2002). Hits to the left, flops to the right. different emotions during listening to music are reflected in cortical lateralisation patterns. *Neuropsychologica, 40*, 2242-2256.

Anders, S., Lotze, M., Erb, M., Grodd, W. & Birbaumer, N. (2004). Brain activity underlying emotional valence and arousal: a response-related fMRI study. *Human Brain Mapping, 23*, 200-209.

Apkarian, A.V., Bushnell, M.C., Treede, R.D. & Zbieta, J.K. (2005). Human brain mechanisms of pain perception and regulation in health a disease. *European Journal for Pain, 9*, 463-484.

Bartlett, D. L. (1996). Physiological Responses to Music and Sound Stimuli. In D. Hodges (Ed.) *Handbook of Music Psychology* (S. 343-385). San Antonio: IMR Press.

Baumgartner, T., Esslen, M., & Jäncke, L. (2005). From emotion perception to emotion experience: Emotions evoked by pictures and classical music. *International Journal of Psychophysiology, 60,* 34-43.

Bigand, E., Filipic, S. & Lalitte, F. (2005). The Time Course of Emotional Response to Music. *Annals of the New York Academy of Sciences*, 1060, 429-437.

Blood, A., Zatorre, R., Bermudez, P. & Evans, A.C. (1999). Emotional responses to pleasant and unpleasant music correlate with activity in paralimbic brain regions. *Nature Neuroscience, 2*, 382-387.

Blood, A. & Zatorre, R. (2001). Intensely pleasurable responses to music correlate with activity in brain regions implicated in reward and emotion. *Proceedings of the National Academy of Science, 98*, 11818-11823.

Bostanov, V. & Kotchoubey, B. (2004). Recognition or affective prosody: Continuous wavelet measures of event-related brain potentials to emotional exclamations. *Psychophysiology, 41*, 259-268.

Bradley, M.M. & Vrana, S.R. (1993). The startle probe in the study of emotion and emotional disorders. In: Birbaumer, N. & Öhman, A. (Eds.), *The structure of emotion* (pp.21-50). Toronto: Hogrefe & Huber Publishers.

Brown, S., Martinez, M. & Parsons, L. (2004). Passive music listening spontaneously engages limbic and paralimbic systems. *NeuroReport, 15*, 2033-2037.

Bush, G., Luu, P. & Posner, M.I. (2000). Cognitive and emotional influences in anterior cingulate gyrus. *Trends in Cognitive Sciences, 4*, 215-222.

Calder, A.J., Lawrence, A.D. & Young, A.W. (2001). Neuropsychology of fear and loathing. *Nature Review of Neurosciences, 2*, 352-363.

Cohen, L.G., Celnik, P., Pascual-Leone, A., Corwell, B., Falz, L., Dambrosia, J., Honda, M., Sadato, N., Gerloff, C., Catalan, M.D. & Hallet, M. (1997). Functional relevance of cross-modal plasticity in blind humans. *Nature, 389*, 180-183.

Critchley, H.D., Elliott, R., Mathias, C.J. & Dolan, R.J. (2000). Neural activity relating to generation and representation of galvanic skin conductance responses: a functional magnetic resonance imaging study. *Journal of Neuroscience, 20*, 3033-40,

Csikszentmihalyi, M. (1990). *Flow: the psychology of optimal experience*. New York: Harper Collins.

Damasio, A.R. (1996). The somatic marker hypothesis and the possible functions of the prefrontal cortex. *Philosophical Transactions of the Royal Society of London, 351,* 1413-1420.

Damasio, A.R. (1998). Emotion in the perspective of an integrated nervous system. *Brain Research Reviews, 26*, 83-86.

Davidson, R.J. (1984) Affect, cognition and hemispheric specialization in: C.E. Izard, J. Kagan, & R. Zajonc (Eds.), *Emotion, cognition and behavior* (pp. 320-365). New York: Cambridge University Press.

deCharms, R.C., Maeda, F., Glover, G.H., Ludlow, D., Pauly, J.M., Soneji, D., Gabrieli, J.D. & Mackey, S.C. (2005). Control over brain activation and pain learned by using real-time functional MRI. *Proceedings of the National Academy of Sciences, 102*, 18626-18633.

Devinsky, O., Morrell, M.J. & Vogt, B.A. (1995). Contributions of anterior cingulate cortex to behaviour. *Brain, 118*, 279-306.

Dowling, W.J. & Harwood, D. (1986). *Music cognition*. San Diego, CA: Academic Press.

Drevets, W.C. & Raichle, M.E. (1998). Reciprocal Suppression of Regional Cerebral Blood Flow during Emotional versus Higher Cognitive Processes: Implications for Interactions between Emotion and Cognition. *Cognition and Emotion, 12*, 353-385.

Ekman, P. & Friesen, W. V. (1982). Felt, false, and miserable smiles. *Journal of Nonverbal Behavior, 6,* 238–252.

Ekman, P., Levenson, R.W. & Friesen, W.V. (1983). Autonomic nervous system activity distinguishes among emotions. *Science, 221*, 1208-10.

Freeman, W. J. (2000). A neurobiological role of music in social bonding. In N. Wallin, B. Merker & S. Brown (Eds), *The origins of music* (pp. 411-424). Cambridge, MA: MIT-Press.

Foxton, J.M., Dean, J.L., Gee, R., Peretz, I. & Griffiths, T. (2004). Characterization of deficits in pitch perception underlying tone deafness. *Brain, 127*, 801-810.

Gabrielsson, A (2002). Perceived emotion and felt emotion: Same or different? *Musicae Scientiae, Special Issue 2001-2002*, 123-149.

Gabrielsson, A. (2001). Emotions in strong experiences with music. In P. Juslin & J. Sloboda (Eds), *Music and emotion. Theory and research* (pp. 431-449). Oxford: Oxford University Press.

Gosselin, N., Peretz, I., Noulhiane, M., Hasboun, D., Beckett, C., Baulac, M. & Samson, S. (2005). Impaired recognition of scary music following unilateral temporal lobe excision. *Brain, 128*, 628-640.

Gottfried, J.A., O'Doherty, J. & Dolan, R.J. (2002). Appetitive and aversive olfactory learning in humans studied using event-related functional magnetic resonance imaging. *Journal of Neuroscience, 22*, 10829-37.

Goydke, K.N., Altenmueller, E., Moeller, J. & Muente, T. (2004). Changes in emotional tone and instrumental timbre are reflected by the mismatch negativity. *Cognitive Brain Research, 21*, 351-359.

Greenwald, M.K., Cook, E.W. III & Lang, P.J. (1989). Affective judgement and psychophysiological response: dimensional covariation in the evaluation of pictorial stimuli. *Journal of Psychophysiology, 3*, 51-64.

Grewe, O., Nagel, F., Kopiez, R. & Altenmueller, E. (2005). How does music arouse "chills"? Investigating strong emotions, combining psychological, physiological, and psychoacoustical methods. *Annals of the New York Academy of Sciences, 1060*, 446-449.

Griffiths, T. D., Warren, J. D., Dean, J. L. & Howard, D. (2004). 'When the feeling's gone': a selective loss of musical emotion. *Journal of Neurology, Neurosurgery & Psychiatry, 75,* 344-345.

Heilman, K.M. (1997). The neurobiology of emotional experience. *Journal of Neuropsychiatry and Clinical Neuroscience, 9*, 439–48.

Hitchcock, J. & Davis, M. (1986). Lesions of the amygdala, but not of the cerebellum or red nucleus, block conditioned fear as measured with the potentiated startle paradigm. *Behavioral Neuroscience, 100,* 11-22.

Hitchcock, J.M. & Davis, M. (1991). Efferent pathway of the amygdala involved in conditioned fear as measured with the fear-potentiated startle paradigm. *Behavioral Neuroscience, 105*, 826-842.

James, W. (1890). *The Principles of Psychology* (2 vols.). New York: Henry Holt (reprinted by Thoemmes Press, Bristol, 1999).

Johnsen, E. L. (2004). *Neuroanatomical correlates of emotional experiences from music.* Doctoral Dissertation. University of Iowa.

Juslin, N.P. (2001). Communicating emotion in music performance. A review and a theoretical framework. In N.P. Juslin, & J.A. Sloboda (Eds), *Music and emotion. Theory and research* (pp. 309-340). Oxford: Oxford University Press.

Juslin, N.P. & Sloboda, J. A. (2001) (Eds.). *Music and emotion. Theorie and research.* Oxford: Oxford University Press.

Kaplan, J.A. & Gardner H. (1990). Artistry after unilateral brain disease. In: H. Goodglass and A.R. Damasio (Eds.), *Handbook of neuropsychology*, Vol. 2: Language, Aphasia and related Disorders (pp 141-155). Amsterdam: Elsevier.

Khalfa, S., Peretz, I., Blondin, J.-P., & Manon, R. (2002). Event-related skin conductance to musical emotions in humans. *Neuroscience Letters, 328*, 145-149.

Koelsch, S. & Fritz, T. (2003). Untersuchungen von Emotion mit Musik: eine funktionell-bildgebende Studie [Investigating emotions in music: a functional imaging study]. *Sprache - Stimme - Gehör, 27*, 62-65.

Koelsch, S. (2005). Investigating Emotion with Music. Neuroscientific Approaches. *Annals of the New York Academy of Sciences, 1060*, 412 – 418.

Koelsch, S., Fritz, T., Schulze, K., Alsop, D. & Schlaug, G. (2005). Adults and children processing music: an fMRI study. *NeuroImage, 25,* 1068-1076.

Koelsch, S., Fritz, T., v.Cramon, D.Y., Müller, K. & Friederici, A. (2006). Investigating Emotion with Music: An fMRI-study. *Human Brain Mapping, 27*(3):239-50.

Kreutz, G. (*in press*). Musik und Emotion. In H. Bruhn, R. Kopiez, A. Lehmann & R. Oerter (Eds.). *Neues Handbuch Musikpsychologie*, 5. Auflage [New Handbook of Music Psychology, 5th edition].

Kreutz, G., Bongard, S. & von Jussis, J. (2002). Kardiovaskuläre Wirkungen beim Musikhören. Zur Bedeutung von musikalischer Expertise und Emotion [Cardiovascular responses to music listening: Effects of musical expertise and emotion]. *Musicae Scientiae, 6*, 257-278.

Kreutz, G., Bongard, S., Grebe, D., Rohrmann, S. & Hodapp, V. (2004). Effects of choir singing or listening on secretory IgA, cortisol, and emotional state. *Journal of Behavioral Medicine, 27*, 623-634.

Kreutz, G., Ott, U. & Wehrum, S. (2006). Cerebral correlates of musically-induced emotions: an fMRI-study. In M. Baroni et al. (Eds.) *Proceedings of the 9th International Conference on Music Perception and Cognition (ICMPC)*, Bologna, Aug 22-26, 2006.

Kreutz, G., Russ, M. O., Bongard, S. & Lanfermann, H. (2003). Zerebrale Korrelate des Musikhörens. Eine fMRT-Studie zur Wirkung "fröhlicher" und "trauriger" klassischer Musik [Cerebral correlates of music listening. An fMRI-study on the effects of 'happy' and 'sad' classical music]. *Nervenheilkunde, 3*, 56-64.

Krumhansl, C. L. (1997). An exploratory study of musical emotions and physiology. *Canadian Journal of Psychology, 51*, 336-352.

Lang, P.J., Bradley, M.M., Fitzsimmons, J.R., Cuthbert, B.N., Scott, J.D., Moulder, B. & Nangia, V. (1998). Emotional arousal and activation of the visual cortex: an fMRI analysis. *Psychophysiology, 35*,199-210.

Lazarus, R. S. (1991). Cognition and motivation in emotion. *Amercian Psychologist, 46*, 352-367.

Leichner, R. & Bröscher, N. (1998). Hemisphärenasymmetrien bei der Beurteilung von Musik [Hemispheric differences in judging music]. *Musikpsychologie, 14*, 69-86.

LeDoux, J. (1996). *The Emotional Brain. The Mysterious Underpinnings of Emotional Life.* N.Y.: Basic Books.

Lerdahl, F. & Jackendoff, R. (1983). *A Generative Theory of Tonal Music.* Cambridge, MA: MIT-Press.

Lotze, M., Wietek, B., Birbaumer, N., Grodd W. & Enck P. (2001). Cerebral activation maps during anal and rectal stimulation. *NeuroImage 14*, 1024-1034.

Lotze, M., Scheler, G., Tan, HRM., Braun, C. & Birbaumer, N. (2003). The musician's brain: functional imaging of amateurs and professionals during performance and imagery. *NeuroImage 20*, 1817-1829.

Lotze, M., Markert, J., Sauseng, P., Hoppe, J., Plewnia, C. & Gerloff, C. (2006a). The role of multiple contralesional motor areas for complex hand movements after internal capsular lesion. *Journal of Neuroscience, 26*: 6096-102.

Lotze, M., Heymans, U., Birbaumer, N., Veit, R., Erb, M., Flor, H. & Halsband, U. (2006b). Differential cerebral activation during observation of expressive gestures and motor acts. *Neuropsychologia*, PMID: 16730755.

MacLean, P.D. (1949). Psychosomatic disease and the visceral brain; recent developments bearing on the Papez theory of emotion. *Psychosomatic Medicine, 11*, 338-353.

Mathiak, K. & Weber, R. (2006). Towards brain correlates of natural behavior: fMRI during Violent Video Games; *Human Brain Mapping* (in press).

Mayberg, H.S., Brannan, S.K., Mahurin, R.K., Jerabek, P.A., Brickman, J.S., Tekell, J.L., Silva, J.A., McGinnis, S., Glass, T.G., Martin, C.C. & Fox, P.T. (1997). Cingulate function in depression: a potential predictor of treatment response. *Neuroreport, 8*, 1057-1061.

McKinney, C. H., Tims, F. C., Kumar, A. M. & Kumar, M. (1997). The effect of selected classical music and spontaneous imagery on plasma-endorphin. *Journal of Behavioural Medicine, 20*, 85-99.

Mehrabian A. & Russel, J.A. (1974). *An approach to environmental psychology*. Cambridge, MA: MIT Press.

Meyer, L.B. (1956). *Emotion and meaning in music*. Chicago. University of Chicago Press.

Mills, C.K. (1912). The cortical representation of emotion with a discussion of some points in the general nervous system mechanism of expression in its relation to organic nervous disease and insanity. *Proceedings of the American Medical Psychiatric Association, 19*, 297-300.

Mitchell, R.L.C., Elliott, R., Barry, M., Cruttenden, A. & Woodruff, P.W.R. (2003). The neural response to emotional prosody, as revealed by functional magnetic resonance imaging. *Neuropsychologia, 41*, 1410-1421.

Murayama, J., Kashiwagi, T., Kashiwagi, A. & Mimura, M. (2004). Impaired pitch production and preserved rhythm production in a right brain-damaged patient with amusia. *Brain and Cognition, 56*, 36-42.

Murphy, F.C., Nimmo-Smith, I. & Lawrence, A.D. (2003). Functional neuroanatomy of emotions: a meta-analysis. *Cognitive and Affective Behavioral Neuroscience, 3*, 207-233.

Nyklicek, I., Thayer, J.F. & van Doornen, L.J.P. (1997). Cardiorespiratory differentiation of musically-induced emotions. *Journal of Psychophysiology, 11*, 304-321.

Ochsner, K.N., Ray, R.D., Cooper, J.C., Robertson, E.R., Chopra, S., Gabrieli, J.D. & Gross, J.J. (2004a). For better or for worse: neural systems supporting the cognitive down- and up-regulation of negative emotion. *NeuroImage, 23*, 483-499.

Ochsner, K.N., Knierim, K., Ludlow, D.H., Hanelin, J., Ramachandran, T., Glover, G. & Mackey, S.C. (2004b) Reflecting upon feelings: an fMRI study of neural systems supporting the attribution of emotion to self and other. *Journal of Cognitive Neuroscience, 16*, 1746-1772.

Pallesen, K.J., Brattico, E., Bailey, C., Korvenoja, A., Koistovo, J., Gjedde, A. & Carlson, S. (2005). Emotion processing of major, minor, and dissonant chords. A functional magnetic resonance imaging study. *Annals of the New York Academy of Sciences, 1060*, 450-453.

Panksepp. J. & Bernatzky, G. (2002). Emotional sounds and the brain: the neuro-affective foundations of musical appreciation. *Behavioral Processes, 60*, 133-155.

Panksepp. J. (1995). The emotional sources of "chills" induced by music. *Music Perception, 13*, 171-207.

Peper, M. & Lüken, U. (2002). Persönlichkeitsforschung im Wandel: Neuropsychophysiologie der Emotionalität [Personality research is changing: Neuropsychophysiology of emotion]. In M. Myrtek (Ed.). *Die Person im biologischen und sozialen Kontext* (pp. 85-114) [The person in biological and social contexts]. Göttingen: Hogrefe

Peretz I. (1990) Processing of local and global musical information by unilateral brain-damaged patients. *Brain, 113*, 1185-205.

Peretz, I., Gagnon, L. & Bouchard, B. (1998). Music and emotion: perceptual determinants, immediacy, and isolation after brain damage. *Cognition, 68*, 111-141.

Peretz, I (2001). Listen to the brain: a biological perspective on musical emotions. In P. Juslin & J. Sloboda (Eds.), *Music and emotion. Theory and research* (pp. 105-134). Oxford: Oxford University Press.

Peretz, I. & Sloboda, J.A. (2005). Part VII: Music and the emotional brain. Introduction. *Annals of the New York Academy of Sciences, 1060*, 409-411.

Peretz, I. & Zatorre, R. (2005). Brain organization for music processing. *Annual Review of Psychology, 56*, 89-114.

Phan, K.L., Wager, T.D., Taylor, S.F. & Liberzon, I. (2002) Functional neuroanatomy of emotion: a meta-analysis of emotion activation studies in PET and fMRI. *NeuroImage, 16*, 331-48.

Phan, K.L., Wager, T.D., Taylor, S.F. & Liberzon, I. (2004a) Functional neuroimaging studies of human emotions. *CNS Spectrum, 9*, 258-66. Review.

Phan, K.L., Taylor, S.F., Welsh, R.C., Ho, S.H., Britton, J.C. & Liberzon, I. (2004b). Neural correlates of individual ratings of emotional salience: a trial-related fMRI study. *NeuroImage 21*, 768-780.

Robinson, S., Windischberger, C., Rauscher, A. & Moser, E. (2004). Optimized 3 T EPI of the amygdalae. *NeuroImage, 22*, 203-10.

Rolls, E.T., Critchley, H.D. & Treves A. (1996). Representation of olfactory information in the primate orbitofrontal cortex. *Journal of Neurophysiology, 75*, 1982-96.

Royet, J.P., Zald, D., Versace, R., Costes, N., Lavenne, F., Koenig, O. & Gervais, R. (2000). Emotional responses to pleasant and unpleasant olfactory, visual, and auditory stimuli: a positron emission tomography study. *Journal of Neuroscience, 20*, 7752-7759.

Russell, J.A. (1980). A circumplex model of affect. *Journal of Personality and Social Psychology, 39*, 1161-1178.

Russell, J.A. & Barrett, L.F. (1999). The structure of current affect: Controversies and emerging consensus. *Current Directions in Psychological Science, 8*, 10-14.

Samson, S. & Peretz, I. (2005). Effects of prior exposure on music liking and recognition in patients with temporal lobe lesions. *Annals of the New York Academy of Sciences, 1060*, 419-428.

Satoh, M., Takeda, K., Murakami, Y., Onouchi, K., Inoue, K. & Kuzuhara, S. (2005). A case of amusia caused by the infarction of anterior portion of bilateral temporal lobes. *Cortex, 41*, 77-83.

Schachter, S. & Singer, J.E. (1962). Cognitive, social, and physiological determinants of emotional stress. *Psychological Review, 69*, 379-399.

Schmid, L.A., Trainor, L.J. & Santesso, D.L. (2003). Development of frontal encephalogram (EEG) and heart rate (ECG) responses to affective musical stimuli during the first 12 months of post-natal-life. *Brain and Cognition, 52*, 27-32.

Schön, D., Lorber, B., Spacal, M. & Semenza, C. (2003). Singing: a selective deficit in the retrieval of musical intervals. *Annals of the New York Academy of Sciences, 999*, 189-192.

Schubert, E. (2001). Contiuous measurement of self-report emotional response to music. In N.P. Juslin & J.A. Sloboda (Eds.), *Music and emotion. Theory and research* (pp. 393-414). Oxford: Oxford University Press.

Singer, T., Seymour, B., O'Doherty, J.P., Stephan, K.E., Dolan, R.J. & Frith, C.D. (2006). Empathic neural responses are modulated by the perceived fairness of others. *Nature, 439,* 466-469.

Sloboda, J.A., Wise, K. & Peretz, I. (2005). Quantifying tone deafness in the general population. *Annals of the New York Academy Sciences, 1060*, 255–261.

Spreckelmeyer, K.N., Kutas, M., Urbach, T.P., Altenmueller, E. & Muente, T.F. (2006). Combined perception of emotion in pictures and musical sounds. *Brain Research, 1070,* 160-170.

Steptoe, A. (2001). Negative emotions in music making: the problem of performance anxiety. In N.P. Juslin & J.A. Sloboda (Eds.), *Music and emotion. Theory and research* (pp. 291-308). Oxford: Oxford University Press.

Veit, R., Lotze, M., Caria, A., Gaber, T., Sitaram, R. & Birbaumer, N. (2006). Real-time fMRI of human anterior insula during emotional processing. *NeuroImage 12*; Suppl.

Wager, T. D., & Feldman Barrett, L. (2004). From affect to control: functional specialization of the insula in motivation and regulation. Published online at PsycExtra: http://www.columbia.edu/cu/psychology/tor/ (access date: 01 June 2006).

Wildgruber, D., Hertrich, I., Riecker, A., Erb, M., Anders, S., Grodd ,W., & Ackermann H. (2004). Distinct frontal regions subserve evaluation of linguistic and emotional aspects of speech intonation. *Cerebral Cortex, 14*, 1384-1389.

In: Neurosciences in Music Pedagogy
Editors: W. Gruhn and F. Rauscher, pp. 171-201
ISBN: 978-1-60021-834-7

Chapter VII

UNPACKING THE IMPACT OF MUSIC ON INTELLIGENCE

James S. Catterall and Frances H. Rauscher

ABSTRACT

Reports that exposure to music causes benefits in nonmusical domains have received widespread attention in the mainstream media. Despite the fact that the term "Mozart effect" has been grossly misapplied and over-exaggerated, the effects of music on intelligence have caught the interest of scientists concerned with cognitive, neuroscientific, and educational approaches to the study of music. This chapter addresses the question, "How does music learning affect other intellectual capacities?" We review research on the effects of music instruction on spatial abilities and arithmetic, followed by an examination of its effects on other cognitive skills, including visual-motor integration and verbal/reading performance. After suggesting some possible mechanisms for these effects, we then discuss a widely-publicized paper suggesting that music instruction enhances general intelligence (IQ). We revisit the data from this study to investigate if the reported IQ gains were driven by one of the two sub-domains measured, i.e., visual-spatial reasoning or verbal reasoning. We conclude from our analyses that the gains in general intelligence derive from gains in visual-spatial reasoning skills more than from gains in verbal skills—for all children and particularly for children who scored low on the IQ test prior to musical instruction.

INTRODUCTION

This chapter extends a tradition of research on music and cognition that has buoyed the careers of psychologists, sent parents to the school board demanding support for an orchestra, and left music teachers and musicians scratching their heads over their professional and social purposes. We refer of course to the popularized notion that "music makes you smarter" as well as to responding convictions that music is about skill, appreciation, affect, emotion, and beauty. Propelled by events catalogued below, the past fifteen years have seen cognitive specialists, neuroscientists, music educators, and an economist or two train their lenses on learning in music to see how music experience affects individuals. Of overwhelming and primary interest in this tradition are the increasingly documented influences of music on how people think. The quest plays out in behavior broadly conceived as well as the inferences we can derive from behavior about the contours of cognition. In recent years, in part because of the accelerating quality of imaging techniques, scientists have examined the manifestations of behavior and experience, including the musical, on brain function and structure.

In this chapter we try to push forward our understanding of music learning and intelligence. Very few music studies have enlisted full-blown standardized tests of intelligence, but a recently published and widely heralded report (Schellenberg, 2004) supports a broad claim that music enhances general intelligence. We revisit the impressive and detailed data from this study to investigate a question growing from a broad current in cognitive music research: Can measured gains in general intelligence be driven disproportionately by gains in a broad sub-domain of thinking capacities, namely spatial-temporal or visual-spatial thinking? We raise this question with hopes that any answers will advance our knowledge of music's impacts on cognition and how such impacts materialize.

Historical Roots

The study of the extra-musical benefits of music exposure has a long history in Western intellectual culture. Early Greek education emphasized rhetoric and mathematics, athleticism, the visual arts, and music. As the Greek curriculum expanded to include other subjects, there ensued a debate regarding the role of music in the educational system. The debate continues today. With the continuing cuts to school budgets, music teachers throughout the world are continually forced to justify their existence. A brief examination of the history of public education in the United States provides some insight into why music teachers have to fight for their right to teach music in the public schools.

The primary goal of education for the Puritans was the ability to read the Bible. Developing basic literacy remained the principle purpose of both private and public (government-funded) education throughout the nineteenth century. However, on August 28, 1838 the School Board of Boston Massachusetts made music a part of the regular curriculum. It was determined that music should not occupy more than two hours a week, that it should be provided at a fixed time throughout the city, and that the classroom teacher must be present during all lessons in order to maintain order. The music teacher was paid $100/year, out of which he or she was expected to provide the piano. This marked the first time in the U.S. that music education was provided in an organized fashion with a teacher paid by public funds. But the original justification for including music in the curriculum showed no interest in artistic goals. Instead, it emphasized the broad educational goals supported by the Boston school board at that time. The development of reading, writing, and arithmetic skills remains a preemptive goal for public education. The failure of the public schools in recent years to develop basic abilities in these skills has resulted in a school environment in which arts education is not a high priority.

Music in Today's Curriculum

Today, music instruction is usually considered an educational add-on, and music education programs are being scaled down or eliminated. Although most music teachers value artistic goals above all, these values unfortunately are not shared where they matter, and music teachers are often expected to justify their employment by resorting to claims that music has extra-musical benefits. Although we believe it is myopic to justify music programs based on their possible non-musical benefits, we believe the study of the non-artistic effects of music instruction is an important contribution to the music cognition literature.

Chapter Overview

After dispelling a common myth regarding the effects of listening to music on cognitive performance, this chapter outlines some of the cognitive effects of music instruction. A discussion of the effects of music training on spatial-temporal abilities and arithmetic is followed by a brief outline of some of the other cognitive benefits of music instruction. Possible mechanisms for these effects are then considered, followed by a description of a recent study exploring the effects of music instruction on general intelligence (i.e., IQ) (Schellenberg, 2004). The remainder of this chapter focuses on "unpacking" some of the cognitive benefits described by Schellenberg (2004), with the emphasis on *Verbal IQ*, *Performance IQ*, and arithmetic skills. The chapter concludes with a discussion of the scientific and educational implications of this new research.

The Myth of the Mozart Effect

The announcement of the "Mozart effect," a term originally applied by the media (Knox, 1993), has recently spawned countless articles, books and recordings making assertions that are largely unsubstantiated. In the original study college students who listened to the first 10 min of Mozart Sonata K. 448 (including the first movement and a small portion of the second movement) scored higher on a spatial-temporal reasoning task than after they listened to relaxation instructions or silence (Rauscher, Shaw, & Ky, 1993; see also Rauscher & Shaw, 1998). The effect was short-lived and small. Nevertheless, the report was touted by various news media, popular magazines, and entrepreneurs as an effortless way to boost children's intelligence. The study's authors have not supported these claims (Rauscher, 2002; Rauscher & Hinton, 2006). The original research did not utilize children as subjects, and the initial finding was specific to one aspect of intelligence only (spatial-temporal reasoning). Furthermore, current research on the "Mozart effect" in adults suggests it is due to arousal, mood, and/or preference rather than to music in general or Mozart in particular. Several studies support this hypothesis (for review, see Schellenberg, 2005), and further suggest the effect is not limited to spatial-temporal abilities (Schellenberg et al., 2007). To our knowledge, only six published studies have searched for a "Mozart effect" in children. One study found an effect (Ivanov & Geake, 2003); one study did not find an effect (McKelvie & Lowe, 2002); two studies found a musical preference effect rather than a Mozart effect (Schellenberg & Hallam, 2005; Schellenberg et al., 2007); and two studies found neither a Mozart effect nor a musical preference effect (Crncec, Wilson, & Prior, 2006; Hui, 2006). The existence of a "Mozart effect" in children is thus highly debatable and certainly not what early enthusiasts had hoped.

The Effects of Music Instruction on Spatial Reasoning and Arithmetic

Despite the controversy surrounding the "Mozart effect," its discovery has prompted a renewed interest in the effects of music on cognition. Recently, researchers have attempted to understand the relationships between music exposure and cognition by providing children with music *instruction* and then testing their performance using a variety of cognitive tasks. Piano instruction, voice instruction, and rhythmic training have been found to improve cognitive performance. For example, Rauscher et al. (1997) assigned students from three preschools to music, computer, or no instruction groups. The instruction groups received several months of individual instruction in either piano keyboard coupled with group singing lessons or computer usage. The children were tested using one spatial-temporal reasoning task and three spatial recognition tasks taken from the Wechsler Preschool and Primary Scale of Intelligence-Revised (WPPSI-R) (Wechsler, 1989) before and after instruction began. Results indicated that children in the music group scored higher on the spatial-temporal task only. There were no differences in the children's pre- and post-test scores on the spatial recognition tasks. The computer and no instruction groups did not improve significantly on any of the tests administered. This study should be interpreted with caution, however, because only one of the three preschools permitted random assignment to groups.

A study with older (9-year-old) children yielded similar effects, although the enhancements did not persist into the third year of the three-year study (Costa-Giomi, 1999). One group of children received weekly piano lessons whereas another group received no lessons. All children were pre- and post-tested using verbal, quantitative, and spatial subtests. An overall score of intelligence was derived from the subtest scores. The children in the music group's overall intelligence scores were higher after the second year of the study. They also scored higher on the spatial subtest after both the first and second years of the study, but there were no differences between the groups on the overall intelligence measure or on any subtest, including the spatial subtest, by the end of the third and final year. This study is somewhat consistent with research finding that the spatial-temporal scores of children who received keyboard lessons in second grade (at approximately age 7) did not differ from children who received no lessons, whereas children enrolled in the same school who received lessons starting in kindergarten (age 5) or first grade (age 6) did significantly differ from controls (Rauscher, 2002; Rauscher & Zupan, 2000). No differences were found between any groups on a visual memory test. It is therefore possible that early instruction is necessary in order for musical learning to transfer to spatial-temporal performance.

Keyboard instruction has also been found to improve the spatial-temporal and arithmetic skills of kindergarten children in an experiment conducted with Greek children (Zafranas, 2004). Sixty-one children received two thirty-min keyboard lessons for seven months. In addition to examining the effect of the lessons on spatial-temporal abilities, a major goal of the study was to determine if effects could also be found for other tasks. Accordingly, the researcher administered six subtests of the Kaufman Assessment Battery for Children (KABC) (Kaufman & Kaufman, 1983) before and after the instruction. The subtests included Hand Movements, Gestalt Closure, Triangles, Spatial Memory, Arithmetic, and Matrix Analogies. Comparisons were made between the children's pre- and post-test scores and those of age-standardized norms. Children showed significant improvement on all but the Matrix Analogies subtest, described as an "analogic thinking task" (p. 203), with the greatest improvement found for the Hand Movements task. Although these findings are intriguing, it is unfortunate that the author did not include any tasks that do not have spatial attributes.

Three studies conducted over a period of five years also utilized the K-ABC, in addition to several other cognitive tests, to explore the impact of music instruction on a variety of skills (Rauscher, LeMieux, & Hinton, 2005). For the first study, Head Start[4] preschool children were randomly assigned to one of three conditions: Keyboard instruction (n=33), computer instruction (n=28), or no instruction (n=26). The children were pre- and post-tested using the KABC (Kaufman & Kaufman, 1983), the Developmental Test of Visual Perception (DTVP-2) (Hammill, Pearson, & Voress, 1993), the Test of Auditory Perceptual Skills-Revised (TAPS-R) (Gardner, 1996), and the WPPSI-R (Wechsler, 1989). These tests measure spatial-temporal, visual-spatial, auditory, verbal, and arithmetic skills. Overall, the authors administered twenty-six subtests. The children in the keyboard group received 48 weeks of individual weekly lessons distributed over two academic years. The computer group received an equal number of individual lessons on a laptop computer. The no-lessons group received no individualized instruction. Results indicated that the children in the keyboard group performed higher than the computer or no lessons groups on the tasks containing spatial or temporal content and arithmetic, but not higher on the Matrix Analogies subtest of the K-ABC (replicating Zafranas, 2004) or any of the verbal measures.

The second of the three studies (Rauscher, LeMieux, & Hinton, 2005) was designed to determine if the type of music instruction children received had measurably different effects on cognition. Most researchers agree that musical skill is an alliance of a number of separate and relatively independent abilities. The authors proposed that early music instruction emphasizing different musical skills would produce correspondingly differential effects on cognitive performance. Head Start preschoolers were randomly assigned to four groups: keyboard instruction (n=34), singing instruction (n=28), rhythm instruction (n=35), or no lessons (n=26). Lessons were again provided for 48 weeks over two years. The same subtests administered in the first study were given before and after instruction. The researchers

[4] Head Start is a federally-funded enrichment program designed to help underpriveledged preschool children begin primary school on an equal academic footing as their middle-income peers.

predicted improvement in spatial-temporal tasks following all types of music instruction, greater improvement in mental imagery tasks following singing instruction (due to singing's strong reliance on auditory imagery), and greater improvement in temporal tasks following rhythm instruction (due to rhythm training's emphasis on the temporal qualities of music). As in the first study, the children in the three music groups scored higher than the control group on the spatial, temporal, and arithmetic tests. As predicted, the rhythm group scored significantly higher than the keyboard, singing, or control groups on the temporal and arithmetic tests. Scores on the verbal tests did not differ from controls. The prediction regarding the effects of singing instruction on mental imagery tasks was not supported.

The third of the three studies (Rauscher, LeMieux, & Hinton, 2005) was conducted to test the durability of the enhancements found in the first two studies, as well as to compare the scores of the Head Start children who participated in Studies 1 and 2 (now in second grade and kindergarten) to those of randomly chosen age-matched middle-income children and disadvantaged children who were not enrolled in a Head Start program. The researchers thus compared the scores of the children who received music lessons in Studies 1 and 2 to three groups of age-matched children: (1) Head Start children who did not receive music instruction; (2) at-risk children who were not enrolled in Head Start, and (3) middle-income children. The children who participated in the control (n = 24) and piano (n = 31) groups in Study 1 were re-tested, as were children from the music groups who participated in Study 2 (piano, n = 27; singing, n = 20; rhythm, n = 29, following attrition). Twenty-seven at-risk kindergartners, 24 at-risk second-graders, 32 middle-income kindergartners, and 28 middle-income second graders were also tested. All children were administered the K-ABC Hand Movements, Number Recall, Gestalt Closure, Faces and Places, Arithmetic, and Riddles subtests. Although interviews revealed no differences in the extent of music instruction between the three disadvantaged groups, 12% of children in the middle-income group received individual instrumental instruction outside of school. Results indicated that the Study 1 music groups' scores on tasks previously enhanced by the instruction (Hand Movements, Gestalt Closure, and Arithmetic) remained higher than age-matched at-risk children who were not in Head Start and Head Start children who did not receive keyboard instruction. The Study 1 music groups' scores did not differ from those of the middle-income group on any test. All music groups from Study 2 scored significantly higher than age-matched at-risk and Head Start groups on the Arithmetic, Hand Movements, Number Recall, and Gestalt Closure tasks. However, the arithmetic scores of the children who received rhythm instruction in Study 2 were significantly higher than age-matched middle-income children. It is important to note that the children from Study 1 were tested two years after instruction was terminated, suggesting that the effects of music instruction on spatial abilities may persist.

Another study provides additional support for the notion that music instruction enhances young children's spatial-temporal task performance (Gromko & Poorman, 1998). Seventeen preschool children received seven months of weekly music lessons consisting of songbell instruction along with singing, rhythm, and kinesthetic activities. The children were encouraged to practice at home between lessons. Seventeen children received no instrumental instruction. The performance subtest of the WPPSI (consisting of five tasks) was administered at the beginning and end of the seven-month period. The children's scores on the five tasks were used to compute their "*Performance IQ*" The pre-test *Performance IQ* scores were then subtracted from the post-test *Performance IQ* scores to yield gain scores. The researchers reported that the gain in *Performance IQ* of the music group was greater than the gain in *Performance IQ* of the control group, and concluded that "...we believe music training can have a positive effect on the development of spatial intelligence in preschool children" (p. 180).

The Stanford-Binet Intelligence Scale (Thorndike, Hagen, & Sattler, 1986) also has been used to test the generalizability of the effects of music instruction (Bilhartz, Bruhn, & Olson, 2000). Six of the eight subtests were administered to four- to six-year-old children: Vocabulary, Memory for Sentences, Bead Memory, Pattern Analysis, Quantitative, and Copying[5]. Thirty-six children participated in a Kindermusik program that included vocal exploration, pitch matching, learning to read music, and instruction in percussion instruments and glockenspiel. The 75-min weekly lessons were provided for 30 weeks. Thirty-five children served as the control group. Gain scores were calculated by subtracting the children's age-standardized scores on the pre-test from their scores on the post-test. The authors found no differences in gain scores between the music and control groups on any measures other than Bead Memory, an abstract reasoning subtest. Furthermore, only those children who met the minimum compliance criteria for the music instruction (e.g., completed take-home assignments, had parental assistance, and attended classes regularly) showed the improvement, with the degree of improvement increasing linearly with the amount of music instruction. The authors suggest that the Bead Memory subtest "measures both visual imagery and sequencing strategies" (p. 629) and concluded that "the results of this study lend support to the hypothesis that there is a significant link between early music instruction and cognitive growth in specific nonmusic abilities" (p. 629).

A study conducted in four lower-income Milwaukee public schools found improved spatial-temporal and arithmetic scores for children who received classroom keyboard instruction (Rauscher, 2005). Kindergarten through 5th-grade children (n=627) were assigned to one of two conditions: keyboard or no lessons. All children were tested at the beginning and end of each of three years using three verbal tests, two mathematics tests, and one spatial-temporal test. There were several problems with the instruction during the first year of the study. Music teachers appointed by the school district did not receive classroom assignments until one month into the school year. They were unfamiliar with the keyboard curriculum and had not taught piano in the classroom before. In addition, there were high rates of absenteeism, lateness, and conduct disorders among the children, and classroom space was inadequate. These problems persisted through the second year of the study. During the third year, a music specialist with proficiency teaching keyboard to elementary school children at a different school district volunteered to assist the teachers in the classroom and share her expertise, greatly improving the quality of the instruction. No significant differences were found between groups for the cognitive measures during the first two years of the study. Differences between the piano and control groups emerged for the arithmetic and spatial-temporal tests during the third year of the study only, after the keyboard instruction had improved. These effects were found only for the kindergarten and first grade children. There

[5] The Copying subtest was later eliminated from the analysis due to poor inter-scorer reliability.

were no differences between any groups on the verbal measures. This research emphasizes the importance of quality music instruction for cognitive transfer effects, and supports previous research reporting differential effects of piano instruction on cognition.

Differences in the visual-spatial skills of adult musicians compared to non-musicians have been found using a reaction time task (Brochard, Dufour, &Després, 2004). To test their vertical discrimination, subjects were presented with a small target dot either above or below a horizontal reference line. Horizontal discrimination was tested by presenting the dot either to the left or to the right of a vertical reference line. Two experimental conditions were employed. In one condition ("line on"), the reference line was present during the presentation of the dot. In the other condition ("line off"), the reference line was absent. The subjects' task was to indicate which side of the reference line the dot was flashed. The researchers predicted that "...if musical expertise has a long-term influence on visual-spatial abilities, ...musicians' performance on both perceptual "line on" and imagery "line off" conditions [would] be significantly better than non-musicians'. Moreover, if this effect relies on a more efficient use of visual representations, an advantage of musical expertise should be greater in the imaging ("line off") conditions" (p. 104). The data supported these predictions. The authors concluded that "such perceptual and imagery advantages partly explain why music instruction generally increases children's scoring in visuospatial tasks (such as paper folding, mental rotation, and tridimensional reasoning) which all involve the mental manipulation of visual representations on several dimensions" (p. 106).

An additional study compared four classrooms of first-grade children who received seven months of Kodàly music instruction along with visual arts training (experimental group) with four classrooms of children who received the school's standard arts curriculum (control group) (Gardiner et al., 1996). Although more students in the experimental group scored below grade averages for reading and math than the standard curriculum children in kindergarten, after the special arts training in first grade they scored equal to the standard group in reading and above them in math. When tested again in the second grade, following an additional 7 months of special arts training, the experimental group's scores again equaled those of the control group in reading and exceeded their scores in math. Unfortunately, because the music and arts instruction were provided together it is not possible to determine if the results were due to the music curriculum.

Graziano, Peterson, and Shaw (1999) explored the effects of keyboard instruction coupled with a video game designed to train spatial-temporal skills and proportional math concepts. There were 6 groups in the study, three of which are discussed here. One group received the keyboard lessons together with the video game, a second group interacted with the video game and received lessons in English, and a third group received no special training. Children were pre-and post-tested with three spatial subtests of the Wechsler Intelligence Scale for Children (WISC-III) (Wechsler, 1991), and were also post-tested with the Spatial-Temporal Math Video Game Evaluation Program, a software program designed by the study's authors to assess the children's proportional math reasoning. Results indicated that the keyboard/video game group scored significantly higher on the proportional reasoning test than the English/video game group. Both groups scored significantly higher than the no lessons group. The scores of the children in the two video game groups did not differ from each other on the WISC-III subtests, although both groups scored higher than the no lessons group on these measures as well. This study suggests a link between music, spatial reasoning, and the spatial aspects of mathematics.

Two meta-analytic studies are relevant to the above discussion on the effects of music instruction on spatial and mathematical abilities. Hetland (2000) performed three meta-analyses on fifteen published and unpublished studies examining the relationship between music and non-musical outcomes. The age of the subjects ranged from 3 to 15 years, and the duration of the music instruction ranged from four weeks to two years. The goal of Hetland's first analysis was to test the hypothesis that music instruction improves spatial-temporal reasoning. Her second analysis sought to examine the effects of music instruction on general intelligence as measured by Raven's Standard Progressive Matrices (Raven, 1986), and the third analysis included studies that used spatial tests other than spatial-temporal reasoning tests. From her first analysis, Hetland concluded that "active music instruction lasting two years or less leads to dramatic improvements on spatial-temporal measures" (p. 203). The second analysis revealed a small but insignificant relationship between music instruction and subjects' scores on Raven's Standard Progressive Matrices test, and the third analysis suggested that the effects of music instruction are not limited to spatial-temporal tasks only, but rather other spatial tasks may be affected as well. However, the small number of studies used in the third analysis makes these results difficult to interpret.

A second meta-analytic study explored the effects of music instruction on mathematical abilities (Vaughn, 2000). Twenty-five studies were included and assigned to one of three groups: correlational, experimental-music instruction, and experimental-music listening. Separate meta-analyses were performed on each group. The analysis performed on the first group of studies yielded a modest relationship between music instruction and mathematics achievement. However, the researcher points out that, "...while correlation is a necessary condition for causality, it is not sufficient" (p. 154). One cannot conclude from this analysis that music instruction *caused* higher math achievement. The analysis of the experimental-music instruction studies showed that music instruction provided to children does indeed appear to cause increases in mathematics achievement. Finally, the author concluded from her third analysis on the music listening studies that "playing music in the background while students are taking math tests has only a small, positive effect at best" (p. 163).

In addition to the experimental and quasi-experimental research, a number of correlational studies have also explored the link between music and spatial intelligence. It is important to note here that correlational studies differ from experimental and quasi-experimental studies in that causality cannot be established due to lack of manipulation of independent variables. However, correlational studies can help determine the relationship between variables. In general, it appears that spatial and musical skills are related (Hassler & Birbaumer, 1988; Hassler, Birbaumer, & Feil, 1985; Hassler, Birbaumer, & Feil, 1987; Hassler & Feil, 1986; Hassler, Nieschlag, & de la Motte, 1990; Karma, 1982; Orsmond & Miller, 1995).

The Effects of Music Instruction on other Cognitive Abilities

Although the data supporting the effects of music instruction on spatial abilities is compelling, a series of other studies supports the notion that music training may enhance other skills as well. In particular, music instruction has been found to enhance visual-motor integration. A study examining the effects of Kodàly training on first-grade children found that children who received the training scored higher on a visual-motor integration task than children who did not receive the lessons (Hurwitz et al., 1975). The task required the children to copy complex figures, some portraying three-dimensional spatial relationships. Other spatial tasks were also administered, but the effects appeared to be found only for the boys.

Orsmond and Miller (1999) assessed three- to six-year-old children using one verbal task measuring receptive vocabulary, a melody recognition task, and three types of spatial tasks, including a visual-motor integration task. Twenty-nine children were enrolled in Suzuki music programs and 29 children received no music lessons. The researchers predicted improvement for the music group on the melody task and the three spatial tasks following four months of music instruction. No improvement was predicted for the verbal task. Results indicated that the children in the music group improved significantly on the melody recognition and visual-motor integration tasks. Little improvement was found for the other spatial tasks. Receptive vocabulary was not affected by the music training. The authors concluded that "music instruction improves fine motor skill" (p. 35).

A provocative study by Lamb and Gregory (1993) suggests that the aural skills involved in learning music can affect phonemic awareness, an important skill in learning to read. The authors tested 18 five-year-old children using a variety of reading, phonemic, and musical tasks. A test of general nonverbal ability, the *Coloured Progressive Matrices* test (Raven, 1956) was administered as well. The musical test was designed by the authors, and included measures of pitch and timbre discrimination. Results indicated that pitch discrimination was significantly correlated with both phonemic awareness and reading scores, which also were correlated with each other. Timbre discrimination and the matrices test did not correlate with any measures. Although this was a correlational study and precludes causal inferences, the data suggest that musical awareness may be a reliable predictor of phonemic awareness and reading ability.

Music training has also been found to affect verbal memory (Chan, Ho, & Cheung, 1998; Ho, Cheung, & Chan, 2003). In one study, the researchers compared the verbal and visual memory scores of children ages 6 to 15 who had received 0 to 5 years of music instruction (Ho et al., 2003). Based on lesion studies showing that verbal memory is processed by the left temporal lobe and that visual memory is processed by the right temporal lobe, the authors hypothesized that music training would influence only verbal memory due to the assumption that musicians' "...left but not right temporal lobe is...better developed" (p. 439). The musically trained children were compared to a group of demographically-matched children who had received no instruction in music. The verbal memory test required the children to recall as many previously spoken words as possible following a 10- or 30-min delay period. The visual memory test was similar to the verbal memory test, only the items to be recalled were designs rather than words. The results were as predicted: The children who had received music training scored higher than those who had not on the verbal memory test during both delay periods. There was no difference between the two groups on the visual memory test. There was also a significant positive correlation between the duration of music training and the children's scores on the verbal memory test, but not on the visual memory test. This study is consistent with research by Kilgour, Jakobson, and Cuddy (2000), which found that Canadian children with music instruction could recall more verbal material than children who had not received music instruction. It is also consistent with studies that found no differences between music and control groups on a test of pictorial memory (Rauscher, 2002; Rauscher & Zupan, 2000).

NEUROSCIENTIFIC EXPLANATIONS

The relationship of music exposure to spatial-temporal abilities is strengthened by the few studies employing neurophysiological measures. For example, an electroencephalographic (EEG) study found gamma band phase synchronization was significantly increased between the frontal cortex and the right parietal cortex during mental rotation tasks, with musicians showing greater synchronization than non-musicians (Bhattacharya et al., 2001). These findings suggest a link between music instruction and mental rotation.

A further EEG study suggests a link between music exposure and spatial-temporal reasoning. Subjects performed a spatial-temporal task immediately after listening to either a composition by Mozart or a spoken story (Sarnthein et al., 1997). The researchers found increased temporo-parietal coherence and prefrontal amplitude both during performance of the spatial-temporal task and while they were listening to the music. The parietal cortex has been implicated in the performance of a number of spatial tasks, including mirror-reading tasks (Dong et al., 2000) and visuomotor behaviors (Jeannerod, Decety, & Michel, 1994). Prefrontal cortex has been associated with working memory and with the temporal sequencing of patterns (Fuster, 1995).

EEG studies have consistently shown that oscillations in the gamma band frequency play an important role in music perception (Bhattacharya & Petsche, 2001). A recent study found that subjects who listened to music displayed increased activity in the gamma band while performing a spatial task than subjects who performed the task in silence (Jausovec & Habe, 2004). The authors concluded that listening to the music influenced visual as well as auditory brain activity. A further study by the same authors found similar results (Jausovec & Habe, 2005). The authors concluded that "...listening to a certain type of music...increases the activity of specific brain areas and in that way facilitates the selection and 'binding" together of pertinent aspects of sensory stimulus into a perceived whole" (p. 215).

A mathematical neural model of higher brain function, developed by Gordon Shaw and Xiaodan Leng, takes a slightly different approach to understanding the relationship between music exposure and spatial reasoning (Leng & Shaw, 1991). The theory is based upon Vernon Mountcastle's (1978) theory of cortical columnar organization. Mountcastle proposed that the cortex is composed of blocks of cells (about 300-500 micrometers in diameter), called columns. Adjacent blocks have very different properties of place (i.e., location in the animal's body which, when stimulated, evokes a response in the cortical tissue) and modality (i.e., the nature of the stimuli that evoke a response and the rate of adaptation to the stimuli). The vertical groupings of neurons with similar properties were called "minicolumns." Mountcastle proposed that these minicolumns are the irreducible basic functioning units of the neocortex. They are the primary processing and distributing units in the brain, and are able to both give and receive information. He hypothesized that each column connects with 10-30 other columns in various areas of the brain, suggesting that the brain functions as a distributed system. This was a very different view compared to previous thought, which depicted brain function as varying by location. It is now generally accepted that the entire cortex is composed of a repetitive and pervasive internal structure.

Leng and Shaw (1991) created a mathematical realization of Mountcastle's theory to try to understand how the brain processes information. They combined Moutcastle's (1978) theory with William Little's (1974) analogy of neural networking and physical spin models. Their goal was to test Mountcastle's theory using various computer simulations. Leng and Shaw named their model the "trion model," with the word "trion" referring to the minicolumns, which they postulated to have three possible levels of firing activity (high, average, low). Leng and Shaw proposed that the columns can be excited into complex spatial-temporal firing patterns. (In this case, the spatial scales are the minicolumns, columns, and cortex as a whole, and the temporal scales are the various timescales that exist in the brain, such as the circadian rhythm, which operates on a 24-hour cycle). Leng and Shaw suggested that the brain has the ability to recognize these trion firing patterns when they repeat, and can also recognize symmetries among these patterns. These patterns can be learned and strengthened based on Hebb learning principles. Thus, the highly structured and organized cortex is constantly searching for patterns and symmetry in the environment, including music.

Leng and Shaw (1991) argued that the trion model explains the coding of musical structure in humans, suggesting that we are biologically wired to create and understand music—particularly music that is high in "periodicity" (i.e., the change in relative amplitude of a piece of music over time) and/or symmetry (i.e., repeating notes, intervals, durations, reversed notes, and sequences). The finding that listening to a brief passage by Mozart can temporarily improve spatial-temporal reasoning in college students—the phenomenon popularly known as the "Mozart effect" — has been cited to support this theory (e.g., Rauscher, Shaw, & Ky, 1993).

In an attempt to define and measure characteristics of periodicity and symmetry in music by different composers, Hughes (2001) analyzed over 600 compositions, representing five composers. His results indicated the highest periodicity index scores for Mozart, J.S. Bach, and J.C. Bach, all of whom are known to have produced music with highly organized musical architecture. Mozart's music also displayed greater symmetry than did the other composer's compositions. Hughes concluded that "it is likely that the superorganization of the cerebral cortex resonates with [the] great organization found in Mozart['s] music." He suggested that Mozart's music may enhance certain abilities (e.g., spatial-temporal abilities) through Hebb learning principles.

It is unfortunate that the studies examining brain function relevant to music exposure and spatial task performance all involved listening to music rather than making music. To our knowledge, there exist no studies that have imaged people while actually performing a musical instrument, probably due to methodological difficulties involving movement artifact. However, given that playing a musical instrument typically involves listening we suggest that the brain areas involved in listening to music are also activated while making music. Given the decades of research showing that early experiences can functionally and structurally affect the developing brain (e.g., Hebb, 1949), it seems possible that early music exposure may affect related brain areas relevant to spatial task performance. It seems that among the music-related brain functions neuroscientists have imaged, spatial reasoning and understanding hold an important place.

Cognitive Explanations

Given the variety of stimuli music can present and the wide range of behaviors and capacities researchers have investigated, it is no surprise that the field has not gravitated toward unifying theories. We suggest that perhaps the strongest theoretical strain running through research on music learning is a cluster of effects centering on music's reported effects on spatial reasoning and arithmetic. Similarities between music, mathematics, and spatial reasoning have been suggested for decades (e.g., Cranberg & Albert, 1988). The argument in support of such theory derives from common and vital attributes of both music and spatial ability. Perhaps the strongest of such attributes is proportional reasoning. For example, the part-whole concept is a very important construct for many spatial and mathematical problems. This concept requires understanding the relationship between parts to wholes, such as when learning percents, decimals, and fractions. In music, proportions play out in musical notes, tempo, and pitch and involve both visual and auditory perception. The understanding of rhythmic patterns, in particular, seems reliant on the part-whole concept. To process rhythm, for example, the musician must repeatedly mentally subdivide the pulse in such a way that whole notes can be conceptualized as two half notes, four quarter notes, eight eighth notes, etc., or vice versa. The task is essentially the same as other part-whole problems posed spatially or mathematically. Similar computations are required for conceptualizing intervallic pitch relationships in music. Certain musical instruments, especially the keyboard and also the string instruments, are physically configured to reinforce the importance of proportion for effective musical production and performance.

Transfer is defined as the ability to extend what has been learned in one context to new contexts (e.g., Byrnes, 1996). The notion of transfer of learning from one domain to another has proved to be extremely controversial and difficult to demonstrate (Perkins & Salomon, 1989). Researchers interested in transfer were initially guided by theories that emphasized the similarities between the initial learning experience and later learning. Thorndike (1913) proposed that the amount of transfer that could occur between two domains was dependent upon the similarity of the elements of the domains. The more equivalent the elements of the two domains, the greater the likelihood of positive transfer. To our knowledge, only one study has specifically examined the transfer between mathematical skill and musical abilities (Bahr & Christensen, 2000). The researchers examined the mathematical skills of students using tasks that were deemed to be either structurally or not structurally related to music.

Structural learning analysis (Scandura, 1984) was performed to determine the degree of overlap between mathematical tasks and musicianship. (In this study, musicianship was defined as requiring formal training in music.) This form of analysis proposes that both mathematics and musicianship require the abstraction of patterned relationships over time. Music notation was likened to the use of graphs in mathematics, with the y-axis in music representing the frequency of a pitch and the x-axis representing time. "This symbolic and pattern comprehension facility is central to all tasks of literate musicianship, and is a common 'rule' for mathematical problem solving. It would seem then that structural analysis would confirm the likelihood that music and maths may 'overlap' for symbol and pattern usage" (Bahr & Christensen, 2000, pp. 192-193).

The procedure used to test this hypothesis was quite straightforward. Eighty-five students were given a test of musicianship designed to assess their knowledge of pitch and tonality notation, keys and scales, intervals and harmony, time and rhythm, and terminology. The students were also administered a mathematics test assessing a variety of different mathematical concepts including number handling, algebraic equations, three-dimensional shape visualization, rotation, graphing, etc. Items that required pattern analysis and symbol usage were deemed structurally similar to the domain of knowledge utilized by musicians. Results indicated that the students who had formal training in music performed better on the numerical tasks identified as overlapping with musicianship skills than the students who did not have formal music instruction. The groups' scores on tasks that did not overlap were not significantly different.

Although this study is intriguing, it leaves much to the imagination of the reader as the authors did not specifically identify which mathematical tasks were identified as overlapping with musicianship, other than to say they involved pattern analysis and symbol usage. They also did not discuss the nature or duration of the subjects' musical training. It appears that the presence of musical training was inferred from the students' scores on the musicality test. It is also important to note that the students enrolled in the study were chosen based on their proficiency in mathematics. All students were recruited from math classes that catered to students who excel in mathematics. It is therefore impossible to determine in which direction the transfer occurred. Perhaps rather than musical knowledge having transferred to mathematical knowledge, mathematical knowledge transferred to musical knowledge.

We believe that the diverse tasks involved in playing a musical instrument strengthen a number of cognitive skills, including auditory, visual, and motor abilities. Measuring the overlap between the original domain of learning and the novel one requires a theory of how knowledge is represented and conceptually mapped across the domains. Transfer is thus difficult to test experimentally until the task components are identified. Although the transfer of musical knowledge to spatial-temporal or mathematical understanding may appear obvious to the musician, there is little research that has directly pursued these mechanisms. Furthermore, several other aspects of intelligence may be affected by music instruction. The latter portion of this chapter will examine this possibility.

Unpacking the Impact of Music on Intelligence

Theory and evidence about the effects of music on spatial-temporal reasoning offer researchers a good framework for exploring diverse outcomes of music learning and experience. As illustrated above, many effects of music learning show well-reasoned and close connections to spatial relations. A strong and well-documented evidentiary base suggests learning and playing music on a keyboard can touch a multitude of spatial relationships measured by an assortment of instruments. We argued above that the configuration of the keyboard, along with the nature of written music, present logical suggestions for why such experiences engage spatial-temporal reasoning. Most researchers and musicians would agree that music is a rich symbol system indicating myriad qualities of tone, representations of distances in space, and organization around time.

Despite emphasis in music research on transfer to spatial-temporal reasoning skills, and in some cases on transfer to specific skills such as mathematics or verbal understanding, it is reasonable to wonder whether or not the main effect of music learning and experience on cognition is largely an effect on general performance capacity, or general intelligence. It is conceivable that general intelligence gains are responsible for performance gains in widely differing areas researchers have tested. We discussed above the ways that research has shown transfer of abilities from music experiences to a spectrum of capacities: literacy, general intelligence, verbal memory, spatial ability, reading ability, selective attention, and mathematics achievement[6]. As Schellenberg (2004) summed things up, "...the most parsimonious explanation of these diffuse associations is in that they stem from a common component, such as general intelligence"[7]. About three years ago, Schellenberg then set out to test the impact of music learning on general intelligence in a high-power and rigorous experimental design. As we describe this seminal research, it is clear that the study maintained classical conditions of learning experiments, the treatments were of high quality and long duration, and the reasoning system for drawing conclusions was sophisticated. In short, it is one of the most well designed studies of music and cognition we have seen. But the study left open a question vital to our purposes here: Just what does music contribute to intelligence, and how? Based on years of research, and especially its concentration on spatial reasoning skills, we wondered if any demonstrated effect of music on intelligence might not manifest disproportionately through gains in visual-spatial skills (essentially half of the leading intelligence test batteries) as opposed to gains in verbal skills (essentially the other half). As it turned out, data from Schellenberg's (2004) experiment could support an examination of this question.

[6] Cites from Schellenberg, 2004, p. 1.

[7] Schellenberg, 2004, p. 511.

The Schellenberg Data: Reviewing Contributions to Intelligence

What follows is a description of the essential components of Schellenberg's (2004) study. (Additional description and detail are available in the published work.) The experiment used random assignment of six-year-olds (N=144) into four groups: two music lesson groups (keyboard or voice) and two no-music groups (drama lessons or no lessons of any sort). Music and drama lessons were taught over a period of 36 weeks by high-level certified teachers employed at the Royal Conservatory of Music in Toronto, Canada. Children received their instruction in classes of six. An IQ test, the WISC-III (Wechsler, 1991) was administered before and after the 36-week period.

Compared to children in the drama and no-lesson control groups combined, the keyboard and voice groups showed greater increases in *Full Scale IQ*. The differences in effects favoring the music groups were small but statistically significant. The effect was greatest for the voice group. For this main effect assessment, Schellenberg argued that combining the music groups made sense because the keyboard and voice groups experienced similar IQ score gains, and collapsing groups was also desirable because the increased group size would make for more powerful statistical tests. Schellenberg probed some differences across groups in individual WISC-III subtests, such as verbal comprehension and perceptual organization, and found that a significant advantage showed for the consolidated music group on 10 of 12 scales.

Schellenberg's main findings are displayed in Table 1. This table shows the pre- and post- IQ scores for each of the four groups.

Table 1. Schellenberg 2004 Results
Comparisons of WISC III Full Scale IQ Score Gains:
Keyboard, Voice, Drama, and No Lessons Groups

	Group	N	Pre-Score	Post-score	Gain-score	Sig. p<__*	Pct. Gain	Effect Size
Music	**Keyboard**	30	102.6 8.8	108.7 12.5	6.1	0.000	5.9%	**0.44**
	Voice	*32*	*103.8* *10.9*	*111.4* *12.6*	7.6	*0.000*	*7.3%*	***0.55***
No-Music	**Drama**	34	102.6 *13.6*	107.7 *10.19*	5.1	0.000	5.0%	**0.37**
	No Lessons	*36*	*99.40* *13.8*	*103.3* *9.9*	3.9	*0.005*	*3.9%*	***0.28***

* t-tests for sig. of pre- to post-score gains;
Effect size = gain score / std. deviation of No Lessons group; simple standardized gain for No Lessons group. Difference in gain scores across groups is reliable (p=.05);
Gain scores of music groups are similar (p<.08);
Gain scores of No-Music groups are similar (p<.07).

The advantage for the music groups (keyboard and voice) appears here as greater average gains in IQ scores. The voice group outpaced the keyboard group; the no-lessons group showed the smallest gain. That all groups gained to some degree is not surprising since the general pattern of WISC-III IQ scores shows increases over time for six-year-olds. That the effect of the drama group fell short of that of the music groups addresses an important concern in this and many studies. As pointed out above, it is possible that more general qualities of the disciplined study of music might characterize disciplined specialized study in other domains, for example, or that a productive arousal of interest in music lesson groups might be no different than arousal witnessed in drama education. This study's data suggest that something small but significant transfers from music – more than from drama. The most aggregated expression of the implication of this study appears in its published title: *Music Lessons Enhance IQ* (Schellenberg, 2004).

Can this Study Tell us More?

Long before the publication of Schellenberg's work, the "music raises IQ" results of his study were transmitted widely in the popular media. This dissemination bore a rough similarity to the phrase "Mozart makes you smarter" that was popularly celebrated more than a decade earlier. The latter ushered in a period of active studies of transfer from music and artistic learning, many discussed above. This field has been propelled in recent years by the interests of cognitive neuroscientists exploring the impacts of music on brain function.

Cognitive researchers who focused on music, including ourselves, naturally took an interest in the news of Schellenberg's study. Schellenberg graciously furnished both of us with a proof of his upcoming publication. After considering the study, several questions not addressed in the article stood out to us on the basis of prior research in music and learning. The most important of these was the potential in the data for investigating the possibility that spatial reasoning gains for the music subjects may have disproportionately accounted for the reported gains in general intelligence. While the study concluded that keyboard and voice lessons affect intellectual capacity generally, prevailing theory in music learning and transfer suggest that developments in spatial-temporal reasoning ability stand out among measured effects of music on cognitive function. A more fine-tuned curiosity pointed to the keyboard group in this study. Based on what research has reported to date, there is good reason to believe that music learning involving the keyboard might advance spatial abilities more than would learning in other types of music or on other musical instruments. The data from Schellenberg's study seemed to offer ways to sort these things out.

Unpacking the Impact of Music on Intelligence

We conducted an independent analysis of Schellenberg's data to explore the possibility that differential gains within the construct of general intelligence accounted for the *Full Scale IQ* results. More specifically, we theorized that music students in the study might have experienced larger gains in the underlying WISC-III intelligence scales related to spatial ability than in other scales. Our procedure was to revisit Schellenberg's WISC-III results to see if the embedded scales could tell us more about the cognitive development of the subjects during the course of the experiment. Results on a wide range of measures were available to us. The WISC-III derives its *Full Scale IQ* measure by combining a six-item *Verbal IQ Scale* and a five-item *Performance IQ Scale*. The verbal scale contains measures of these items: *Information* (general "trivia-like" questions), *Comprehension* (common sense reasoning, judgment), *Vocabulary* (knowledge of words and grouping words by meaning), *Similarities* (categorizing information, abstract reasoning), and *Arithmetic* (attention, concentration, and numerical reasoning (oral)). The *Performance IQ Scale* contains measures of these items: *Object Assembly* (visual analysis), *Block Design* (spatial problem solving with puzzles), *Picture Arrangement* (sequential, logical thinking), *Picture Completion* (identifying missing parts of pictures; mastery of essential detail), *Symbol Search* (locating symbols in a grid of symbols; rate of processing new information), and *Coding* (visual-motor skills; marking rows of shapes with appropriate codes).

Spatial Reasoning in the WISC-III

There is ample precedent for enlisting the WISC-III for assessing spatial reasoning ability, but a scan of our review shows modest use of WISC-III scales in studies of music. Apart from Schellenberg, we are not aware of any study in this field that has used the *Full Scale Intelligence* measure, and only two that exploit a significant subset of its major scales[8]. One reason for its limited use is that the WISC-III test battery is time consuming, painstaking to administer, and expensive by norms of research in education. Its publisher, the Psychological Corporation, restricts use of the test to Ph.D. psychologists who have been trained to administer and interpret it. The entire battery requires more than two hours of test time for each subject, and two hours of the trained administrator's time. (Thus Schellenberg's study retaining 132 subjects through the post-test required 528 hours of test administrator time beyond set-up times and testing intervals between subjects.)

Rauscher's early studies (Rauscher, Shaw, Levine, Ky, & Wright, 1994; Rauscher et al., 1997) enlisted the WPPSI-R rather than the WISC-III, and administered four of the Performance sub-test tasks: Object Assembly, Block Design, Geometric Design, and Animal Pegs. Her studies found enhancement only for the *Object Assembly* task as an indicator of spatial reasoning skills in very young children. More than half of the studies assessed by Hetland (2000) in her meta-analysis of 39 music and spatial ability experiments used the Paper Folding and Cutting task from the Stanford-Binet battery (pp. 120-122)[9].

The full range of WISC-III scales presents analytical opportunities generally not enlisted in the music field. If we look at studies using the WISC-III more generally to assess cognitive functioning, and in particular studies that focus on spatial skills, a tradition of indicators is

---

8 Hetland's (2000) meta-analysis of 35 music and spatial reasoning studies generally conducted in the late 1990s reports no results from the WISC. Sixty-one percent of the reported studies used the Paper Folding and Cutting task from the Stanford-Binet battery.

9 The Object Assembly task requires children to assemble cardboard cut-out pieces of a familiar object into a unified whole in the absence of a physical model; the Paper Folding and Cutting task presents subjects with a drawing of a piece of paper that has been folded and cut in several places. The subjects' task is to mentally unfold the paper in order to choose an illustration depicting what the paper would like after it has been unfolded.

visible. First, professionals involved in the assessment of learning disabilities find useful the distinction between visual-spatial skills on the one hand and verbal skills on the other hand. The Learning Disabilities Association of Ontario advises its members to consider the *Performance IQ* scale of the WISC-III to top the inventory of tests for visual-spatial processing[10]. In an excerpt from her book, Bonnell (n.d.) claims that, "...poor *Performance IQ* means a general visual-spatial disability."

Riverside Publishing, publisher of a rival test, the Woodcock-Johnson (WJ) III, aligns spatial-relations measures of the WJ III to three sub-tests in the WISC-III: Block Design, Object Assembly, and Picture Completion. These are three of the five sub-tests in the WISC-III *Performance IQ* scale (excluding Symbol Search and Coding). The Psychological Corporation, publisher of the Differential Abilities Scales (1990), states in its handbook for the DAS that, the DAS spatial cluster and the WISC III Performance (IQ) scale correlated the highest (r=.82). Geneticist H. Hirota (2003) referred to the *Performance IQ* scale of the WISC-III as a measure of visual-spatial intellectual ability. Hirota used *Performance IQ* in a recent study of spatial ability markers located on human chromosomes. And in their very recent review of "types of intelligence" enlisted by cognitive psychologists, Phelps et al. (2005) associate the WISC-III's *Performance IQ* with *Gv*, or visual intelligence, and *Verbal IQ* more with generalized intelligence[11]. To provide a concrete illustration, Gabel (2001), in the Psychological Corporation's WISC Interpretive Guide, discussed *Gv* (visual intelligence) and its mapping onto the Block Design, Symbol Search, and Coding items of the WISC-III – another three of five subscales comprising *Performance IQ*. Gabel suggested that G*v* is used by physicists, engineers, auto mechanics, architects, carpenters, sculptors, and parts department managers in their jobs. It could also be that musicians use *Gv* to deal with the demands of their instruments and music notation.

Taken together, professional and scholarly usage of the WISC-III *Performance IQ* scale supports its potential value in sifting through the "types of intelligence" that may be influenced by music lessons. In traditional usage, *Performance IQ* is thought to involve capacities largely distinct from the capacities measured by the *Verbal IQ* scale[12]. Since *Full Scale IQ* is solely derived from *Performance IQ* and *Verbal IQ*, an imbalance of experimental score gains between *Performance IQ* and *Verbal IQ* might point to specific building blocks of measured general intelligence gains. In the case of music, there is a possibility that IQ gains such as those reported by Schellenberg (2004) came more through advances in visual-spatial processing skills than through gains in verbal skills. In addition, an inquiry into this possibility might suggest mechanisms underlying the increase in *Full Scale IQ* accompanying music lessons in the Schellenberg study.

---

[10] Learning Disabilities Society of Ontario. Idao.caa/pei/assessment/supporting.php.

[11] General intelligence is represented by Gc, crystallized intelligence (facts), and Gf, fluid intelligence (reasoning with facts).

[12] We should note that Verbal IQ and Performance IQ are not completely unrelated. In our database, the correlation of pre-test measures of the two scales is about 0.3. As administered to 6 year olds, the WISC-III is administered orally, and all scales would be impacted by children's aural information processing skills.

Mathematics in the WISC-III

The *Full Scale IQ* and *Performance IQ* scales do not include a test of mathematics proficiency. An item labeled *Arithmetic* is included in the WISC-III *Verbal IQ* battery, ostensibly because the items depend more on decoding of oral language than success in solving mathematics problems. The student assessments in the Schellenberg experiment went beyond the WISC-III to include the Kaufman Test of Educational Achievement (K-TEA), Comprehensive Form (Kaufman & Kaufman, 1985). This test includes items scaled for mathematics computation (calculation and operations) and mathematics applications (math reasoning and application to problems). These scales present an opportunity to explore another reported impact of music instruction, i.e., its effects on mathematics skills.

Music and Verbal Versus Spatial Skills

For our first analyses we compared the effects of music participation on *Verbal IQ* and *Performance IQ* respectively. We first combined the keyboard and voice groups into a single "music" group and the drama and no lessons groups into a single "no-music" control group. We then examined the differences between the combined music group and the combined control group. We next assessed the effects of the keyboard only group on verbal and performance intelligence and compared the keyboard group to the no-lessons control group. The results are shown in Table 2.

Table 2. Comparisons of WISC III Scale Score Gains: Music Groups vs. No-music Groups Verbal IQ, Performance IQ

Music (Keyboard or voice) vs. No Music (Drama or no lessons)								
Measure	**Group**	**N**	**Pre-Score**	**Post-score**	**Gain-score**	**Sig. p<___*****	**Pct. Gain**	**Effect Size**
Verbal IQ	**music**	62	114.02 12.29*	119.76 14.35	5.74	0.000	5.0%	**0.45**
	no-music	*70*	*112.9* *12.84*	*115.97* *12.24*	*3.07*	*0.011*	*2.7%*	
Performance IQ	**music**	62	51.74 *6.87*	57.08 *10.19*	5.95	0.009	10.3%	**0.55**
	no-music	*70*	49.99 *9.7*	53.57 *10.71*	2.06	0.226	7.1%	
Keyboard vs. No Lessons								
Verbal IQ	**keyboard**	34	113.2 12.11	117.63 15.56	4.43	0.021	3.9%	**0.42**
	no-lessons	*42*	*112.56* *10.62*	*115.47* *11.34*	*2.91*	*0.053*	*2.6%*	
Performance IQ	**keyboard**	34	50.05 7.13	56.33 9.35	5.8	0	11.5%	**0.68**
	no-lessons	*42*	*47.89* *9.51*	*51.17* *9.01*	*3.28*	*0.004*	*6.8%*	

* Standard deviation;
* t-tests for sig. of pre- to post-score gains; Data derived from Schellenberg, 2004.

The upper half of Table 2 displays the results for the music/no-music comparison. There were 62 students in the music group and 70 students in the no-music group – accounting for all subjects for which the study obtained pre- and post-surveys. The second and third numerical columns show the actual pre-scores and post-scores on the *Verbal IQ* and *Performance IQ* scales, along with standard deviations. Pre-test to post-test gain scores are shown along with the statistical significance of score gains.

What do we see in Table 2? First, both groups made significant gains on the *Verbal IQ* and *Performance IQ* scales over the course of their first grade year. This seems indicative of an expected gain on these tests over a school year at the age of six. When we look at the performance of the music vs. no-music groups, raw gain scores cannot be directly compared because *Verbal IQ* and *Performance IQ* are built on different scales – medians of about 100 points for *Verbal IQ* and 50 points for *Performance IQ*. One way to compare score gains is by simply looking at the percentage gains represented by the gain-scores. In this case, it can be seen that both groups gained in percentage terms on both scales. The music group outpaced the no-music group in *Verbal IQ* gains, 5.74 percent as opposed to 3.07 percent. The music group gained 10.3 percent in *Performance IQ* while the no-music group gained 7.1 percent. Based on percentage gain score changes, it appears that the music group made comparatively larger gains in *Performance IQ* than in *Verbal IQ*.

Since the numerical scales differ, the best test for differences in *Verbal IQ* versus *Performance IQ* gain scores is to standardize the gains and to calculate the effect sizes associated with participating in music. This is accomplished by dividing the music group's score gains on the measures by the respective standard deviations of comparison group pre-scores on the same measure. For example the effect size of participating in music on *Verbal IQ* (in comparison to the control group or no-music status) equals the music group's Verbal IQ gain (5.74 points) divided by the standard deviation of the no-music group's Verbal IQ pre-score (12.84). The effect size is 0.44. An effect size of 0.30 is considered moderate but significant. Effect sizes above 0.50 are considered robust.

The critical comparison for our analysis shows up in relative effect sizes that gauge music's effect on verbal intelligence versus performance (or visual-spatial) intelligence. As shown in Table 2, music participation in this experiment shows an effect size of 0.45 on *Verbal IQ* and a somewhat (but not greatly) larger effect size (0.55) on *Performance IQ*. This supports the contention that gains in *Full Scale IQ* associated with music lessons benefited more from gains in visual-spatial intelligence than in verbal intelligence. This suggests that a specific effect of music lessons on visual-spatial intelligence was important to the outcome of the *Full Scale IQ* analysis, and that spatial reasoning might be considered a mechanism through which music impacts intelligence.

The bottom portion of Table 2 shows an analysis parallel to the one just described exploring music versus no-music in this study: a comparison of the keyboard group, one of the two music sub-groups, to the no-lessons control group (one of the no-music sub-groups). Our interest in the keyboard group grew from research described above often finding that instruction on the keyboard contributes to spatial reasoning skills, possibility due to the spatial layout of the keyboard. The data shown in Table 2 support a claim that the keyboard group experienced more growth in *Performance IQ* than any group studied – including the no-lessons group. Percentage gains in both verbal and visual-spatial scores for the keyboard group outpaced those of the control group, especially in *Performance IQ*. The effect size on *Performance IQ* (0.68) is larger than the effect size on *Verbal IQ* (0.42). Overall, the disparity favoring visual-spatial development over verbal development is greater for the keyboard group than for the combined music (singing and keyboard) group (a 0.26 point difference versus a 0.10 point difference, for want of a better metric). These data support a contention that the keyboard students in the study contributed significantly to the *Full Scale IQ* gain through disproportionate learning in *Performance IQ*, or the visual-spatial domain.

Visual-Spatial Gains for Whom?

Studies of learning such as the ones we reported in this chapter rely mainly on average measures for constructed groups. We decided to explore how music instruction in general and keyboard lessons in particular impacted children with low *Full Scale IQ* pre-test scores. We thus performed tests analogous to the ones shown above in Table 2, but we included only those subjects who had scored at the 50th percentile or below on the *Full Scale IQ* pre-test. These children are likely to suffer more difficulties in school than the average participant in this study. We chose to call this group an *at-risk* sub-sample.

Table 3 shows the results of these analyses. In percentage terms, the music group outscored the no-music group in both *Verbal IQ* and *Performance IQ*. The percent increase in scale scores was higher for *Performance IQ* than for *Verbal IQ*, but the effect size associated with music participation is about the same for verbal and visual-spatial intelligence.

As we did for the whole sample, we also examined the effects of keyboard lessons, as opposed to no lessons at all, for the at-risk sub-sample. As shown in the lower half of Table 3, we found the largest disparity favoring the effects of music instruction on visual-spatial development compared with the effects on *Verbal IQ*. The keyboard group showed 12 percent score gains in *Performance IQ* in contrast to 4.1 percent score gains in *Verbal IQ*. The effect sizes of the keyboard group for both verbal and spatial intelligence are substantial, but a large difference favors effects on visual-spatial intelligence (effect size 0.88 on *Performance IQ* versus 0.55 on *Verbal IQ*).

Table 3. Keyboard Group versus No-Lessons Control Group: Comparisons of WISC III Score Gains: Verbal IQ, Performance IQ, Verbal Composite, and Arithmetic

Music (Keyboard or voice) vs. No Music (Drama or no lessons)								
Measure	**Group**	**N**	**Pre-Score**	**Post-score**	**Gain-score**	**Sig. p<__****	**Pct. Gain**	**Effect Size**
Verbal IQ	**music**	4	106.8 9.23**	114.2 12.71	7.4	0.08	6.5%	**0.80**
	no-music	*42*	*105.8* *9.28*	*110.3* *9.68*	*4.5*	*0.034*	4.1%	
Performance IQ	**music**	34	48.41 5.68	54.5 9.05	6.09	0.000	11.2%	**0.82**
	no-music	*42*	*44.79* *7.46*	*49.35* *9.82*	4.56	*0.000*	*9.1%*	
Keyboard vs. No-lessons								
Verbal IQ	**keyboard**	18	106.78 10.45	111.33 15.26	4.55	0.080	4.1%	**0.55**
	no-lessons	*26*	*108.27* *8.29*	*111.58* *9.11*	3.31	*0.066*	3.0%	
Performance IQ	**keyboard**	18	47.78 6.06	54.28 8.64	6.5	0.000	12.0%	**0.88**
	no-lessons	*26*	*45.04* *7.39*	*49.54* *8.43*	4.5	*0.002*	9.1%	

Subsample scoring at or below 50th percentile in Full Scale IQ on pre-test (56.7 percent of subjects due to ties).

** standard deviation

*** t-tests for sig. of pre- to post-score gains.

Data derive from Schellenberg, 2004.

AT-RISK VERSUS ALL STUDENTS

It appears that the effects of music on both *Verbal IQ* and *Performance IQ* are higher for the at-risk group (Table 3) than for the consolidated music group. A portion of this general effect is probably due to a natural regression of scores toward the mean. Students in the lowest reaches of a performance distribution on a pre-test are likely to improve their scores more than the average student. Another possible explanation for the disparity is that children with lower spatial ability tend to benefit more from keyboard lessons than other children. For some reason their visual-spatial learning curve is steeper. In any case, the relatively strong effect of keyboard lessons on *Performance IQ* in this sub-sample is an important finding and an issue that has not been attended to in research to date.

MUSIC AND MATHEMATICS

In this final section we turn to our analysis of the effects of music instruction in general as well as the effects of keyboard instruction in particular on mathematics skills. Our approach parallels our assessment of music instruction's effects on *Visual* and *Performance IQ*. We examined standardized gains in mathematics computation and math applications indicated by student performance on pre- and post-test measures from the K-TEA.

We first examined relationships between the music lesson and no-music lesson groups in the experiment. About half of the children were assigned to either keyboard or voice education (music lesson group). The others were assigned to drama lessons or no lessons at all (no-music lesson group). Our comparisons in mathematics learning are based, as above, on standardized gains in average group test scores and effect sizes.

Our main comparisons for music versus no-music groups are shown in the upper half of Table 4. Here it can be seen that all groups made significant gains in math computation and math applications scores. Percentage score gains were in the 9.8 percent to 15.8 percent range, and effect sizes were in the 0.62 to 0.93 range. These gains are generally larger than any effects we saw for treatment and comparison groups on *Visual* or *Performance* IQ. Disproportionate gains in mathematics achievement for all groups probably speak to the fact that math is generally taught every day in the elementary school curriculum and few students fail to progress at some level. The data in Table 4 show that effect sizes in mathematics for the consolidated music groups marginally exceed gains for the no-music group – effect sizes of 0.80 versus 0.62 in math computation and of 0.93 versus 0.87 in math applications. These margins do not attest to dramatic differences in math scores between music and no-music groups, but some link across music learning, spatial skills development, and mathematics learning may be at work. The data more firmly reveal that all groups on average made significant progress between pre- and post- mathematics tests.

Table 4 also displays the results of our analysis of mathematics impacts for keyboard students versus the no-lesson control group. The results resemble what we found for all students[13]. We show 11 to 15 percentage point gains in computation and application scale scores across all groups and very modest differences in music instruction's effect sizes favoring the all-music and keyboard groups. If something is propelling math scores through gains in spatial-temporal skills or through other mechanisms, the valence seems positive. However, additional research is needed to say much more.

A similar story emerges for our final analysis, the impacts of music on mathematics for students who started this experiment scoring in the lower half of the *Full Scale IQ* distribution. We characterized this group above as having higher risk of difficulties in school. Indeed, a comparison between Table 4 (all students) and Table 5 (at-risk sub-sample) shows uniformly lower average math scores for at-risk students, as one indicator of relative preparedness for what lies ahead. As we saw in Table 4, percent score gains in mathematics shown in Table 5 are large for all groups. A very small advantage generally accrues to the music and keyboard groups. An exception shows in the bottom lines of Table 5 where we see

---

[13] The keyboard students made up half of the consolidated music groups; thus their respective performance outcomes will be significantly correlated.

that the no-lessons control group gained slightly more than the keyboard group in math applications scores, albeit with nearly identical gains.

Table 4. Comparisons of Scale Score Gains: Music Groups vs. No-music Groups Math Computation: Math Applications

Measure	Group	N	Pre-Score	Post-score	Gain-score	Sig. p<__***	Pct. Gain	Effect Size
Music (Keyboard or voice) vs. No Music (Drama or no lessons)								
Math Computation	music	62	89.55 16.68	103.69 11.09	14.14	0.000	*15.8%*	**0.80**
	no-music	*70*	90.29 *17.63*	*100.44* *10.13*	*10.89*	*0.000*	*11.2%*	*0.62*
Math Applications	music	62	90.27 *10.13*	99.74 *10.22*	*9.45*	0.000	*10.5%*	**0.93**
	no-music	*70*	90.70 *10.15*	99.56 *11.1*	*8.86*	0.000	*9.8%*	*0.87*
Keyboard vs No Lessons								
Math Computation	keyboard	30	90.93 17.29	104.43 13.2	*13.50*	0.000	*14.8%*	**0.80**
	no-lessons	*36*	*86.5* *16.87*	*97.75* *10.65*	*11.25*	0.000	*13.0%*	*0.67*
Math Applications	keyboard	30	90.97 11.68	101.8 11.68	*10.83*	0.000	*11.9%*	**1.06**
	no-lessons	*36*	*88.89* *10.22*	*98.42* *10.97*	*9.53*	0.000	*10.7%*	*0.93*

* Standard deviation.
** Standardized gains for control groups.
*** t-tests for sig. of pre- to post-score gains.
Data derive from Schellenberg, 2004.
Mathematics scales from the Kaufman Test of Educational Achievement (K-TEA; Kaufman & Kaufman, 1985).

Table 5. Comparisons of Scale Score Gains:
Music Groups vs. No-music Groups
Math Computation: Math Applications
Subjects w/ Low Pre-Full Scale IQ****

Music (Keyboard or voice) vs. No Music (Drama or no lessons)								
Measure	**Group**	**N**	**Pre-Score**	**Post-score**	**Gain-score**	**Sig. p<__*****	**Pct. Gain**	**Effect Size**
Math Computation	**music**	34	85.18 17.64	101.5 12.21	16.32	0.000	*19.2%*	**0.93**
	no-music	*42*	83.05 *17.63*	*96.67* *10.13*	*13.62*	*0.000*	*16.4%*	*0.77*
Math Applications	**music**	34	86.24 *10.53*	95.76 *9.09*	*9.52*	0.000	*11.0%*	**0.94**
	no-music	*42*	86.74 *10.15*	95.36 *11.1*	*8.62*	0.000	*9.9%*	*0.85*
Keyboard vs No Lessons								
Math Computation	**keyboard**	18	88.67 18.54	102.44 15.43	*13.77*	0.002	*15.5%*	**0.82**
	no-lessons	*26*	*83.05* *16.87*	*95.35* *10.65*	*12.30*	0.001	*14.8%*	*0.73*
Math Applications	**keyboard**	18	87.78 10.13	97.28 10.22	*9.50*	0.000	*10.8%*	**0.97**
	no-lessons	*26*	*86.73* *9.825*	*96.46* *11.1*	*9.73*	0.000	*11.2%*	*0.99*

* Standard deviation.
** Effect size for treatment groups. Standardized gains for control groups.
*** t-tests for sig. of pre- to post-score gains.
**** Subsample scoring at or below 50th percentile on WISC III Full Scale IQ pre-test.
Data derive from Schellenberg, 2004.

CONCLUSION

In this chapter, we reviewed leading research on the extra-musical effects of music learning. This body of research focuses on the cognitive manifestations of music experiences. It consequently also focuses on the possible cognitive effects of music and especially our capacities to think. A preponderance of cognitive studies in music suggest that music instruction's most commonly seen effect is on visual-spatial thinking skills. This conclusion is based on studies of varying music modalities using a variety of research instruments designed to measure visual-spatial skills. Studies also indicate that among music experiences, learning to play the keyboard is prominent among the musical instruments showing effects on spatial thinking. This may be because of the geometric/spatial layout of the keyboard paired with the corresponding geometry and proportionality of written music.

In the latter half of this chapter, we assessed data from a recent high-quality study that concluded that music lessons enhance intelligence (Schellenberg, 2004). Our purpose was to examine this popularized and scientifically justified claim to see if spatial reasoning development was disproportionately responsible for measured growth in general intelligence. This curiosity was of course propelled by the traditions in cognitive music research just described. We also capitalized on an opportunity to explore possible relations between music lessons and student mathematics achievement.

Our instincts were rewarded for the most part. Using *Verbal IQ* and *Performance IQ* measures from the WISC-III—the two scales that account completely for *Full Scale IQ*—we assessed developments of music versus no-music students. In one comparison, we showed that for the consolidated music group (both the keyboard and voice lessons groups together), music participation or learning boosted *Performance IQ* (visual-spatial) more than *Verbal IQ*, with effect sizes 0.55 versus 0.45, respectively. Thus music instruction in this experiment not only boosted general intelligence, but also the overall measure showing increased intelligence was comprised of modestly stronger developments in visual-spatial skills than in verbal skills.

When we turned to keyboard lessons versus no lessons at all, our instincts led us to hypothesize that any contributions to increased intelligence from the keyboard group would favor growth in visual-spatial skills perhaps even more than the keyboard and voice groups combined. Keyboard instruction's effect on visual-spatial skills (effect size 0.68) was in fact stronger than its effect on verbal skills (effect size 0.42). Of the two musical experiences in this experiment, keyboard lessons showed greater influence on spatial skills development. Furthermore, for students scoring low in *Full Scale IQ* at the start of the study, the components of general intelligence gains were balanced between visual IQ and verbal IQ. For the lower IQ keyboard group, however, gains in visual skills outpaced gains in verbal skills by the highest proportion of all or our comparisons.

A general finding of this analysis is that comparative gains in general intelligence reported for music students may be caused more by gains in visual-spatial reasoning skills than by gains in verbal skills. We could stretch this point with an assertion that we did not test: That the importance of music instruction's contribution to spatial skills may be masked though the analytical techniques we used in this research. It is possible that gains in visual-spatial skills for music students contributed to increases in their verbal skills, based on supported contentions that language has a significant spatial-relations component. (Of course things could work the other way around: children may gain more on spatial tests if their language skills have improved.) At any rate, this discussion seems to touch on some underlying mechanisms through which music bears on thinking skills.

We then explored possible effects of music lessons on proficiency in mathematics. In this analysis, we found generally large effect sizes (proficiency gains for all groups). These gains favored all-music and keyboard students over comparison groups, but the differences were small and probably not educationally important. We also performed the same analysis with an at-risk sub-sample of participating students. Contrary to what we observed in comparisons involving all students, the at-risk music groups – all music and keyboard – showed gains in math skills approximating the gains found for the whole sample. We did not find that music lessons improve mathematics achievement. Another relationship in our data presents another challenge to music-mathematics hypotheses. In our first pass through the data, music showed stronger effects on visual-spatial skills for at-risk students than for the average music student. At the same time, music showed no greater effects on mathematics proficiency for at-risk students than for all music students. Thus we did not manage to detect an advantage in mathematics for music students that might have been mediated by the development of visual-spatial skills.

Finally, we need to remind ourselves of an issue we broached at the start. If one intends to boost IQ scores, verbal or visual, or mathematics proficiency, music would not be the tool of choice. There are certainly learning programs that would reach such goals more efficiently. Nonetheless, it is important to recognize that music has positive cognitive implications. We do not know the duration of the effects of music exposure on children, at least not the effects that have occupied the learning researchers discussed here (i.e., cognitive effects). However, based on present indications, the valence of music experience in the long run is probably positive for many participants and in ways that have not been fully unpacked. The cellist, guitarist, and trumpet player with ten years of lessons can speak to us directly about many more important effects.

Acknowledgment

We are extremely grateful to Dr. E. Glenn Schellenberg for providing us with the data we used in this analysis.

References

Bahr, N., & Christensen, C.A. (2000). Inter-domain transfer between mathematical skill and musicianship. *Journal of Structural Learning and Intelligent Systems, 14*, 187-197.

Bhattacharya, J., & Petsche, H. (2001). Universality in the brain while listening to music. *Proceedings of the Royal Society of London, 268*, 2423-2433.

Bhattacharya, J., Petsche, H., Feldmann, U., & Rescher, B. (2001). EEG gamma band phase synchronization between posterior and frontal cortex during mental rotation in humans. *Neuroscience Letters, 311*, 29-32.

Bilhartz, T.D., Bruhn, R.A., & Olson, J.E. (2000). The effects of early music training on child cognitive development. *Journal of Applied Developmental Psychology, 20*, 615-636.

Brochard, R., Dufour, A., & Després, O. (2004). Effect of musical expertise on visuospatial abilities: evidence from reaction times and mental imagery. *Brain and Cognition, 54*, 103-109.

Byrnes, J.P. (1996). *Cognitive development and learning in instructional contexts*. Boston: Allyn and Bacon.

Chan, A.S., Ho, Y.C., & Cheung, M.C. (1998). Music training improves verbal memory. *Nature, 396*, 128.

Costa-Giomi, E. (1999). The effects of three years of piano instruction on children's cognitive development. *Journal of Research in Music Education, 47*, 198-212.

Cranberg, L.D., & Albert, M.L. (1988). The chess mind. In K.L. Obler & D. Fein (Eds.), *The exceptional brain* (pp. 156-190). New York: The Guilford Press.

Crncec, R., Wilson, S.J., & Prior, M. (2006). No evidence for the Mozart effect in children. *Music Perception, 23*, 305-317.

Dong, Y., Fukuyama, H., Honda, M., Okada, T., Hanakawa, T., Nakamura, K. et al. (1995). Essential role of the right superior parietal cortex in Japanese kana mirror reading: an fMRI study. *Brain, 123*, 790-799.

Fuster, J.M. (1995). Temporal processing. *Annals of the New York Academy of Sciences, 769*, 173-181.

Gardiner, M.F., Fox, A., Knowles, F., & Jeffrey, D. (1996) Learning improved by arts training. *Nature, 381*, 284.

Gardner, M.F. (1996). *Test of auditory-perceptual skills*. Hydesville, CA: Psychological and Educational Publications.

Graziano, A.B., Peterson, M., & Shaw, G.L. (1999). Enhanced learning of proportional math through music training and spatial-temporal training. *Neurological Research, 21*, 139-152.

Gromko, J.E., & Poorman, A.S. (1998). The effect of music training on preschoolers' spatial-temporal task performance. *Journal of Research in Music Education, 46*, 173-181.

Hammill, D.D., Pearson, N.A., & Voress, J.K. (1993). *Developmental test of visual perception*. Austin, TX: Pro-ed.

Hassler, M., Birbaumer, N., & Feil, A. (1985). Musical talent and visual-spatial abilities: a longitudinal study. *Psychology of Music, 13*, 99-113.

Hassler, M., & Feil, A. (1986). A study of the relationship of composition/improvisation to selected personal variables. *Bulletin of the Council for Research in Music Education, 87*, 26-34.

Hassler, M., Birbaumer, N., & Feil, A. (1987). Musical talent and visual-spatial ability: onset of puberty. *Psychology of Music*, 15, 141-151.

Hassler, M., & Birbaumer, N. (1988). Handedness, musical abilities and dichaptic and dichotic performance in adolescents: a longitudinal study. *Developmental Neuropsychology, 4*, 129-145.

Hassler, M., Nieschlag, E., & de la Motte, D. (1990). Creative musical talent, cognitive functioning, and gender: Psychological aspects. *Music Perception, 8*, 35-48.

Hebb, D.O. (1949). The organization of behavior: a neuropsychological theory. New York: Wiley.

Hetland, L. (2000). Learning to make music enhances spatial reasoning. *Journal of Aesthetic Education, 34*, 179-238.

Ho, Y.C., Cheung, M.C., & Chan, A.S. (2003). Music training improves verbal but not visual memory: Cross sectional and longitudinal explorations in children. *Neuropsychology, 17*, 439-450.

Hui, K. (2006). Mozart effect in preschool children? *Early Childhood Development and Care, 176*, 411-419.

Hurwitz, I., Wolff, P.H., Bortnick, B.D., & Kokas, K. (1975). Nonmusical effects of the Kodàly music curriculum in primary grade children. *Journal of Learning Disabilities, 8*, 167-174.

Ivanov, V.K., & Geake, J.G. (2003). The Mozart effect and primary school children. *Psychology of Music, 31*, 405-413.

Jausovec, N., & Habe, K. (2004). The influence of auditory background stimulation (Mozart's sonata K. 448) on visual brain activity. *International Journal of Psychophysics, 51*,261-271.

Jausovec, N., & Habe, K. (2005). The influence of Mozart's sonata K. 448 on brain activity during the performance of spatial rotation and numerical tasks. *Brain Topography, 17*, 207-218.

Jeannerod. M., Decety, J., Michel, F. (1994). Impairment of grasping movements following a bilateral posterior parietal lesion. *Neuropsychologia, 32*, 369-380.

Karma, K. (1982). Musical, spatial, and verbal abilities: a progress report. *Psychology of Music*, Special Issue, 69-71.

Kaufman, A.S., & Kaufman, N.L. (1983). *Kaufman assessment battery for children*. Circle Pines, MN: American Guidance Service.

Kilgour, A.R., Jakobson, L.S., & Cuddy, L.L. (2000). Music training and rate of presentations as mediator of text and son recall. *Memory and Cognition, 28*, 700-710.

Lamb, S.J., & Gregory, A.H. (1993). The relationship between music and reading in beginning readers. *Educational Psychology, 13*, 19-27.

Leng, X., & Shaw, G.L. (1991). Toward a neural theory of higher brain function using music as a window. *Concepts in Neuroscience, 2,* 229-258.

Little, W.A. (1974). The existence of persistent states in the brain. *Mathematical Bioscience, 19*, 101-120.

McKelvie, P., & Low, J. (2002). Listening to Mozart does not improve children's spatial ability: final curtains for the Mozart effect. *British Journal of Developmental Psychology, 20*, 241-258.

Mountcastle, V.B. (1978). An organizing principle for cerebral function: the unit module and the distributed system. In G.M. Edelman & V.B. Mountcastle (Eds.), *The mindful brain* (pp. 1-50). Cambridge, MA: MIT Press.

Orsmond, G.I., & Miller, L.K. (1995). Correlates of musical improvisations in children with disabilities. *Journal of Music Therapy, 32*, 152-166.

Orsmond, G.I., & Miller, L.K. (1999). Cognitive, musical and environmental correlates of early music instruction. *Psychology of Music, 27*, 18-37.0

Perkins, D.N., & Salomon, G. (1989). Are cognitive skills context-bound? *Educational Researcher, 18*, 16-25.

Rauscher, F.H. (2002). Mozart and the mind: Factual and fictional effects of musical enrichment. In J. Aronson (Ed.), *Improving academic achievement: Impact of psychological factors on education* (pp. 269-278). New York: Academic Press.

Rauscher, F.H., Shaw, G.L., & Ky, K.N. (1993). Music and spatial task performance. *Nature, 365*, 611.

Rauscher, F.H., Shaw, G.L., Levine, L.J., Wright, E.L., Dennis, W.R., & Newcomb, R. (1997). Music training causes long-term enhancement of preschool children's spatial-temporal reasoning abilities. *Neurological Research, 19*, 1-8.

Rauscher, F.H., & Shaw, G.L. (1998). Key components of the "Mozart Effect." *Perceptual and Motor Skills, 86*, 835-841.

Rauscher, F.H., & Zupan, M.A.(2000). Classroom keyboard instruction improves kindergarten chldren's spatial-temporal performance: A field experiment. *Early Childhood Research Quarterly, 15*, 215-228.

Rauscher, F.H., LeMieux, M., & Hinton, S.C. (2005). *Selective effects of music instruction on cognitive performance of at-risk children.* Paper presented at the bi-annual meeting of the European Conference on Developmental Psychology, Tenerife, Canary Islands.

Rauscher, F.H., & Hinton, S.C. (2006). The Mozart effect: Music listening is not music instruction. *Educational Psychologist, 41,* 233-238.

Rauscher, F.H., LeMieux, M.M., & Hinton, S.C. (2006). *Quality piano instruction affects at-risk elementary school children's cognitive abilities and self-esteem.* Paper presented at the Ninth International Conference on Music Perception and Cognition, Bologna, Italy.

Raven, J.C. (1956). *Coloured progressive matrices.* London: H.K. Lewis.

Raven, J. C. (1986). *Standard progressive matrices.* San Antonio, TX: The Psychological Corporation.

Sarnthein, J., von Stein, A., Rappelsberger, P., Petsche, H., Rauscher, F.H., & Shaw, G.L. (1997). Persistent patterns of brain activity: an EEG coherence study of the positive effect of music on spatial-temporal reasoning. *Neurological Research, 19*, 107-116.

Scandura, J.M. (1984). Structural (cognitive task) analysis: a method for analyzing content: Part 2. Toward precision, objectivity and systematization. *Journal of Structural Learning, 8*, 1-28.

Schellenberg, E.G. (2004). Music lessons enhance IQ. *Psychological Science, 15*, 511-514.

Schellenberg, E. G. (2005). Music and cognitive abilities. *Current Directions in Psychological Science*, 14, 322-325.

Schellenberg, E. G., & Hallam, S. (2005). Music listening and cognitive abilities in 10- and 11-year-olds: The Blur effect. *Annals of the New York Academy of Sciences*, 1060, 202-209.

Schellenberg, E. G.; Nakata, T., Hunter, P.G. (2007). Exposure to music and cognitive performance: Tests of children and adults. *Psychology of Music, 35*(1), 5-19.

Thorndike, E.L. (1913). *Educational psychology*. New York: Columbia University Press.

Thorndike, E.L., Hagen, E.P., & Sattler, J.M. (1986). *The Stanford-Binet intelligence scale* (4th ed.). Chicago: Riverside.

Vaughn, K. (2000). Music and mathematics: Modest support for the oft-claimed relationship. *Journal of Aesthetic Education, 34*, 149-166.

Wechsler, D. (1989). *Wechsler preschool and primary scale of intelligence-revised.* San Antonio, TX: The Psychological Corporation.

Wechsler, D. (1991). *Wechsler intelligence scale for children—third edition.* San Antonio, TX: Psychological Corporation.

Zafranas, N. (2004). Piano keyboard training and the spatial-temporal development of young children attending kindergarten in Greece. *Early Childhood Development and Care, 174*, 199-211.

In: Neurosciences in Music Pedagogy
Editors: W. Gruhn and F. Rauscher, pp. 203-231
ISBN: 978-1-60021-834-7

Chapter VIII

NEUROSCIENTIFIC ASPECTS OF GIFTEDNESS AND MUSICAL TALENT

Marianne Hassler and Leon K. Miller

ABSTRACT

This chapter reviews neuroscientific research on musical giftedness from two different perspectives: At first, experimental findings on musicians and musical talent in normal subjects are presented.. After a short look at past research on the brains of famous musicians, on definitions of musical talent, and on basic assumptions in neuroscience, recent studies on brain structure and function in musicians as compared to non-musicians are reviewed. Differences between the two groups are discussed regarding musical talent and the whole individual´s living condition. Additionally, theoretical models and experimental data are presented on how musical giftedness might develop, and on the possible influences of gonadal hormones on the respective brain processes during prenatal development, in puberty, and in adulthood. A further focus is on developmental changes in musical giftedness occurring in the course of puberty, when many of those who have shown exceptional musical talent in childhood loose their musical excellence. Finally, differences between male and female musicians in brain structure and function are discussed and findings are referred to showing that sex differences in brain structure and function do not mirror differences in performance levels.

In the second part discusses research on musical savants. This research suggests that several component skills contribute to the emergence of exceptional musical skill in those with disabilities: pitch memory, sensitivity to musical structure, and the coordination of music perception and production. It is argued that in each case the component skills are normative in the sense that similar skills are also present in those without disability, though typically at a much less developed level. Musical savants usually exhibit their special skills early in development. Moreover, several types of developmental disability with early onset, specifically congenital blindness and autism, are prominent in savant case histories. The review concludes with a discussion of hypotheses linking these types of disability to early language development, cerebral lateralization of function, and the enhancement of musical skills.

A Neuroscientific Approach to Musical Talent in Normal Subjects

Introduction

The history of experimental neuroscientific research on giftedness and musical talent goes back at least to the turn of the 19th to the 20th century. In a historical review, Alfred Meyer (1977) reminded us for example of Auerbach´s (1906 ff) pioneering work on the "Beitrag zur Lokalisation des musikalischen Talentes im Gehirn und am Schädel" (localization of musical talent in the brain and scalp). Auerbach, in his search for a morphological substrate of musical talent, was able to study the brains of famous musicians, among them the conductors Felix Mottl and Hans von Bülow, the music teacher Naret Koning, the singer Julius Stockhausen, and Bernhard Cossmann, the "Joachim of the violoncello" (Meyer, 1977, p. 264). Compared to non-musicians, he found the middle and posterior thirds of the superior temporal gyrus were strongly developed and showed great width. He described them as being intimately connected with the equally well developed gyrus supramarginalis. Auerbach tentatively suggested that these brain structures might be related to exceptional musical ability.

To investigate musical talent, Auerbach studied the brains of persons who were well known for their exceptional musical capacities. Thus, he avoided taking part in the vivid and controversial debates among musicologists, music teachers, and musicians about the nature of musical giftedness. Which are the constituents of musical talent? The history of music psychology shows that the way musical talent was defined has changed over time. Gembris (1999) has recently distinguished three historical phases that were not clearly separated but overlapping: 1. The *phenomenological phase* for which three famous researchers were referred to, Michaelis (1805), Billroth (1895) and Kries (1926), 2. the *psychometric phase* with Seashore (1919), Wing (1939/61) and Gordon (1989) as exponents, and 3. the *musical meaning phase* whose advocates are Stefani (1987), Sloboda (1993), and Blacking (1990). In Auerbach´s times, the view on musical talent was in its phenomenological phase, i.e., musical talent was described comprehensively and all-embracingly in a phenomenological way. While the philosopher Michaelis published a list of characteristic features of musical talent, such as the ability to recognize tones and melodies, attention to music, pleasure in music, a strong hearing ability, musical memory, and a facility in singing a heard melody, the surgeon Theodor Billroth, in his book "Wer ist musikalisch" (who is musically gifted), argued for the ability to perceive musical forms to be at the core of musical education (Bildung) (Gembris, 1999).

In their experimental work, neuroscientists of our days use to follow Auerbach´s tradition. To test their hypotheses on a possible relationship of musical giftedness to brain structure and function, they look for subjects who have shown exceptional musical capacities by playing in orchestras, performing as soloists, or at least by playing musical instruments regularly and having had at least several years of formal music training. In some cases, psychometric music tests are used, in addition.

In the seventies of the past century, neuroscientific research on musical ability was spurred on by split brain research. Investigations were concerned with the relative participation of the two hemispheres in musical, verbal and non-verbal performance after brain surgery. Questions about localization in patients and in healthy subjects were tested with techniques like dichotic listening, EEG, and tachistostcopic measurement. Results gave insight into brain activity only in a very undifferentiated way, i.e., informations were restricted to participation of the left or right hemisphere in the respective processes. However, despite methodological limitations, the first differences in function could be reported between musicians and non-musicians. While, in non-musicians, musical materials were found to be processed predominantly with right hemisphere functions, musicians showed a relative superiority of the left hemisphere for the respective processes. Speech, on the other hand, which is, in non-musicians, predominantly processed by functions of the left cerebral hemisphere showed tendencies for bilateral or even right lateral dominance in musicians (Bever & Chiarello, 1974; Gordon, 1983). Such atypical processing was accompanied by enhanced occurance of non-righthandedness, i.e., of ambidextrous or left handedness in musicians (Witelson, 1985; Hassler & Birbaumer, 1988).

In the nineties of the past century, new imaging techniques allowed the study of the working brain. The above outcome was confirmed and specified. This was the beginning of continuous and fruitful research activities in the field of music and neuroscience.

The Neuroscientific Approach

There are some fundamental assumptions in neuroscience that also apply to research on musical talent. First of all, and in accord with Auerbach, the neuroscientific point of view looks at all behaviors as a reflection of brain function. What we commonly call mind is a range of functions carried out by the brain. The action of the brain underlies not only relatively simple motor behaviors such as walking, breathing, smiling or, very important for musicians, movement of the arms and fingers, but also elaborate affective and cognitive behaviors such as feeling, learning, thinking, and composing a symphony (Kandel, 1993, p. 1).

In addition, neurobiological research has shown that human life should be considered as an indivisible continuum where each of the developmental stages is equally important, all stages are interdependent and not separable from the whole individual life´s continuum. In this continuum, the individual represents an indivisible entity of all functions both physiological or physical, psychological and social. The physical, biochemical, endocrinological and psychological processes represent a whole which cannot be divided (Fedor-Freybergh, 1999). With respect to musical talent, this view means that musical giftedness cannot been seen in isolation but as part of a complex organization of the whole individual.

Other important points of interest are questions about the nature and nurture of talent. Neuroscientific research has shown that, from the beginning of life, there is an intimate interaction between genetic and environmental influences on the developing brain. Environmental stimulation is absolutely necessary to allow the premature brain of a newborn to reach its final structure and function. Learning alters brain structure. Thus, we have to add the understanding of how the interactive process of learning and inborn markers works in future musicians to the question whether the talent – or its components –is influenced by genetic or environmental agents.

Musical Talent and the Brain of Musicians

In the nineties with new imaging techniques, two brain structures, the corpus callosum and the planum temporale, were of special interest when the neural basis of musical talent was investigated. From results of the earlier functional studies mentioned above, both structures were expected to show differences in size between musicians and non-musicians. These differences may explain findings of atypical processing of verbal and musical material in musicians. The first structure, the corpus callosum, is the most prominent fiber bundle connecting the right und the left cerebral hemispheres. Results were expected to show differences between musicians and non-musicians in the posterior part of the structure. However, it was found that the anterior part was larger in musicians than in non-musicians (Schlaug, 2001). Data were interpreted as indication of a difference between musicians and non-musicians in interhemispheric communication of sensorimotor areas. It was assumed that playing an instrument with daily training would result in enlargement of the respective parts. Since the differences in this area of the corpus callosum were due to musicians who had started music training before the age of seven it was assumed that differences use to emerge with practice beginning at an early age when brain development that has not yet come to its end.

The second structure of interest is the planum temporale, a brain area containing auditory association cortex which is usually larger on the left side (Schlaug, 2001). Schlaug has argued that outstanding musical ability is associated with increased leftward asymmetry of cortex subserving music-related functions. It was also found that the left planum temporale was larger in those musicians who possessed absolute pitch compared to musicians without absolute pitch and to non-musicians.

Because of the possible influence of training on sensorimotor brain areas, the cerebellum was also measured in musicians and non-musicians. A higher relative cerebellar volume was found in male musicians compared to male non-musicians, but not in female musicians compared to female non-musicians (Hutchinson, Lee, Gaab & Schlaug, 2003).

With the extension of research in this field, more and more differences between musicians and non-musicians at different places of the brain were detected (see Baeck, 2002, and Norton et al., 2005, for review). To give some examples: Using magnetic encephalography, Elbert et al. (1995) found that fingers of the left hand of violinists have greater representation in the primary somato-sensory cortex than those in a control group. Increased gray matter density in Broca´s area in male symphony orchestra musicians was found with voxel-based morphometry (Sluming et al., 2002). A study from Pantev et al. (1998) has investigated the representation of tones in the cortex. Using functional magnetic source imaging technique, they measured cortical representations in highly skilled musicians. Dipole moments for piano tones, but not for pure tones of similar fundamental frequency (matched in loudness) were found to be enlarged by about 25 % in musicians compared to control subjects who had never played an instrument. Enlargement was correlated with the age at which musicians began to practice and did not differ between musicians with absolute or relative pitch.

Other findings were also related to hearing ability. Micheyl and colleagues (1997) have reported musicians' auditory systems track sound levels more accurately than non-musicians'. The activity of the MOBC (medial olivocochlear bundle), an auditory efferent subsystem, was measured in musicians and non-musicians and a significant difference between the two groups was found. Musicians had greater ability to maintain the perceived loudness of a sound, that is, to hear more closely the actual real levels of sound. When, for instance, a musician and a non-musician standing side by side listen to birds' singing, church bells′ ringing, or to the the sound of a violin, the musician will hear more exactly than the non-musician the loudness of these acoustic events. While it seems likely that this is learned unconciously, the authors also discuss the possibility that if innate, this ability predisposes people to become musicians.

There is also evidence for pre-attentive superior auditory processing in musicians. This was revealed by the brain′s automatic change-detection response, which is reflected electrically as the mismatch negativity (MMN) and generated by the operation of sensoric memory, the earliest cognitive memory system (Koelsch et al., 1999). Major chords and single tones were presented to both professional violinists and non-musicians. Slightly impure chords, presented among perfect major chords elicited a distinct MMN in professional musicians, but not in non-musicians. According to Koelsch et al., the outcome demonstrates that compared to non-musicians, musicians are superior in pre-attentively extracting more information out of musically relevant stimuli. Sensory memory systems, they argue, have been modulated by training.

Is it because musicians have a finer auditory system that they react to music more sensitively than non-musicians? The auditory system is a highly complex one. Relations between music and emotions are well known (LeDoux et al., 1984). Already in the seventies of the past century, studies with musicians and non-musicians pointed to a greater sensitivity of the autonomic nervous system of musicians to music (Harrer & Harrer, 1977). New studies have confirmed early findings (Evers et al., 1999). If musicians are exposed to music – the pieces may be vivid or calm – they react with an increase of heart rate or of other indicators of the autonomic nervous system. If one measures cortisol, a stress hormone, an increase is sometimes found in musicians even when they hear peaceful music. In non-musicians, on the other hand, peaceful music tends to initiate a decrease in heart rate or in stress hormones (Weinberger, 1997).

Hormones and Musical Talent - Theoretical Considerations and Experimental Findings

By referring to studies about effects of music on stress hormones, we have turned the interest towards a possible relationship of musical talent to the hormone system. Here, my own research team has been working for about 15 years looking, among other things, for differences between musicians and non-musicians in hormone parameters. They found young adult musicians differ from non-musicians of the same age in the gonadal hormone testosterone, in the pineal hormone melatonin, and, at some developmental stages, in ACTH (adrenocorticotropic hormone), a neuroactive peptide (Hassler, 1992; Hassler & Gupta, 1998; Hassler, Gupta & Wollmann, 1992). As a first step, testosterone was measured in saliva, i.e., non-invasively. Testosterone is usually much higher in males than in females. Results showed that male musicians (composers) had significantly lower mean testosterone levels than male non-musicians and female musicians (composers) had significantly higher mean testosterone levels than female non-musicians (Hassler & Nieschlag, 1989). The relatively small differences between male and female composers in the gonadal hormone amount were intepreted as physiological androgyny.

Some years later another group of creative musicians was tested for several hormones and neurotransmitters as well as for immune parameters.This time, we measured the substances in blood. In addition to the results of former studies, we found the pineal hormone melatonin was higher in musicians than in non-musicians of both sexes. Melatonin and testosterone are parts of a complex hormone system, and our results seemed to give some indications of differences between the musicians and non-musicians in this system.

Our research in this field was guided by theoretical considerations from Doerner (1985), and Geschwind and Galaburda (1985). These experts have stressed the importance of gonadal hormones´ influence on prenatal brain development and on later behavior. Doerner´s findings showed prenatal influences of gonadal hormones on hypothalamic structures that are responsible for adult male and female aspects of behaviors and allow for predictions about androgyny. Androgyny is likely to occur when relatively small amounts of testostereone in males and relatively high amount of testosterone in females organize the respective hypothalamic nuclei arount prenatal weeks 12 to 14.

Geschwind and Galaburda (1985) have summarized experimental findings on cerebral lateralization and have discussed biological mechanisms, associations and pathology. Based on the long-lasting and in genious research and vision of the late Norman Geschwind, Geschwind and Galaburda (1985) argued that relatively high prenatal testosterone after week 20 of gestation causes a developmental slowing of some left hemisphere structures and a compensatory growth of parts of the right and/or left hemisphere resulting in a somewhat atypical brain organization. This atypical organization has consequences for brain structure and function. Doerner´s and the GBG (Geschwind-Behan-Galaburda) model would suggest a somewhat atypical time course of gonadal hormone release during critical prenatal periods.

The GBG model has been inspiring though it was also critisized for vagueness (MacManus & Bryden, 1991). In addition, one has to accept that, for ethical reasons, prenatal hormonal influences on brain development cannot be directly investigated and research is restricted to testing predictions from the model.

With respect to musical talent, the GBG model offers a hypothesis about how this kind of giftedness might develop. It was suggested that in about 1/3 of people prenatal development of the two cerebral hemispheres is somewhat atypical because the steroid hormone testosterone´s stronger impact on the premature brain. The hormone is, prenatally, produced by the child´s own hormone system and, in addition, comes from the mother´s. Whether the postulated strong testosterone effect is due to enhanced amount or enhanced effect, consequences for brain development are far-reaching. Atypical hemispheric lateralization will occur in the affected subjects, and this may be accompanied by non-righthandedness.

According to Geschwind and Galaburda (1985), the atypical brain organization may favor, among others, the occurrence of exceptional musical and spatial abilities. One could argue this model views musical talent as one out of several side-effects of atypical brain anatomy. It stresses the point of view that musical talent has to be seen as part of the whole individual´s condition.

The GBG model also makes predictions for immunity. Since immune structures develop together with the left cerebral hemisphere, the above-mentioned retardation of left hemisphere development also affects immune structures. They develop more slowly in above-mentioned fetuses and influence postnatal functioning. As a consequence, in people with atypical brain anatomy, among them musicians, one can expect vulnerability towards autoimmune diseases and towards atopic diseases, i.e., these persons will be vulnerable towards allergies, asthma, eczemas.

We tested the predicitions of the Doerner and the BGB model twice. The first study was cross-sectional, with groups of adult composers, painters, instrumentalists, and non-musicians. Each group contained 20 males and 20 females, except for the composers, which contained 14 females. The second study was longitudinal with 60 male and 60 female children. Boys and girls were 11,5 years on average when we started the study and we followed our participants for 8 years throughout puberty. We then were able to retest them five years after our 8-year longitudinal study had ended. Children differed with respect to musical capacities. When we started, one group (20 males, 20 females) played one or more musical instruments and demonstrated creative musical talent, the second group (20 males, 20 females) played musical instruments but did not compose/improvise, and the third group (20 males, 20 females) consisted of non-musicians who scored between B and E on the Wing music tests which we used to measure musical talent. Non-musicians did not play musical instruments.

Musical, verbal and spatial psychometric tests were used to assess abilities, a dichotic listening test was assumed to investigate hemispheric lateralization for verbal materials, and testosterone was measured in saliva to assess physiological androgyny in addition to Bem´s Sex Role Inventory (Bem, 1974) which was administered to assess psychological androgyny. Compositions and improvisations were provided by the creative (experimental) group. Musicians surpassed non-musicians on spatial tests. They showed atypical lateralization for language processing and, if male, they were more often non-righthanded compared with non-musicians. Creative musicians were psychologically androgynous even at the age of twelve and, as adults, they were also physiologically androgynous. These results confirmed predictions from the GBG model and Doerner´s hypothesis (Hassler & Birbaumer, 1988; Hassler, 1992).

To investigate immunological parameters, we administered questionnaires to our subjects three times and, when they were about 19 years old, we measured immunoglobulin E, the total amount of immunoglobulins, and ß-endorphin in blood serum. According to Geschwind, we expected musicians would suffer more often than non-musicians from atopic diseases and would be more vulnerabel towards atopic diseases as reflected by enhanced immunoglobulins E values. Our data have confirmed these expectations for males, only. In addition to be muscially talented, males had also to be left-handedness (Hassler & Birbaumer, 1988; Hassler & Gupta, 1993).

When we summarize the results of our own and others' neuroscientific studies on musicians and musical talent and try to imagine a person described by the outcome, we envision a man or a woman with exceptional musical capacities who shows atypical brain organization for verbal and non-verbal materials, who possesses a very fine ear for music and attends pre-consciously to musical materials; he/she reacts to music as if music is a stressor, i.e., with increased heart rate and/or increased stress hormone production. The person is likely to be non-righthanded, i.e., a lefthanded or ambidextrous, and may be vulnerable to atopic diseases, especially if male. At the same time he/she is psychologically and physiologically androgynous, the latter means testosterone levels are lower (males) or higher (females) in creative musicians than in sex-typed individuals; in addition the pineal hormone melatonin is higher in musicians of both sexes compared to non-musicians. The brain reflects early music training by enlarged structures in the corpus callosum, planum temporale and cerebellum, and representations for instance for piano tones and for the left thumb and little finger in string players are larger. In addition, musical talent is accompanied by above average spatial abilities. Male and female musicians have successfully maintained musical talent during the critical period of puberty.

Development of Musical Capacities in the Course of Puberty

Many of the child prodigies whose performances on stage we enjoy, disappear from our horizon nearly as quickly as they came up. This phenomenon has been well known for centuries and was noticed in the 19th century for example by the famous music critic Eduard Hanslick in Vienna (Hanslick, 1869/70). In the 20th century, Révész (1925), Scott and Moffett (1977), Bamberger (1982), Manturzewska (1990, 1996) and Hassler (1992) have contributed to our knowledge. Assessing neurobiological parameters possibly influencing musical excellence in the course of puberty is a relatively new approach to research in this field.

From a neurobiological point of view, puberty is the time when the brain reaches its final structural appearance. New MRI data coming from a research group at the UCLA Laboratory of Neural Imaging led by Paul Thompson show how, exactly, the brain matures: Overall, total grey matter (GM) volume was found to increase at earlier ages, followed by sustained loss starting around puberty. The process of GM loss (maturation) begins first in dorsal parietal cortices, particularly the primary sensorimotor areas, and then spreads rostrally over the frontal cortex and caudally and laterally over the parietal, occipital, and finally the temporal cortex. Frontal and occipital poles lose GM early, and in the frontal lobe, GM maturation ultimately involves the dorsolateral prefrontal cortex which loses GM only at the end of puberty (Gogtay et al., 2004). With respect to behavior, Gogtay et al. have argued that the sequence by which the cortex matures coincides with regionally relevant milestones in cognitive and functional development. Parts of the brain associated with more basic functions mature early: motor and sensory brain areas mature first, followed by areas involved in spatial orientation, speech and language development. Later to mature are areas involved in executive functions, attention, and motor coordination (Gogtay et al., 2004).

Puberty is also marked by dramatic changes in neuroendocrine functioning (Cameron, 2004). These changes have fundamental effects on structure and function of the maturing nervous system resulting in altered physiological and behavioral potentials in the adult organisms. As Romeo (2005) has stressed, the changes in neurobehavioral development during puberty rival those occurring during neonatal development. Increased hypothalamic secretion of the gonadotropin-releasing hormone (GnRH) is essential for the activation of the pituitary-gonadal axis to initiate puberty. The GnRH secretory network initially develops and is temporarily active during periods of fetal/neonatal development, so puberty is the secondary reactivation of an existing system (Ebling, 2005).

Gonadal steroid hormones, among them testosterone, play an important role in regulating plastic changes in neuronal functioning. The respective receptors are distributed throughout the brain allowing the steroids to influence information processing in many places. We know from EEG studies that, during listening and while creating compositions, many brain areas of both hemispheres are activated in musicians and, while composing, subcortical structures are also involved (Petsche & Etlinger, 1998, p. 329).

It has been proposed that testosterone is an important agent with respect to music (Hassler & Nieschlag, 1989, Fukui, 2001). Our own data seem to suggest that testosterone amount and/or effect plays a role in adult musicians with respect to creative musical activities, and changes in testosterone production in the course of puberty may affect musical creativity in this important developmental period. We found that an optimal testosterone range may exist for the creative aspect of musical talent. This range is characterized by low levels in males and in high levels in females. When, in the course of puberty, this optimal range was exeeded in males, musical creativity deteriorated. We tentatively suggested that rising testosterone level may be one component out of a complex system that influences the functional network subserving creative musical capacities (Hassler, 1992; Hassler & Gupta, 1993). This network is reorganized when the brain matures in the course of puberty.

In our study, out of 20 male and 20 female young participants who were able to compose and improvise at the mean age of 12 years, only two males and one female provided compositions or improvisations at the age of 18 years. When we retested our subjects five years after our 8-year-longitudinal study had ended, subjects had already chosen their professions. We found four male participants from the experimental group (with the ability to compose and/or improvise) had become professional musicians – one was a conductor, one a pianist, one a drummer and one was a violinist playing in a symphony orchestra. One female from group 2 (children with musical talent but without the ability to compose and/or improvise) had become a singer. All of the other participants played their instruments from time to time, but all of them had become expert listeners with a deep affinity towards music.

Experimental findings prior to ours have also shown that the creative aspect disappeared in many of those who had been able to compose and/or improvise in their early teens (Hanslick, 1869; Revesz, 1925; Scott & Moffett, 1977). Bamberger (1982) found young instrumentalists have to overcome a critical period and have to cope with new ways of thinking und appreciating their own performance. It is, however, important to note that adolescent brain development is a period of vulnerabilities and opportunities (Dahl, 2004). Puberty is also the time when new interests and abilities arise that have to be acknowledged.

Sex Differences in Brain Structure and Function

Many of the studies on music in relation to brain structure and function were conducted with male musicians only. However, as soon as females were included in the assessment, sex differences were likely to occur. To give some examples:

A larger anterior corpus callosum in musicians compared to non-musicians is one of the well established findings (Schlaug, 2001). However, when female musicians matched with male musicians in all relevant variables were included in the assessment female musicians did not differ from female non-musicians in the anterior corpus callosum (Lee et al., 2003). When the cerebellum was measured, a higher relative cerebellar volume was found in male musicians but not in female musicians compared to non-musicians (Hutchinson et al., 2003).

Testing the Geschwind and Galaburda hypothesis about atypical brain organization and special talents with female subjects, talent in a female population was found to be related to the strength of non-righthandedness but not to atypical lateralization or to immune disorders as expected from the model (McNamara et al., 1994). Our own study showed atypical hemispheric dominance for verbal material in composers of both sexes. This was in line with predictions. However, male creative musicians but not female creative musicians were more often non-righthanded and suffered more often than non-musicians from atopic deseases (Hassler & Birbaumer, 1988; Hassler & Gupta, 1993) indicating the model is only partly able to make predictions for females.

Koelsch et al. (2003a), using event-related brain potentials (ERPs), found sex differences in the functional organization of the brain for music processing in adults. Results demonstrated that an electrophysiological correlate of music-syntactic processing is generated bilaterally in females, and with right hemispheric predominance in males. According to the authors, their findings indicate that sex differences for the analysis of auditory information are not restricted to processes in the linguistic domain such as syntax, semantics, and phonology. Koelsch et al. (2003b) also assessed 5- and 9-year old children in order to understand how the physiological correlates of the processing of musical information develop during childhood. Subjects listened to (major-minor tonal) music while event-related brain potentials were measured. Stimuli were chord sequences, infrequently containing inappropriate chords. The results demonstrated that the degree of (in)appropriateness of the chords modified the brain response in both groups according to music-theoretical principles. The authors argued that their results suggest already 5-year old children process music according to a well-established cognitive representation of the major-minor tonal system and according to music-syntactic regularities. An early negative brain response was left-predominant in boys, whereas it was bilateral in girls, indicating a sex difference also in children processing music, and suggesting that children process music with a hemispheric weightening different from that of adults. This was especially true for males.

An earlier ERP-study (Hantz et al., 1996) had reported sex differences in memory for timbre in adults. Male and female highly trained musicians performed a memory task for musical timbre modeled after the missing-displaced visual object test known to favor female performance, and the missing-displaced visual object test. In the missing-displaced object test subjects were tested on memory for visual objects. They had to detect missing or displaced objects from a previouly presented set of objects. Female and male musicians performed equally well on both tests. However, females showed a greater differentiation than males for the auditory test, i.e., a greater sensitivity to changes in intensity.

Sex differences in brain structure and function are not restricted to musical talent (Njemanze, 2005). From the neuobiological point of view sex differences in brain organization are not surprising. The basic brain development of all mammals, including humans, is female. Several intervening steps are necessary to enable the embryo and the fetus to develop in the male direction. Gonadal hormones play an important role in this respect. If the right amount of the right hormone is not available to the developing organism at the right time, or when the tissue is not sensitive towards the hormone, a female body and/or brain will develop despite XY chromosomes (Doerner, 1985; Money, 1986). With respect to puberty, the Thompson team from UCLA found female brains mature about one year earlier than male brains, though there are considerable interindividual variations (Gogtay et al., 2004). The outcome is in line with data about the general onset of puberty in boys and girls (Tanner et al., 1966). The Thompson team (Luders et al., 2004) has also reported sex differences in cortical complexity in adults. Cortical complexity is a measure that quantifies the spatial frequency of gyrification and fissuration of the brain surface. They found a greater gyrification in women than in men in frontal and parietal regions. Increased complexity implies more cortical surface area.

Sex differences in brain structure and function do not imply, however, that sex differences exist in competence, in performance levels, or in the degree of musical giftedness in musically talented males and females. They simply mean the brains of males and females work differently to reach comparable results. An early example for contributions of male and female musicians to music performances on stage comes from Eduard Hanslick (1869/70). He reported that, in Vienna around the year 1792, more than fifty percent of all musicians in concert were younger than 14 years, and he counted and named nine girls and six boys, i.e. more young female than male musically talented artists. In adulthood, there seem to be more male than female members of orchestras, more male conductors and more male composers, and it seems likely that developmental processes in the course of puberty contribute to the respective changes as do environmental conditions. However, there have been females among creative musicians in all centuries though they were less well known and less accepted in society than males; and it seems possible that they were fewer in number. To name just two:

In 2005, we celebrated the 200th birthday of Fanny Hensel-Mendelssohn, the sister of Felix Mendelssohn-Bartholdy, a female composer of excellence and originality who has come to our knowledge only late and incompletely. Her exceptional musical talent and musical creativity was obvious early in childhood and she got over the time of puberty without negative influences on musical giftedness. She was a composer, but she was also an excellent pianist, well known in her home town of Berlin for her "Sonntagskonzerte" (Sunday concerts) in the beautiful garden of the Mendelssohn Palais. She shared the early onset of musical giftedness with the second female composer to whom I want to refer. When, in 1985, Dora Pejacevic became "Composer of the Year" in the former Yugoslavia, it was exactly one century after her birth. She showed exceptional musical talent at preschool age. She played the violin and piano publicly from a very young age and published her first compositions in her teens. During her life she was a successful composer whose compositions were performed by famous orchestras (Kos, 1982).

Other findings also seem to suggest that differences between male and female musicians in brain structure and function are compatible with fewer sex-typed characteristics and behaviors. Psychological research has shown musicians have “a broader range of temperaments” (Kemp, 1985); male and female characteristics coexist in a musician independently of the biological sex. With respect to cognitive functioning, many of the tests reliably showing sex differences in favor of either males or females revealed no differences between male and female musicians. In our studies, the difference between males and females in the gonadal hormone testosterone – testosterone is usually at least ten times higher in males than in females – was minimized between male and female composers. Hemispheric lateralization as measured with a dichotic listening task showed a “male” pattern – a left lateralization - in female composers and a “female” pattern – a bilateral activation - in male composers. Jonsson and Carlsson (2000) reported a relationship of creativity to androgyny and, with respect to the neurobiology of creativity, differences in frontal activity between high and low creative subjects as measured by regional cerebral blood flow were found (Carlsson, Wendt, & Risberg, 2000).

Data in previous sections, though they are necessarily selective, show the broad range of approaches of neuroscientific research to musical giftedness. However, our knowledge is still like a puzzle with many missing pieces. Therefore, a great deal of further interdisciplinary effort is needed to detect the pattern of brain structure and function underlying musical giftedness. Moreover, the correct interpretation of existing data is not an easy undertaking. It seems unclear, for instance, what the structural enlargements in those who have started music training before age seven mean for musical giftedness. Musicians who started music training after age seven have also reached high levels of musical competence, allowing them to play in the same prominent orchestras or to have a solo career comparable to that of their early-beginning colleagues. Another important question still unanswered was formulated by Norton et al. only recently (2005): Are there pre-existing neural, cognitive, or motor markers for musical ability? This group has started a longitudinal study with 5-7-year-old children to answer the question. Finally, the biochemical processes in the brain that are related to musical activities, and the compounds that contribute to the respective brain processes are still largely unknown. This field belongs to the most challenging responsibilities for future research.

Although the neurobiological findings are still incomplete, they help us to understand musical giftedness in a new way. We have to realize that the experimental data show a musician differs from a non-musician not only with respect to his or her musical talent but also in brain structure and function, i.e., in the whole individual´s condition. This seems to be true, for example, for androgynous personality characteristics and behaviors, for some cognitive abilities, and also with respect to health or illness, the latter aspect is subject of music medicine.

We know from experimental studies and from history that exceptional musical abilities go through a critical period in the course of puberty, and many of the highly talented children loose their remarkable musical potentials in their teens despite supreme efforts of children, teachers, and parents. Neurobiology has shown that changes in brain functioning and in capacities occur during this developmental period when the brain matures. It is a biological process which also offers new opportunities and chances (M.H.).

Musical Savants

Occasionally, exceptional musical skill emerges in those with mental disability. Usually these are classified as instances of the savant syndrome (Miller, 1999). Other manifestations of savant behavior occur in the visual arts, primarily drawing, certain kinds of arithmetic (e.g. "calendar calculating") and mechanical skills. Musical savants have been the subject of interest and controversy for many years. Much of this controversy originally concerned aesthetic evaluations of savants' musical products. Southall (1979) for example, examined the controversy surrounding the career of Blind Tom, a savant pianist-composer who toured the United States in the 1800's. Reactions to Blind Tom's recitals and compositions ranged from astonished admiration to charges of fraud or quick dismissal. A similar debate can be found in reactions to the contemporary savant composer, Hikare Oe. (Cameron, 1998).

Recent research on musical savants has been less concerned with aesthetic evaluation of their musical output. Rather, the work has focused on the nature of musical savant skill and the circumstances under which it emerges. This research suggests several respects in which musical savants may provide a special perspective on the nature and enhancement of musical skill. Typically, musical giftedness and more general intellectual achievement are at least moderately correlated (Lynn, Wilson, & Gault, 1989). The stark contrast provided by musical savants' behavior in the musical domain and their (often severe) intellectual disability (e.g. Howe, 1989) represents a rare separation of musical talent from the general intellectual matrix in which it is usually found. Thus, analysis of musical savants may provide important clues about the substrate of musical exceptionality.

The relevance of musical savants to discussions of musical giftedness in a more general sense depends upon the satisfaction of two criteria. First, that the musical "products" of savants are sufficiently like those of gifted musicians without disability to justify their inclusion in discussions of musical giftedness. As noted above, historically the answer to this question has been equivocal. Characterizations of savants' performance as being "rote" in nature, without nuance or communicative intent (e.g. O'Connell, 1974) suggested little relevance to musical expression in those without disability. However, upon closer examination (Sloboda, Hermelin & O'Connor, 1985; Miller, 1989), far from being "rote," savants' musical performances display the same sorts of sensitivity to musical structure found in gifted musicians. Moreover, musical savants can be strongly emotionally invested in their music (Howe, 1989; Newman, 1989). At this point, savant musical behavior seems sufficiently comparable to the musical behavior of those without impairment. A second issue is perhaps less easily resolved. Possibly, either the context in which musical savant behavior occurs, or the disability that accompanies it, offers few points of contact with more normative development Rather than deciding this issue on an *a priori* basis, it seems that a more fruitful approach is to consider such factors as representative of continua of experience, a common approach in the field of developmental disability (e.g. Bates, 2004).

The Nature of Savant Skill

What, then, might savants tell us about emerging musical exceptionality? Musical savants are not equally represented across musical domains (Miller, 1989). The vast majority are performers. While vocalists appear occasionally, most savants are instrumentalists, the favorite instrument being the piano. (In this respect, the unequal distribution within the domain resembles that of savants in other areas. For example, "visual artist" savants are by and large restricted to landscape drawing/portraiture, with depiction of layout and depth a particular strength). This means that the implications to draw from examining the structure of savant skill may be limited to certain kinds of musical expression.

Three component behaviors are emerging as noteworthy in descriptions of savant musical skill; absolute pitch, memory for melodic and harmonic structure, and perceptual-motor integration of musical information. Absolute pitch refers to the ability to give the class name (e.g. note name) of given tones, independent of context (e.g. an accompanying scale, a tonal anchor, and the like) (Takeuchi & Hulse, 1993). It might be more appropriate to enclose the term in quotes since the note naming skill is usually not error-free and is frequently constrained by such factors as register and timbre. Absolute pitch may be more prevalent in those with certain types of disability, particularly autism (Frith, 1989) and in the congenitally blind (Welch, 1988). However, phenomena usually associated with absolute pitch may be normative early in development. There is evidence of memory for specific pitch information over time in the first year of life (Kessen, Levine & Wendrick, 1978; Saffran & Griepentrog, 2001). Note that appearance of specific pitch memory so early in life suggests that the actual learning of note names and their subsequent accurate assignment to tones is probably secondary to the basic skill. Indeed, among savants, there can be indications of exceptional specific pitch memory well before note names are known (e.g. Newman, 1989). Although there is some evidence of a genetic component to the presence of absolute pitch (Barharloo, et al., 2000), it is amenable to modification either through direct training or possibly as an indirect consequence of early musical instruction (Crozier, 1997). Finally, there is some evidence to suggest that those with absolute pitch utilize somewhat different memory routes than those usually used in processing pitch information (Schlaug, 2001).

The value of absolute pitch, musically, has been contested by some and it may interfere with performance on certain tasks (Takeuchi & Hulse, 1993). Nevertheless, experiences with musical savants suggest that it is a powerful mechanism for young children in responding in a sophisticated fashion to their acoustic environment. For example, the young savant "Eddie" noted the pitch given by the elevator motor on the way to a lesson (Miller, 1989). Anecdotes like this suggest that for the savant with absolute pitch relatively well-formed sounds in the environment have at least a potential musical meaning. (It is important that these be coherent sounds, acoustically. This same savant would cover his ears when exposed to such things as screeching tires, slamming doors, and multiple conversations in a room). (Please explain "coherent sounds." Do you mean sinusoidal sounds?) Having such a vehicle for classifying musical sounds provides an entry to the elementary structure of music. In this respect, those with absolute pitch so early in their musical development may have an advantage in accessing the basic alphabet of their musical environment. Indeed, the structures responsible for processing pitch information in those with absolute pitch may be strongly related to those normally used in processing phonology in language (Schlaug, 2001).

Musical savants are particularly adept at retaining the melodic line of a piece, as well as the harmonic structure implicit in a composition. For example, Miller (1987a) examined the "melody span" of a young savant, or his talent for duplicating strings of notes that he had just heard. His knack for doing so was superior to those with much more musical training. Similarly, the musical savant studied by Sloboda, Hermelin & O'Connor (1985) was able to give a passable rendition of the melodic line of a short piece by composer Edvard Grieg, again after a single hearing. Similarly, when asked to reproduce musical cadences (Miller, 1995), or to recall the harmonic accompaniment to a melodic line (Sloboda, Hermelin, & O'Connor, 1985; Miller, 1987b) savant accuracy is noteworthy. These skills are constrained in an important respect by the structure of the melodic line or harmony. When given material with a relatively unfamiliar structure, or arranged to minimize conventional structure, retention is much less impressive. Miller suggested that this reflects sensitivity to the basic structure of the musical language of the savant's environment. Moreover, music written within the confines of the classical "common practice" period appears to be particularly accessible for the savant. This may, in turn, reflect a combination of circumstances. On the one hand, music reflecting the common practice period has structural elements emphasizing those intervals that are embedded in the overtones of complex sounds, the octave, the fifth and the fourth. Furthermore, these structural elements are likely basic to the processing of pitch information in the auditory cortex (Trano et al., 2001). That is, they may be part of the information routinely extracted when a pitch is heard. For the savant, apparently, they are also easily accessible as tools for retaining and recalling musical information with a high degree of accuracy. And, they may play a role in savant musical preferences. The savant Eddie, for example, was enthusiastic about performances from a variety of genres, including "whole tone" scale compositions, but balked at playing after hearing selections using serial (12 tone) principles (Miller, 1989), and Hikari Oe's compositions are written in the classical style (Cameron, 1998). Finally, Miller (1989) has argued that the musical instrument of choice for savants, the keyboard, represents a propitious choice in that its physical layout may make such structure more apparent. Octaves occur at regular spatial intervals, and intervals such as the fifth and the fourth can be "found" and reproduced relatively quickly. Some untrained children with disability, when confronted by a keyboard for the first time, are able to find these regularities after a brief period of familiarization (Miller & Orsmond, 1994). It should be noted that these are relative strengths, again with counterparts in normative development. Thus, infants show some degree of octave equivalence in their simple discrimination of tones (Demany & Armand, 1984), and are sensitive to the regularities implicit in the conventional scales of their culture within the first year of life (Trehub, 2001). The savant's exceptionality then, seems to reflect the extent to which these nascent sensitivities are transformed into sophisticated memory structures for the retention of musical information.

Perhaps the most impressive and, paradoxically, the least studied feature of savant exceptionality is the musical savants' ability to produce a rendition of what is heard after very little exposure; as noted above, often a single hearing. That is, the savant has an exceptional ability to translate whatever has been retained about a piece of music into a motor program for its reproduction. In fact, the "listen and play" format seems so natural to the savant that it is typically both the research format (Charness, Clifton, & MacDonald, 1988; Miller, 1989) and teaching strategy (e.g. Newman, 1989) for exploring exceptionality.[14] The "listen and play" format can, at times, turn into a shadowing experiment, with the savant following the musical text with a performance as he is listening to it (Miller, 1989). This is the case even when savants have motor impairment (Miller, 1989; Treffert, 1989) suggesting that the exceptionality involves some sophisticated linking of input and output, rather than something particularly noteworthy in the output system itself. Savants' motor behavior is not precocious, even though their musical behavior is. Some indirect evidence that tactile/kinesthetic memory might be particularly important in retention of musical material by savants was obtained by Leopold (1997). A group of musical savants and a group of non-disabled comparison keyboard players, all with absolute pitch, were asked to play sets of chords after a single hearing. On different trials, either an immediate verbal recall or a tactile shape discrimination task was given between presentation and recall of the chords. As expected, the presence of an intervening task was associated with impaired chord recall. However, for the savants, tactile shape discrimination interfered more consistently with musical recall. For the non-disabled musicians, immediate verbal recall more consistently affected subsequent musical recall.

While precocious performance on a musical instrument may be the most salient aspect of this component of the savant's giftedness, anecdotes in case histories suggest a more general proclivity to try to explore and reproduce the sound environment in some fashion. For example, the savant Harriet G. as an infant started vocalizing, with a high degree of accuracy, some of the operatic arias she had been exposed to (Viscott, 1970). The savant CA explored the sounds available in his institution, before being introduced to a musical instrument, by repeated tapping on different items of furniture with utensils, as though to compare the sounds they made (Miller, 1989). The composer Hikare Oe was especially intrigued by bird calls, and devoted himself in his early years to mimicking them. The general descriptions of several savants suggest they were exceptional vocal mimics (Treffert, 1989).

Though striking, the close link between perception and performance seen in musical savants is not singular. Movement in response to music is a prominent feature of early childhood (Moog, 1976). What is, perhaps, remarkable about musical savants is the extent to which content reproduction is an important part of their response to music. However, tendencies to mimic are commonly found in infants, most often with respect to sounds associated with language (Papoušek, 1996). In the musical domain, Howe et al. (1995) found that those infants who were more likely to vocalize musically tended to show superior performance skills after formal musical training later in life. Finally, there is growing evidence that activity in motor cortex accompanies music listening, even when overt performance is not required (Zatorre, Halpern, & Perry, 1996). Taken together, these findings suggest that, as with both absolute pitch and sensitivity to melody and harmony, the savant's precocious perceptual-motor prowess is not a "unique" skill unrelated to those seen in individuals without impairment. Rather, it represents the expression and enhancement of

[14] The "listen and play" format is not exclusive to savants, of course, being a favorite strategy of learning music within the aural tradition (Berliner, 1994) and among the visually impaired (Ockleford, 1998).

tendencies that are inherent, at varying levels of prominence, in normative musical development.

The Context of Savant Skill

Although there may be various developmental disabilities in musical savants, the two most frequently discussed are congenital blindness and autism (Miller, 1999). In each case, research has suggested that conditions associated with the disability may promote the emergence of some of the skills mentioned above. As stated earlier, absolute pitch is more prevalent in those who are congenitally blind (Welch, 1988). The congenitally blind also appear to have a larger capacity for processing auditory information. They excel, for example, in such tasks as enumerating the number of tones in complex chords and in detecting subtle changes in chord structure over time (Pitman, 1965). They also are more adept at consolidating auditory information quickly and efficiently (Stevens & Weaver, 2005). Finally, the tonotopic organization of the auditory cortex appears to be more highly differentiated in the congenitally blind (Elbert, et al, 2002). Traditionally, the enhanced skills of those with congenital blindness were thought to be a simple matter of attention; absence of sight leads to greater attention to other sources of information (Miller, 1992). More recent research has called this interpretation into question, however. This research suggests that when sensory input in one modality is absent, those cortical areas typically dedicated to its processing are involved, instead, in processing information from the remaining, intact sensory modalities (Sur & Leang, 2001). Thus, "visual" cortex to some extent becomes "auditory" and perhaps, "kinesthetic" cortex (Bavelier & Neville, 2002). At this point, it is not clear how extensive this re-dedication might be. Nor is it clear whether the resulting perceptual experience represents a combination of inherent qualities of the sensory systems involved, or merely an expansion of the capacity of the intact sensory system. Imagery reported in the congenitally blind suggests an amalgam of sensory qualities from intact and impaired modalities in perception (Warren, 1984) while the differences in attention span mentioned above suggests fundamental capacity enhancement. In either case, it is likely that the auditory experience of one who is congenitally blind differs in important ways from those who have normal vision, or those whose sight is lost later in life.

There are also possibly links between characteristics associated with autism and musical savant skill. Generally speaking, savant skills are more frequently found in those with autism than those with other developmental disabilities (Heaton & Wallace, 2004). Musical savant skills, in particular, may be closely associated with autism spectrum disorders. Rimland (1978) reported over 50% of the savant cases uncovered in his survey of individuals with autism involved musical skills. This is quite a bit higher than the percentage of musical savants found in savant samples unselected with respect to accompanying developmental disability (Salovilita, 2000). Part of the problem in establishing specific links between autism and savant skill is the heterogeneity in samples diagnosed as having autism as well as controversies surrounding the appropriate diagnostic criteria (Waterhouse, Fein & Modahl, 1996). However, current research suggests several points of contact between musical savant skills and autistic symptoms. Compared to their peers, autistic children have a significant advantage in retaining specific pitch information, and in detecting subtle changes in melodic structure over time (Heaton, Hermelin, & Pring, 1998). Under certain conditions, they are also adept at analysis of individual pitches in chords. (Heaton, 2003). Hermelin (2001) suggests these relative strengths may reflect a tendency on the part of those with autism to attend to the specific details of what is seen or heard, rather than on more global properties or the general context of the sound environment (for example, who or what is producing the sound). Second, variance in emotional reactivity and expressiveness often observed in those with autism may predispose them to develop particularly intense interests. For example, autistic children often exhibit either hypo or hypersensitivity to different types of sensory input (Baranek et al. in press). The sensitivity of musical savants to different types of auditory input has already been mentioned. Motivational systems associated with the presence of obsessive behavior and motor stereotypies in autism might also promote the sorts of single mindedness found in savants when they are exhibiting their skill (Lindsley, 1965). Miller and Monroe (1990) found musical interests to be negatively correlated with the presence of socially maladaptive behaviors such as obsessive behavior and stereotypies in a sample of disabled adults with a variety of diagnoses. Possibly, in this sample the obsessive interests of some had been channeled into more socially appropriate (musical) areas. Finally, the particular language disabilities often found in those with autism point to possible origins of musical skill elaboration in the savant. First, delay or disturbance in early language development is a frequent feature in autism. Moreover, the delay may be uneven, in that semantic and pragmatic aspects of language are often more impaired than syntax and phonology. Indeed, often autistic children show remarkable demonstration of phonological knowledge combined with pragmatic weakness in the form of echolalia (Fay, 1973). In echolalia, one repeats what one has just heard in a conversation, rather than responding more appropriately. Interestingly, this is a characteristic that autistic children may have in common with those who are congenitally blind, particularly if some further intellectual disability is present (Warren, 1984). Note that echolalic responses indicate close attention to the structure of what is heard and an ability (and propensity) to give an accurate rendition of it. Thus, the musical products of savants may represent the operation of some processes common to language (sound structure analysis and reproduction) while not requiring others, for example those concerned with reference (Miller, 1989).

The Emergence of Savant Skill

Two general types of explanation have been advanced to account for the emergence of musical savant skills. One of these suggests that musical savants demonstrate the operation of processes typically associated with the right cerebral hemisphere. For a variety of reasons, in savants these processes are allowed to develop in a relatively unconstrained fashion. A second hypothesis is that savants' precocious music demonstrates the successful musical application of cognitive resources typically devoted to other areas, especially the acquisition of language. Of course, since language processing itself is asymmetrically distributed across the hemispheres (Geschwind & Galaburda, 1985), these two types of explanation are not mutually exclusive. Nevertheless, different characteristics of savant behavior appear to support "language" and "hemispheric" views. First, some savant musical strengths (e.g. sensitivity to melodic and harmonic structure) have been linked to right hemisphere processing while less exceptional skills (e.g. rhythm) have been associated with left hemisphere processes (Zatorre, 2001). More generally, savants are far more adept at performing than conceptual analysis and musical exegesis (Miller, 1989; Treffert, 1989). This procedural, as opposed to declarative nature of savant skill also suggests a predominance of right hemisphere processing (Altenmueller, 2001). However, one prominent savant attribute, absolute pitch, appears to be associated with processes in the left hemisphere closely allied to language (Schlaug, 2001).

The conditions under which savant behavior emerges also suggest mixed support for the two types of explanation. For example, the language delay in autism has already been mentioned. A similar delay occasionally occurs in those who are congenitally blind (Fay, 1973). In both cases pragmatic discourse and semantic understanding are special problems while phonology and syntax may be relatively normal. However, patterns of functional cerebral lateralization in autism (Fein et al., 1984) indicate a right hemisphere predominance more often than is usually the case. Perhaps the strongest evidence for a language-savant skill link comes from the case of Nadia, a precocious visual artist (Selfe, 1995). As her language skills improved, her drawing skills deteriorated. Indeed, some have advocated preventing savants from exercising their skill as a way to encourage more normative development (Lindsley, 1965). On the other hand, Miller (1989) found no deterioration in the musical products of a young savant as language improved. In fact, music provided an important context for the encouragement of conversational skills.

Musical savants typically express their special skills earlier in life than other types of savants do (Miller, 1999). In this context, it is significant that both autism and congenital blindness describe conditions present in early development. Indeed, the most successful model for describing the emergence of savant musical skills will likely include processes associated with early cognitive development and those more generally associated with hemispheric differentiation of function. Perceptual and cognitive functioning during the early sensori-motor period has been characterized as involving primarily right hemisphere functioning (Geschwind & Galaburda, 1985; Goldberg & Costa, 1981). Early configural and procedural learning, for example, are consistent with processing strengths associated with the right hemisphere. Further, in the infant's energetic imitation of what is heard, it is often difficult to distinguish speech from song. The infant's products (and the parents' models) are most often an amalgam of the two (Papoušek, 1996). These early forms of interaction precede the acquisition of declarative communication. As language acquisition becomes a focus of development, processes associated with the left hemisphere become more prominent. The acquisition of declarative knowledge and symbolic reference promoted by language also implicate left hemisphere processes. Moreover, the acquisition of referential systems of knowledge likely represent a new way of organizing and understanding what one hears (Karmiloff-Smith, 1992), in some respects replacing earlier forms of understanding. This may affect how one attends to what one hears. For example, non-disabled preschoolers, when given a new nursery school song, learn the lyrics well before they have mastered the melody (Miller, 1987). For the musical savant, the anecdotal evidence suggests the reverse is true (Miller, 1989).

Savants, then, might represent the consequence of a set of circumstances that are propitious for musical precocity; structural enhancement of the auditory modality, sparing of those mechanisms central to understanding the structure of the acoustic environment, and an extended early developmental period during which the focus of analyzing the sounds one hears is not overshadowed by their reference to objects, people and events in the world. Finally, this type of understanding takes place during a period of development where the interaction with the environment tends to be more sensori-motor than symbolic (Hargreaves, 1996). Certain developmental disabilities may prolong these early modes of interaction and therefore indirectly promote the expansion of skills associated with them. Treffert (2000) argues, for example, that the much higher frequency of males both in samples of savants and in certain types of disability such as autism suggests links between the two.

Clearly, this set of conditions must also include a history of environmental support for musical expression (Sloboda, 2002). As with those who are not disabled, parents and teachers certainly can play an important role in the musical growth of savants (e.g. Monty, 1981; Newman, 1989). It is likely that, for the savant, such external support is significantly augmented by intrinsic motivation. The satisfaction, for example, in being able to model what one has heard, or the feeling of mastery that accompanies successful task completion must be especially important for the savant who has "failed" at many of the developmental tasks faced.

This highly speculative sketch suggests the ultimate explanation for the emergence of musical savant skills will likely be quite complex. It does, however, offer a reason for the relative rarity of the savant phenomenon. If savant behavior requires such a combination of factors, any model of giftedness that hypothesizes multiplicative interactions among factors (e.g. Simonton, 1999) also predicts that exceptional levels of achievement will be rare.

Earlier, we raised the issue whether the particular types of disability associated with the emergence of savant behavior might render these cases irrelevant to more general questions of giftedness. First, though the disabilities of primary concern, autism and congenital blindness, may be atypical, developmentally speaking, the underlying processes are not. Thus, sensitivities to sound structure and the "syntax" of the sounds in one's culture are characteristic of musical development. In one sense, both autism and blindness may represent conditions that simply bring these predispositions to the forefront. Second, mechanisms similar to those we have been discussing in the context of disability may also be operative, or at least implicit, in normal development. For example, Bavelier and Neville (2002) reported that even in some of their subjects with intact senses, cross-modal activity was found (e.g. responses in "visual" cortical areas to sounds). Savant-like behaviors may appear after certain kinds of cortical trauma (Miller, Boone & Cummings, 2000) or even as a consequence of certain temporary experimental manipulations (Snyder & Mandy, 1997). Finally, the models linking savant behavior to the development of language and hemispheric specialization in savants resemble those that were proposed in part I for musical giftedness in general. For these reasons, continued attention to the context of musical savant skill seems warranted (L.M.).

A Comparison of Perspectives

It is clear that these research perspectives on musical giftedness in samples of the disabled and samples of the unimpaired diverge on some points. One of these is methodological. Musical savants are quite rare, and for both ethical and technical reasons, are unlikely candidates for many procedures currently yielding insights about cortical activity associated with music processing. On the other hand, as "experiments of nature" savants often present a kind of extreme manipulation that is unlikely to be duplicated except in paradigms that use animal models. The existence of musical savants has also provided part of the argument for distinctive types of intelligence (Gardner, 1983). As we have seen, however, the "musical intelligence" of musical savants is itself likely a combination of skills and attributes. Both traditions indicate a set of conditions or factors responsible for the emergence of musical exceptionality rather than a core musical faculty. Indeed, at this point in both traditions we are still at the point of trying to identify the ingredients in the recipe for musical exceptionality. In any case, conclusions from examination of savants should probably best be viewed as hypotheses to be tested within the confines of the experimental method.

A second point of divergence concerns the link between musical activity and other aspects of human behavior. One of us (MH) sees these as invariably intimately related, while the other (LM) adopts a more modular approach. These differences probably represent points along several dimensions, for example, differences in the importance of music in one's identity, and differences in the extent to which musical activity is influenced by other aspects of one's life. For example, musical activity may be central to the musical savant but, because of the associated disability, there may be some important constraints on the extent to which this activity can be influenced by conventional instruction, life experiences, etc..

Still, we believe that there are important points of convergence in the two research traditions we have reviewed. First, it is clear that the musical activity of those identified as gifted has important concomitants in cortical organization and development. At this point we are just beginning to explore just what kinds of activity and what kinds of consequences for development these might be. Further, while direct instruction is certainly important as a source of nurturance for musical growth, the research from both traditions suggests important structural contributions as well. There is also suggestive evidence from both the savant literature and that on musical giftedness in general that conditions and/or activities very early in development have especially important consequences for musical exceptionality. However, these factors may turn out to be quite subtle. For example, early sensitivity to musical structure represents a relatively transparent factor with respect to musical giftedness. On the other hand, less obvious factors such as early expansion of auditory cortex or changes in the developmental rate of cerebral lateralization of function may also have important consequences for musical growth, but in neither case can the factor be considered either necessary or sufficient. Rather, they are probably part of a constellation of factors that facilitate the emergence of musical exceptionality.

These different research traditions also emphasize the importance of a developmental perspective in considering musical exceptionality. First, it is clear that the tasks (and the corresponding achievements) for the emerging musician change over development. Contrast, for example, the precocious sensitivity to musical structure characteristic of the savants to the ability to restructure and create new musical expressions found in the musically gifted adolescent. An implication of these differences is that the sensitivity to various types of environmental input also changes. For the young musical savant, the simple opportunity to explore a musical corpus can promote musical growth (e.g. Newman, 1989). For the more mature musician to emerge, however, a quite different set of conditions might be necessary.

A developmental perspective examines the rate as well as the nature of change over time. The research we have reviewed suggests that questions associated with rate of development will continue to be prominent. The association between rate of development and sensitivity to input has already been mentioned. Might slower rates of development provide for a longer period of sensitivity to change agents (Simonton, 1999)? Might relatively rapid development in one domain affect that in another? To what extent might the relatively slow acquisition of language in savants provide an additional opportunity to for growth in non-language areas? More generally, how may the rate of development of skills typically associated with one hemisphere affect the growth of those in the other? Addressing the mystery of musical giftedness may require us to consider these more general questions about human development.

Finally, the research indicates important individual differences that are often overlooked in using more generic terms like musical giftedness or talent. Savants differ considerably in the range of musical interests and favored mode of expression. In those without disability, individual variation in the path to mature musicianship seems to be the rule. At this point we are beginning to trace some of the contributions to this rich variation, gender and pubertal development emerging as important areas for consideration. We believe continued examination of this heterogeneity will enrich our understanding of how exceptional musical skills develop.

References

Altenmueller, E. (2001). How many music centers are there in the brain? In R. Zatorre & I. Peretz (Eds.) *The biological foundations of music* (pp. 273-280). New York: New York Academy of Sciences.

Auerbach, S. (1906). Beitrag zur Lokalisation des musikalischen Talentes im Gehirn und am Schädel. *Virchows Archiv für pathologische Anatomie und Physiologie und für klinische Medizin*, 197-230.

Baeck, E. (2002). The neural networks of music. *The European Journal of Neurology, 9,* 449-56.

Baharloo, S., Service, S., Resch, N., Gitschier, J., & Fremier, N. (2000). Familial aggregation of absolute pitch. *American Journal of Human Genetics, 67*, 755-758.

Bamberger, J. (1982). Growing up prodigies: the midlife crisis. *New Direction of Child Development, 17*, 61-77.

Baranek, G., Fabian, J., Poe, M., Stone, W., & Watson, L. (In press). Sensory experiences questionnaire: discriminating sensory features. In young children with autism, developmental delays and typical development. *Journal of Child Psychology and Psychiatry.*

Bates, E. (2004). Explaining and interpreting deficits in language development across clinical groups. *Brain and Language, 88*, 248-253.

Bavelier, D., & Neville, H. (2002) Cross-modal plasticity: where and how. *Nature Reviews: Neurosciences, 3*, 443-452.

Bem, L.S. (1974) The measurement of psychological androgyny. *Journal of Consultant and Clinical Psychology 42,* 155-162.

Berliner, P. (1994). *Thinking in Jazz.* Chicago: University of Chicago Press.

Bever, T.G. & Chiarello, R.I. (1974). Cerebral dominance in musicians and non-musicians. *Science 185,* 537 – 540.

Billroth, Th. (1895). *Wer ist musikalisch?* Nachgelassene Schrift, hg. Von E. Hanslick. Berlin: Paetel.

Blacking, J. (1990) Music in children´s cognitive and affective development: Problems posed by ethnomusicological research. In: F.R.Wilson and F.L. Roehmann (Eds.) *Music and child development.* St. Louis: MMB Music, pp 68-78

Cameron, L. (1998). *The Music of Light.* New York: Free Press.

Cameron, J.L. (2004). Interrelationships between hormones, behavior, and affect during adolescence: Understanding hormonal, physical, and brain changes occuring in association with pubertal activation of the reproductive axis. *Annals of the New York Academy of Sciences, 1021,* 110-123.

Carlsson, I., Wendt, P.E., & Risberg, J. (2000). On the neurobiology of creativity. Differences in frontal activity between high and low creative subjects. *Neuropsychologia, 38 (6),* 873-85.

Charness, N., Clifton, J. & MacDonald, J. (1988). Case study of a musical "mono-savant" A cognitive-psychological focus. In L. Obler & D. Fein (Eds.) *The exceptional brain.* (pp. 277-294) New York: Guilford Press.

Crozier, J. (1997). Perfect pitch: practice makes perfect, the earlier the better *Psychology of Music, 25*, 110-120.

Dahl, R.E. (2004). Adolescent brain development: a period of vulnerabilities and opportunities. Keynote address. *Annals of the New York Academy of Sciences, 1021,* 1-22.

Demany, L., & Armand, F. (1984). The perceptual reality of tone chroma in early infancy. *Journal of the Acoustical Society of America, 76,* 57-66.

Doerner, G. (1985). Sex-specific gonadotropin secretion, sexual orientation and gender role behavior. *Experimental and Clinical Endocrinology, 86,* 1-6.

Elbert,.T., Pantev, C., Wienbruch, C., Rockstroh, B., & Taub, E. (1995). Increased cortical representation of the fingers of the left hand in string players. *Science, 270,* 305 – 307.

Elbert, T., Sterr, A., Rockstroh, B., Pantev, C., Miller, M,., & Taub, E. (2002). Expansion of the tonotopic area in the auditory cortex of the blind. *Journal of Neuroscience, 22,* 9941-9944.

Ebling, F.J.P. (2005). The neuroendocrine timing of puerty. *Reproduction, 129,* 675-683.

Evers, S., Dannert, J., Rodding, D., Rotter, G. & Ringelstein, E.B. (1999). The cerebral hemodynamics of music perception: A transcranial doppler sonography study. *Brain, 122,* 75-85.

Fay, W. (1973). On the echolalia of the blind and the autistic child. *Journal of Speech and Hearing Research, 11*, 365-371.

Fedor-Freybergh, P.G. (1999). Psychoimmuno-neuroendocrinology: an integrative approach to modern philosophy in medicine and psychology. *Neuroendocrinology Letters, 20,* 205-213.

Fein, D., Humes, M., Kaplan, E., Lucci, D., & Waterhouse, L. (1984). The question of left hemispheric dysfunction in infantile autism. *Psychological Bulletin, 95*, 258-281.

Frith, U. (1989). *Autism: Explaining the Enigma*. Oxford: Blackwell.

Fukui Hajime (2001). Music and testosterone: A new hypothesis for the origin and function of music. *Annals of the New York Academy of Sciences, 930,* 448-51.

Gardner, H. (1983) *Frames of mind.* New York: Basic Books.

Gembris, H. (1999). Historical phases in the definition of musicality. *Psychomusicology, 16,* 17-25.

Geschwind, N. & Galaburda, A.M. (1985). Cerebral lateralization. Biological mechanisms, associations and pathology, I and II,. *Archives of Neurology 42,* 428, 521.

Gogtay, N., Giedd, J.N., Lusk, L., Hayashi, K.M., Greenstein, D., Vaituzis, A.C., Nugent III, T.F., Herman, D.H., Clasen, L.S., Toga, A.W., Rapaport, J.L. & Thompson, P.M. (2004). Dynamic mapping of human cortical development during childhood through early adulthood. *Proceedings of the National Academy of Sciences of the United States of America 101,* 21, 8174-8179.

Goldberg, E., & Costa, L. (1981). Hemispheric differences in the acquisition and use of descriptive systems. *Brain and Language, 14,* 144-173.

Gordon, E. E. (1989). *Advanced measures of music audiation.* Chicago: GIA

Gordon, H.W. (1983). Music and the right hemisphere, in: A. W. Young (Ed.) *Functions of the right cerebral hemisphere,* New York: Academic Press, 65-86.

Hanslick, E. (1869/1870). *Geschichte des Concertwesens in Wien.* Facsimile Reprint Hildesheim - New York 1979.

Hantz, E. C., Marvin, E.W., Kreilick, K.G. & Chapman, R.M. (1996). Sex differences in memory for timbre: an event-related potential study. *International Journal of Neuroscience, 87 (1-2),* 17-40.

Hargreaves, D. (1996) The development of artistic and musical competence. In I. Deliège & J. Sloboda (Eds.) *Musical beginnings.* (pp 145-187) New York: Oxford.

Harrer, G. & Harrer, H. (1977). Music, emotion and autonomic function. In: M. Critchley & R.A. Henson (Eds.) *Music and the brain* (pp 202 – 216), Heinemann Medical Books.

Hassler, M. (1992). Creative musical behavior and sex hormones: musical talent and spatial ability in the two sexes. *Psychoneuroendocrinology, 17(1),* 55-70.

Hassler, M. & Birbaumer, N. (1988). Handedness, musical abilities, and dichotic performance in adolescents: a longitudinal study. *Developmental Neuropsychology, 4(2),* 129-145.

Hassler, M. & Nieschlag, E. (1989). Masculinity, femininity, and musical composition. Psychological and psychoendo-crinological aspects of musical and spatial faculties. *Archives of Psychology, 141,* 71-84.

Hassler, M. & Gupta, D. (1993). Functional brain organisation, handedness, and immune vulnerability in musicians and non-musicians. *Neuropsychologia, 31(7),* 655-660.

Hassler, M. & Gupta, D. (1998). Melatonin is elevated in highly gifted musicians. *Neuroendocrinology Letters, 19*, 87-91.

Hassler, M, Gupta, D. & Wollmann(1992). Testosteron, estradiol, ACTH and musical, spatial and verbal performance. *International Journal of Neuroscience, 65,* 45 – 60.

Heaton, P. (2003). Pitch memory labeling and disembedding in autism. *Journal of Child Psychology and Psychiatry and Allied Disciplines, 44,* 543-541.

Heaton, P., Hermelin, B., & Pring, L. (1998). Autism and pitch processing: a precursor for savant musical ability. *Music Perception, 15,* 291 –305.

Heaton, P., & Wallace, G. (2004). Annotation: the savant syndrome. *Journal of Child Psychology and Psychiatry and Allied Disciplines, 45,* 899-911.

Hermelin, B. (2001). *Bright splinters of the mind.* London: Kingsley Publishers.

Howe, M. (1989). *Fragments of genius: the strange feats of idiots savants.* London: Routledge.

Howe, M., Davidson, J., Moore, D., & Sloboda, J. (1995). Are there early childhood signs of musical ability? *Psychology of Music, 23,* 162-176.

Hutchinson, S., Lee, L.H., Gaab, N. & Schlaug, G. (2003). Cerebellar volume of musicians. *Cerebral Cortex, 13 (9),* 943-949.

Jonsson, P. & Carlsson, I. (2000). Androgyny and creativity: a study of the relationship between a balanced sex-role and creative functioning. *Scandinavian Journal of Psychology, 41 (4),* 269-274.

Kandel, E.R., (1993). Brain and behavior. In Kandel, Schwartz & Jessell (Eds) *Principles of neural science, 3. Ed.*, (pp1-17) Elsevier, New York.

Karmiloff-Smith, A. (1992). *On modularity: a developmental perspective on cognitive science.* Cambridge: MIT Press.

Kemp, A.E. (1985). Psychological androgyny in musicians. *Bulletin of the Council for Research in Music Education, 85,* 102-108.

Kessen, W., Levine, J., & Wendrick, K. (1979). The imitation of pitch in infants. *Infant Behavior and Development, 2,* 93-99.

Koelsch, S., Schröger, E. & Tervaniemi, M. (1999). Superior pre-attentive auditory processing in musicians. *Neuroreport, 10,* 1309-1313.

Koelsch, S., Maess, B, Grossmann, T. & Friederici, A. D. (2003a). Electric brain responses reveal gender differences in music processing. *Neuroreport, 14 (5),* 709-713.

Koelsch, S., Grossmann, T., Gunter, T.C., Hahne, A., Schroger, E. & Friederici, A.D. (2003b). Children processing music: electric brain responses reveal musical competence and gender differences. *Journal of Cognitive Neuroscience, 15 (5),* 683-693.

Kos, K. (1982). *Dora Pejacevic.* Zagreb: Jugoslavenska Academija Znanosti I Umjetnosti.

Kries, J. v. (1926). *Wer ist musikalisch? Gedanken zur Psychologie der Tonkunst.* Berlin: Springer

LeDoux, J. E., Sakaguchi, A. & Reis, D. J. (1984). Subcortical efferent projections of medial geniculate nucleus mediate emotional responses conditioned to acoustic stimuli. *Journal of Neuroscience*; *4 (3),* 683-698.

Lee, D.J., Chen, Y. & Schlaug, G. (2003). Corpus callosum: musician and gender effects. *Neuroreport, 14 (2),* 205-209.

Leopold. P. (1997). Musical savants: implications for the relation between music and language. Ph.D. Dissertation, University Of Illinois, Chicago.

Lindsley, O. (1965). Can deficiency produce specific superiority: the challenge of the idiot savant. *Exceptional Child, 31*, 225-232.

Luders, E., Narr, K.L., Thompson, P.M., Rex, D.E., Jancke, L., Steinmetz, H. & Toga, A.W. (2004). Gender differences in cortical complexity. *Nature Neuroscience, 7 (8),* 799-800.

Lynn, R., Wilson, G., & Gault, A. (1989). Simple musical tests as measures of Spearman's G. *Personality and Individual Differences, 10,* 25-28.

Manturzewska, M. (1990). A biographical study on the lifespan development of professional musicians. *Psychology of Music, 18,* 112-139.

Manturzewska, M. (1996). Identification and promotion of musical talent. In A.J.Cropley & D. Dehn (Eds.): *Fostering the growth of high ability: European perspectives* (pp.271-285).

McGaugh, J.L. & Cahill, L (197). Interaction of neuromodulatory systems in modulating memory storage. *Behavioural Brain Research*, 83, 31-38.

McManus, I.C. & Bryden, M.P. (1991). Geschwind's theory of cerebral lateralization: Developing a formal causal model. *Psychological Bulletin, 110,* 237-253.

McNamara, P., Flannery, K.A., Obler, L.K. & Schachter, S. (1994). Special talents in Geschwind's and Galaburda's theory of cerebral lateralization: an examination in a female population. *International Journal of Neuroscience, 78 (3-4),* 167-76.

Meyer, Alfred (1977). The search for a morphological substrate in the brains of eminent persons including musicians: a historical review. In: M. Critchley & RA Henson (Eds) *Music and the brain,* (pp 255-281). London: Heinemann Medical Books.

Michaelis (1805) Über die Prüfung musikalischer Fähigkeiten. *Berliner Musikalische Zeitung 56-58,* 222-230

Micheyl, C. et al. (1997). Difference in cochlear efferent activity between musicians and non-musicians. *Neuroreport, 8,* 1047-1050.

Miller, B., Boone, K., Cummings, J., Read, S., & Mishkin, F. (2000). Functional correlates of musical and visual ability in frontotemporal dementia. *British Journal of Psychology, 176,* 458-463.

Miller, L. (1987a). Determinants of melody span in a developmentally disabled musical savant. *Psychology of Music, 15*, 76-89.

Miller, L. (1987b). sensitivity to tonal structure in a developmentally disabled musical savant. *American Journal of Mental Deficiency 91,* 467-471.

Miller, L. (1989). *Musical savants: exceptional skill in the mentally retarded.* Hillsdale, NJ.: Erlbaum.

Miller, L. (1992). Diderot reconsidered: Visual impairment and auditory compensation. *Journal of Visual Impairment and Blindness, 86*, 206-211.

Miller, L. (1999). The savant syndrome: intellectual impairment and exceptional skill. *Psychological Bulletin, 125*, 31-46.

Miller, L. & Monroe, M. (1990). Musical aptitude and adaptive behavior in people with mental retardation. *American Journal on Mental Retardation, 95*, 220-227.

Miller, L. & Orsmond, G. (1994). Assessing structure in the musical explorations of children with disabilities. *Journal of MusicTherapy, 31,* 248-265.

Miller, L.B. (1987). Children's musical behaviors in the natural environment In J. Perry, I. Perry and T. Draper (Eds.) *Music and child development.* (pp. 206-224) New York: Springer Verlag.

Money, J. (1986). *Venuses penuses. Sexology, sexosophy and exigency theory.* Buffalo: Prometheus Books.

Monty, S. (1981). *May's Boy.* Nashville, TN: Thomas Nelson.

Moog, H. (1976). *The Musical Experience of the Preschool Child* London: Schott.

Newman, N. (1989). A music teacher's perspective on savant skill. In L.Miller, *Musical savants: exceptional skill in the mentally retarded* (pp.211-242). Hillsdale. NJ. Erlbaum.

Njemanze, P.C. (2005). Cerebral lateralization and general intelligence: gender differences in a transcranial Doppler study. *Brain and Language, 92 (3),* 234-239.

Norton, A., Winner, E., Cronin, K., Overy, K., Lee, D.J. & Schlaug, G. (2005). Are there pre-existing neural, cognitive, or motoric markers for musical ability? *Brain and Cognition, 59 (2),* 124-134.

Ockleford, A. (1998). *Music Moves.* London: Royal National Institute For the Blind.

O'Connell, T. (1974). The musical life of an autistic boy. *Journal of Autism And Childhood Schizophrenia, 3,* 223-229.

Pantev, C., Oostenveld, R., Enngelien, A., Ross, B., Roberts, L.E., Hoke, M (1998). Increased auditory cortical representation in musicians. *Nature 392,* 811-814.

Papoušek, H. (1996) Musicality in infancy research: biological and culturalorigins of early musicality. In I. Deliège & J. Sloboda (Eds.) *Musical beginnings.* (pp 37-55), New York: Oxford.

Petsche, H. & Etlinger, S.C. (1998). *EEG and thinking.* Wien: Verlag der Oesterreichischen Akademie der Wissenschaften.

Pitman, D. (1965). The musical ability of blind children. *Research Bulletin American Foundation for the Blind , 22,* 63-80.

Rauscher, F.H., Shaw, G.L., Levine, L.J., Wright, E.L., Dennis, W.R., & Newcomb, R. (1997). Music training causes long-term enhancement of preschool children's spatial-temporal reasoning abilities. *Neurological Research, 19,* 1-8.

Révész, G. (1925). *The Psychology of a Musical Prodigy.* Freeport: Books for Library Press.

Rimland, B. (1978). Savant capabilities of autistic children and their cognitive implications. In G. Serban (Ed.) *Cognitive deficits in the development of mental illness* (pp. 44-63) New York: Brunner/Mazel.

Romeo, R.D. (2005). Neuroendocrine and behavioral development during puberty: a tale of two axes. *Vitamins and Hormones, 71,* 1-25.

Saffran, J., & Griepentrog, G. (2001). Absolute pitch in infant auditory learning: evidence for developmental reorganization. *Developmental Psychology, 37,* 74-85.

Salovilita, T. (2000). Incidence of the savant syndrome in Finland. *Perceptual and Motor Skills, 91,* 120-122.

Scott, D. & Moffett, A. (1977). The development of early musical talent in famous composers: a biographical review. In: Critchley & Henson (Eds.) *Music and the brain.* (pp. 174-201). William Heinemann Medical Books Lim., London.

Schlaug, G. (2001). The brain of musicians. A Model for Functional and Structural Adaptation. *Annals of the New York Academy of Sciences, 930,* 281-299.

Seashore, C. (1919). *Seashore measurement of musical talent.* New York

Selfe, L. (1995). Nadia reconsidered. In C. Golumb (Ed.) *The development of artistically gifted children* (pp 197-237). Hillsdale, NJ.: Erlbaum.

Simonton, D. (1999) Talent and its development: an emergenic andepigenetic model. *Psychological Review, 105,* 435-457.

Sloboda, J. (1993) Musical ability. In: G.R.Bock & K. Ackrill (Eds.) *The origins and development of high ability* (pp.106 -118),. Chichester: Wiley.

Sloboda, J. (2002). Musical expertise. In D. Levitan (Ed.) *Foundations of cognitive psychology: core readings.*(pp 565-581), Cambridge: MIT Press.

Sloboda, J., Hermelin, B., & O'Connor, N. (1985). An exceptional musical memory. *Music Perception, 3,* 155-170.

Sluming, V., Barrick, T., Howard, M., Cezayirli, E., Mayes, A & Roberts, N. (2002). Voxel-based morphometry reveals increased gray matter density in Broca´s area in male symphony orchestra musicians. *Neuroimage, 17 (3),* 1613-1622.

Snyder, A. & Mandy, T. (1997). Autistic artists give clues to cognition *Perception, 26*, 93-96.

Southall, G. (1979). *Blind Tom: The post-civil war enslavement of a black musical genius.* Minneapolis, Minn: Challenge Productions.

Stefani, G. (1987) A theory of musical competence. *Semiotica 66,* 7-22

Stevens, A., & Weaver, K. (2005) Auditory perceptual consolidation in early onset blindness. *Neuropyschologia, 43*, 1901-1910, *113.*

Sur, M. & Leang, C. (2001). Development and plasticity of cortical areas and networks. Nature Reviews, *Neurosciences, 2,* 251-262.

Takeuchi, A., & Hulse, S. (1993). Absolute pitch. *Psychological Bulletin* 345-362.

Tanner, J. M. (1962). *Growth at adolescence*. Springfield III: Charles C. Thomas.

Trano, M., Cariani, P., Delgutte, B., & Braida, L. (2001). Neurobiological foundations for the theory of harmony in Western tonal music. In R. Zatorre & I. Peretz (Eds.) *The biological foundations of music* (pp. 92-117). New York: New York Academy of Sciences.

Treffert, D. (1989). *Extraordinary People*. New York: Harper & Row.

Treffert, D. (2000). Savant syndrome. In A. Kazdin (Ed.) *Encyclopedia of psychiatry, 7.* (pp. 144-148) Oxford: Oxford University Press.

Trehub, S. (2001). Musical predispositions in infancy In R. Zatorre & I. Peretz (Eds.) *The biological foundations of music* (pp. 1-16). New York: New York Academy of Sciences.

Viscott, D. (1970). A musical idiot savant. *Psychiatry, 33,* 494-515.

Warren, D. (1984). *Blindness and early child development.* New York: American Foundation for the Blind.

Waterhouse, L., Fein, D., & Modahl, C. (1996). Neurofunctional mechanisms in autism. *Psychological Review, 103,* 457-489.

Weinberger N. (1997). The Musical Hormone. *MuSICA Research Notes, IV, 2*, 1-4.

Welch, G. (1988). Observations on the incidence of absolute pitch (AP) ability in the early blind. *Psychology of Music, 16*, 77-80.

Witelson, S.F. (1985). The brain connection: The corpus callosum is larger in left-handers. *Science, 229,* 665-668.

Zatorre, R. (2001). Neural specificity for tonal processing In R. Zatorre & I. Peretz (Eds.) *The biological foundations of music* (pp. 193-211). New York: New York Academy of Sciences.

Zatorre, R., Halpern, A., & Perry, D. (1996). Hearing in the mind's ear: a PET investigation of musical imagery and perception. *Journal of Cognitive Neuroscience, 8,* 29-46.

In: Neurosciences in Music Pedagogy
Editors: W. Gruhn and F. Rauscher, pp. 233-266
ISBN: 978-1-60021-834-7

Chapter IX

MUSICAL LEARNING IN INDIVIDUALS WITH DISABILITIES

Eckart Altenmueller

ABSTRACT

The chapter describes the musical abilities and neurobiological conditions of children with various physical and intellectual disabilities. A short overview of the nature and causes of disabilities in general is presented, including a short discussion on terminology. The impact of physical disabilities, visual and hearing impairment on music making is discussed. Localized brain disorders affecting specifically music perception and production such as amusia or focal dystonia are presented. Developmental disorders like Attention Deficit Hyperactivity Syndrome, Dyslexia, Autism Spectrum Disorder and Savant Syndrome are described. The way in which the symptoms affect musical processing is discussed and suggestions are given as to how music can be beneficial for these individuals. The last part of the chapter focuses on individuals with chromosomal aberrations such as Down Syndrome and Williams Syndrome. The specific musical talent of the latter and its neurological origin will be discussed. In summary, musical activity is an extremely rich experience which can have a positive impact on all of the above mentioned conditions.

INTRODUCTION

The Richness of Musical Experience

Musical experience is probably the richest human emotional, sensorimotor and cognitive experience. It involves listening, watching, feeling, moving and coordinating, remembering and expecting. It is frequently accompanied by profound emotions resulting in joy, happiness, bitter-sweet sadness or even in overwhelming peak experiences that manifest themselves in bodily reactions like tears in the eyes or shivers down the spine. Almost all brain regions contribute to musical experience when considered as a holistic entity (for a review on brain structures and musical functions see also chapters 1 and 5*)*. The various sensory and motor pathways in the brainstem, midbrain, and basal ganglia are the prerequisites of automated processing. Primary and secondary regions in the cerebral cortex are critical for any conscious perception in the auditory, visual, and somatosensory modality. The frontal lobe is involved in guidance of attention, in planning and motor preparation and in complex human specific skills like imitation and empathy, the latter two skills contributing importantly to music learning and emotional expressiveness in music. Brain regions between the sensory fields – commonly termed "association cortices" – are also important structures, where different sensory inputs, for example from the ears and the eyes, or from the ears and the sense of touch are integrated and fused to a single united impression, typical for musical experience. The cerebellum is important in motor coordination and in various cognitive tasks especially when aspects of timing play a role, for example in rhythm processing and tapping in synchrony with an external pacemaker such as a metronome. Finally, the extended emotional network (comprising the basis and the inner surfaces of the two frontal lobes, the cingulate gyrus and brain structures in the evolutionarily old parts of the brain such as the amygdala, the hippocampus and the midbrain) is crucial for emotional perception of music and hitherto for an individual's motivation to listen to or to engage in any musical activity.

Furthermore, the brain is highly dynamically organized, that is, brain structures are subject to changes according to the activities and demands imposed upon them. Musical activity has proven to be a powerful stimulus for this kind of brain adaptation, or brain plasticity, as pointed out by Schlaug and Bangert (see chapter 4). Particularly when music learning starts at an early age, the specific brain regions involved (such as the sensorimotor hand regions in the right hemisphere, controlling the fine motor skills of the left hand in violinists), are enlarged (Bangert & Schlaug, 2006). Other effects include the thickness of nerve fibres and subsequent increased speed of nerve conduction between brain centres communicating during a musical task, such as between the auditory and the motor centres of the brain. These effects of plasticity are not restricted to musical prodigies; they occur in musical amateurs and young adults, albeit to a lesser extent. Thus, with the main topic of our chapter in mind, it is possible to suggest that music-induced brain plasticity may have benefits for disabled children, helping to improve various sensory, motor, coordinative or emotional disabilities.

Due to the multifaceted nature of musical activities, many different types of disability experienced by humans can affect their musical experience. A person with cerebral palsy may not have the motor coordination required to play an instrument easily, while a person with cochlea implants may not hear melodies clearly. A person with severe learning disabilities may find it difficult to keep an auditory signal in memory, while an autistic child may not recognize the sense of melancholy implied by a certain melody. Equally though, and perhaps more importantly, persons with disabilities will always retain a wide range of abilities that will enable them to continue to enjoy musical experiences. An individual with autism may enjoy the beautifully structured patterns and repetitions within music, and may thus extract another quality, valuable and meaningful.

In addition, of course disabled individuals may display extraordinary musical talent, which can sometimes be considered as "islands of genius" that stand in marked, incongruous contrast to an overall deficit in other cognitive domains (Miller, 1989; also see chapter 8). Such musical savants can display extraordinary performance skills, most often piano playing, yet composing in the absence of performing has also been reported. Individuals with Williams Syndrome show an asymmetry of cognitive faculties caused by a genetic defect (microdeletion on chromosome 7). IQs typically in the range of 65-70 usually leave these children with very poor mathematics and spatial reasoning skills, but more adept than might be expected with their language and musical skills, depending on their opportunities for acquisition (Levitin & Bellugi, 1998).

There are remarkable similarities between many of the characteristics of musical savants and child prodigies. Prodigies, of course, do not have a cognitive impairment, but their situations may be similar in that their extraordinary skills appear very early and provide support for the idea of the heritability of musical talent. Révész (1925) intensively studied a musical prodigy and was unable to account for his gifts, other than inherited traits. Recently, Ruthsatz and Detterman (2003) reported a case of a musical prodigy who had no formal tuition on an instrument, did no formal practice and had gained his skills by listening to other performers and improvising his own musical pieces. His family had no particular musical background, although his mother played the piano. He could sing in two languages and had taught himself to play numerous instruments. His musical behaviours seemed self-motivated – he engaged in them spontaneously and with pleasure, in particular in entertaining people. He spent a great deal of time in playful imitation of other musicians and the improvement in his performance was due to this practice. He had extremely high scores on tests of musical ability and intelligence, the latter revealing an extraordinary memory as measured within his cognitive profile.

In this chapter, a short overview of the nature and causes of disabilities in general is presented, including a short discussion of terms like "impairment" and "developmental disability." It is an attempt to systematize the different syndromes and to extract information from the medical and neurobiological literature relevant for music educators. Subsequently, I present a range of the most common hereditary, developmental and acquired disorders that have been found to affect the quality of either musical learning or musical experience, especially in children and adolescents. In each case, I will define and describe the disorder, discuss what is known about the neurobiological basis, discuss the way in which the symptoms affect musical processing and then give suggestions as to how music can be beneficial with these individuals. I do not attempt to be exhaustive in this exercise, since in many conditions scientifically based knowledge concerning the effects of musical activity is still lacking. In some instances, rather selected case vignettes are presented to indicate how children of different ages and needs can benefit from musical activity.

Before going deeper into the details, some selected terms relevant for the topic of this chapter shall be briefly defined. Here, I do not go into details of the specific syndromes, rather I present a concise glossary of conditions related to the general theme of "disability." In addition, I give a general overview of the most common signs of disabilities and their diagnostic criteria.

The broad term *disability* comprises *physical and cognitive disabilities,* defined as an impairment that substantially limits one or more of the major life activities of an individual. A person is defined as disabled if he or she has difficulties in performing certain functions, such as seeing, hearing, talking, walking, climbing stairs, lifting and carrying, or has difficulties in performing the activities of daily living, or has difficulties with certain social roles, such as those required at school or in employment. A person is considered to be severely disabled when she or he is completely unable to perform one or more activities, or uses an assistive device to get around, or needs assistance from another person to perform basic activities of daily life.

An *impairment* is defined as any physiological disorder or condition, cosmetic disfigurement, or anatomical loss affecting neurological, musculoskeletal, respiratory, and cardiovascular or any other bodily system including the sense organs. It also refers to any cognitive or psychological disorder, including developmental disorders, organic brain syndrome, emotional or mental health disorders, and specific learning disabilities.

Of course, there are also social ramifications of having a disability. Very often, it is not the biological condition but the societal barriers that restrict the lives of disabled people. In the past decades, many societies have undertaken great efforts to overcome the restriction of certain social activities in disabled individuals, for example with respect to general mobility or sports.

With respect to the broad term of cognitive disability, three main conditions have to be distinguished: (1) *developmental disabilities*, which manifest before an individual reaches adulthood, (2) *developmental delay* which is frequently used synonymously with developmental disabilities, but usually describes more pronounced intellectual limitations, and (3) *dementia* of various origins, for example Alzheimer's Disease, which are characterized by a progressive decline of cognitive abilities.

Developmental disabilities are characterized by a wide spectrum of conditions such as extremely low IQ, behavioral and emotional disorders and specific learning disabilities, to name a few. It is important to keep in mind that subtle disturbances in a cognitive or emotional domain, for example stuttering, mild dyslexia, or congenital amusia (*see below*) are not subsumed under this term. A developmental disability constitutes a substantial disability for the affected individual and is attributable to a general cognitive deficit, cerebral palsy, epilepsy, autism or other neurological condition resulting in the impairment of general intellectual functioning or adaptive behaviour.

Developmental delay refers to substantial intellectual limitations in a person's present functioning. It is characterized by significantly sub-average intellectual functioning, existing concurrently with related limitations in two or more of the following applicable adaptive skill areas: communication, self-care, home living, social skills, community use, self-direction, health and safety, functional academics, leisure, and work. Developmental delay is diagnosed during the developmental years, usually before the end of puberty. In medical contexts, developmental delay frequently is replaced by the term *mental retardation*. However, the latter acquired pejorative connotations over the last few decades and therefore should not be used in educational contexts. Additionally, developmental delay generally implies that appropriate intervention, for example music therapy, will improve the condition, allowing for "catching up." Importantly, this term carries the emotionally powerful idea that the individual's current difficulties are likely to be temporary.

Dementia refers to symptoms which include changes in memory, personality, and behaviour that result from a change in the functioning of the brain. These declining changes occur typically at older age, are long lasting and involve a progressive loss of brain cell function. There is a broad range of different types of dementia. The most common cause of dementia is Alzheimer's dementia. Here, the progressive death of nerve cells in the brain is associated with the formation of clumps (amyloid plaques) and tangles of protein (neurofibrillary tangles) in the brain. Other types of dementia include vascular dementia, due to multiple strokes and defective blood circulation in the brain of the elderly, and infections, such as Jacob-Creutzfeldt disease. Symptoms of dementia include repeatedly asking the same question, loss of familiarity with surroundings, increasing difficulty in following directions, loss of memory and changes in personality or emotion. Not everyone displays all symptoms since symptoms vary in the different types of dementia.

Behavioural disorders or *emotional disorders* are used interchangeably as terms to classify children who exhibit extreme or unacceptable chronic behaviour problems. These children lag behind their peers in social development and are often isolated from others either because they withdraw from social contact or because they behave in an aggressive, hostile manner. Behaviour disorders frequently result from persistent negative social interactions between the child and the environment. They generally consist of four clusters of traits: conduct disorders, anxiety-withdrawal, immaturity and socialized aggression.

Learning disability is a term used to describe a condition affecting the ability to understand or use spoken or written language, perform mathematical calculations, coordinate movements or direct attention. The term is sometimes used as a means of separating general intellectual limitations from specific, limited minor deficits, as well as indicating that there is no emotional or psychological disability. Learning disabilities comprise such common disorders as Attention Deficit Disorder (ADD) or Attention Deficit Hyperactive Disorder (ADHD), reading and writing difficulties (dyslexia), difficulties in understanding mathematical operations (dyscalculia) and minimal brain dysfunction (MBD) with unspecific lowering of IQ.

Diagnostic Criteria

Although in most instances physical disabilities are recognized early, during the first 12 months, sometimes specific disorders may go undetected. For example, impaired auditory function and even deafness can go unnoticed in infants, unless newborn auditory screening is routinely applied. With developmental disabilities the situation is even more complex and a firm diagnosis is frequently considerably delayed. Early signs of a difficulty are in many instances unspecific. For example, children with developmental disabilities may learn to sit up, to crawl, or to walk later than other children, or they may learn to talk later. However, there is an enormous variability in normal psycho-motor development and thus it can be difficult to distinguish between a normal child developing slowly and a child with a developmental disability. Children with developmental disabilities can also have trouble understanding social rules, discerning cause and effects and solving problems. In early childhood, mild disability (IQ 60–70) may not be obvious and may not be diagnosed until the child begins school. Even when poor academic performance is recognized, it may take expert assessment to distinguish mild developmental disability from a specific learning disability or behavioral problems. Of course, as they become adults, many mildly disabled individuals are able to live and function independently.

Moderate disability (IQ 50–60) is nearly always identifiable within the first few years of life. Children with a moderate disability will encounter difficulties in school, at home, and in the community and in many cases will join special, usually separate, classes in school. However, such individuals can still progress to become functioning members of society. As adults they may live with their parents, in a supportive group home, or even semi-independently with significant supportive services to help them, for example, manage their finances.

Among people with cognitive disabilities, only about one in eight will score below 50 on IQ tests. An individual with such a severe disability will need intensive support and supervision for his or her entire life.

Causes of Disabilities

Physical or cognitive disabilities can have many causes. Genetic conditions are the most frequent causes of inborn cognitive disabilities, with Down Syndrome occurring in one out of 800 children and accounting for about 25% to 40% of severely mentally disabled children as the most common pathological entity. Genetic defects can also cause deafness, hyperlaxity of joints and fibrous tissue (Ehlers-Danlos-Syndrome), impaired motor behaviour such as muscular weakness, or involuntary cramping in dystonia. All of these conditions heavily affect the individual's musical abilities.

Other frequent causes of disability are problems during pregnancy. For example, an infection like rubella during the first trimester of pregnancy can cause deafness or cognitive disability. In industrialized countries, the most frequent cause of mild cognitive disabilities is foetal alcohol syndrome, accounting for up to 8% of the cases and causing defects that can include cognitive deficits, growth deficiencies, central nervous system dysfunction, cranio-facial abnormalities and behavioural maladjustments. Another example, congenital hypothyroidism, is a neurological syndrome that results from severe thyroid hormone deficiency during the foetal period. In the infant, this can translate into deaf mutism, moderate to severe cognitive deficits, spastic paresis of the limbs, and strabismus.

Problems at birth are another cause of various disabilities. If a baby is not getting enough oxygen during labour and birth, he or she may suffer from developmental disabilities, especially cerebral palsy. Cerebral palsy is characterized by muscle weakness or poor control of movement or posture, frequently accompanied by increased muscular tone and stiffness. Other birth stress, for example the use of forceps, can cause mechanical damage to brain tissue.

In infancy and early childhood, infections like whooping cough, measles, or meningitis can cause cognitive disabilities, which can also be caused by exposure to poisons like lead or mercury, usually in polluted environments. Poor or defective nutrition can also be dangerous, for example Iodine deficiency affects approximately 2 billion people worldwide and is the leading preventable cause of cognitive disability in areas of the developing world.

Finally, manifold adverse social conditions, such as institutionalisation and lack of emotional attachment at a young age can cause cognitive deficiencies. Sensory deprivation in the form of severe environmental restrictions, prolonged isolation or severe atypical parent-child interactions can lead to irreversible developmental disabilities, such as those experienced by so called "wolf-children", who have been raised in isolation and in the absence of human language. If these children are not exposed to language before the age of 6, they will never develop adequate language capabilities, due to the fact that language (and other) cognitive capabilities can only be acquired during a sensitive period of the central nervous system, lasting from birth to about age five. Interestingly, the sensitive period for the development of musical faculties seems to be much more extended since music education can yield powerful effects in later childhood and adolescence (for a review see Hodges, 2006). However, scientific investigations on the acquisition of musical faculties in children raised in isolation have not been undertaken to date.

In the following section, selected sensory and physical disabilities that have an impact on musical experience and musical learning, such as motor impairments, blindness and deafness, will be discussed. Then, localized central nervous disorders that have an impact on music perception or production will be presented. Subsequently the section will focus on cognitive disabilities. Finally, I discuss the impact of these disorders on musical development and learning.

Musical Learning with Physical Disabilities

Probably at all times, music-loving individuals with physical disabilities have tried to overcome their condition by searching for and inventing creative solutions. In 1822, Cornelius Ward constructed a flute for Count Rebsomen, who had lost his left arm while fighting as a general for Napoleon. The ingenious mechanism enabled him to play the flute only with his right hand (*Figure 1.,* Brook, 1933).

Musical Learning with Physical Impairments

After the Second World War, makers of woodwind instrument systematically explored possible modifications to instruments by reorganizing the mechanics of the keys or modifying the keys, in order to meet the needs of musicians who had lost single or multiple fingers. In the late1950s, physical deformation caused by thalidomide (Contergan) affected thousands of children, leaving them with abnormally short or even missing limbs. This tragedy motivated instrument makers to modify piano pedals for children with extremely shortened legs, or modify the valves of the French horn for children without hands, allowing them to play the instruments with their feet (Klenerman & Wood, 2006, p.137 -155). Naturally, these creative solutions are more easily implemented in instruments based on mechanical components than in older instruments such as the guitar or the violin. However, stringed instruments can also be adapted to permit bowing with the left hand (rather than the right), or bowing with a prosthesis modification of the bow.

Figure 1. Portrait of count Rebsomen who had lost his left arm when fighting for Napoleon. The mechanism of this "left-handed flute" was invented by the flute maker Cornelius Ward in 1822 (By courtesy of the German collector B.S. in Hannover).

Interestingly, more disabled children play the piano than any other instrument. It is perfectly possible to play it one-handed and there have been some well-known concert artists who played in this way, for example the famous Austrian pianist Paul Wittgenstein. Wittgenstein lost his left arm during the first World War and inspired outstanding composers like Maurice Ravel, Richard Strauss, Benjamin Britten, Sergei Prokoffief and Paul Hindemith to compose concert pieces for the left hand alone, pieces that are now part of the concert repertoire. However, this left hand piano repertoire (to which Johannes Brahms, Joseph Rheinberger, Camille Saint-Saëns and Alexander Skrjabin, among others, contributed) is for advanced pianists. There is little repertoire available for left-, or right-handed players at the beginner level. The beginner therefore is very dependent upon a teacher's willingness to engage in the adaptation of materials.

Thus, many factors have to be taken into account when choosing a conventional instrument for a child with a physical disability. First, of course, one needs to identify an instrument that is preferred by the prospective player, since individuals can overcome enormous obstacles when genuinely motivated and determined. Indeed, motivation and persistence are essential for anyone learning to play a musical instrument. Other factors to be taken into consideration are the physical demands of the instrument, such as the strength required to hold the instrument for a reasonable period of time, and the physical size of the instrument. Of course, it may be possible to adapt the playing position or to obtain stands and other aids. If a stand is required to support the instrument, it may be necessary to obtain advice from an occupational or physical therapist or from a specialist instrument-maker, since playing a badly-positioned instrument can do physical harm and have a negative effect on progress and motivation. Proper positioning is also crucial for the ease of playing, optimisation of technique and continuing motivation.[15] If a large instrument is not possible, many instruments are available in smaller versions that sound at a higher pitch. Flutes, clarinets, and bassoons for example are built in a range of sizes, while stringed instruments are available in quarter and half sizes. Finally, an open-minded teacher who is willing to adapt teaching styles and repertoire according to the complex pattern of strengths and weaknesses of the disabled pupil is essential to the success of instrument playing in physically disabled children.

It should be noted that there are also many novel means to create music and play music with unconventional instruments, based on MIDI-technology and/or electronic devices such as sensors or switches. Such instruments are ideal for sensory and physically disabled children to express their musical creativity and to perform music alone or in groups.[16] The Drake Music Project, established in 1988, provides disabled children and adults who are unable to play conventional musical instruments with opportunities to discover their musical abilities and explore, compose and perform their own music by using novel electronic devices, mapping for examples arm gestures to sounds. In the frame of this project, continued research and development in assistive music technology has created technologies and techniques that provide many, previously unimaginable music-making opportunities for musically talented disabled individuals.

Music Learning with Visual Impairments

The myth of the blind musician is an important touchstone in many cultures. The story of Homer, the blind poet and singer, is probably the most frequently quoted example in Western tradition, even though its historical truth remains uncertain. Poets and painters have been inspired by this topic for centuries. Even today, blind musicians such as the tenor Andrea Bocalli or the guitarist Jeff Healey are sometimes considered to possess a specific charisma and musicality.

There has long been a folk belief that blind people have superior auditory skills compared to sighted people and that therefore the blind have an advantage when performing music. Some recent studies have supported this belief. For example, research has shown that blind musicians are more likely to have perfect pitch than sighted musicians (Hamilton et al., 2004). Similarly, people born blind or who became blind early in childhood are better able to recognize tiny variations in pitch compared to sighted people (Gougoux et al., 2004). These findings are often interpreted as signs of early brain plasticity. Roeder et al. (2002) found that the visual cortices of individuals who were blind from childhood (congenitally blind), was involved in sound and language processing rather than for processing visual input from the retina, as is the case with sighted individuals. A similar effect has also been found for Braille-

[15] In the UK, there is a charity organisation that has designed and fitted stands for disabled instrumentalists (www.remap.org.uk).

[16] An inspirational site introducing several young people with severe muscular disabilities who are able to play an electronic wind instrument is found at www.mybreathmymusic.com, switch users can find original solutions at www.midicreator.co.uk.

reading blind individuals, whose visual cortices have been found to be activated when processing the tactile input from the finger tips (Sadato et al., 1996). Both phenomena demonstrate the adaptability of the brain.

Historically, many blind musicians, including some of the most famous, have performed without the benefit of formal instruction, since frequently such instruction relies on written musical notation. However, today there are many resources available for blind musicians who wish to learn Western music theory and classical notation. Louis Braille, the inventor of the Braille alphabet for the blind, also created a system of classical notation, called Braille music. This system allows the visually impaired to read and write music just as the sighted do. Computer technology and the internet make it possible in theory for blind individuals to be more independent in composing and studying music. In practice, however, most programs rely on graphical user interfaces, which are difficult for the blind to navigate. Lately, there has been some progress in creating screen-reading interfaces for the blind, especially for the Windows operating system.

It is often suggested that music education can bring many benefits to the visually impaired child, such as training them to listen and interpret the sounds they hear, improving auditory memory, developing co-ordination, and enabling them to integrate socially. The quickest way to teach music to visually impaired children is through listening. In this way, children hear the music directly from the teacher and can use tape recordings during practice time at home. However, music literacy, where possible and to whatever appropriate level, should also be encouraged, since it can lead to rewarding participation in music-making and allows more musical independence in later life. Particular attention should be directed to posture with blind musicians. A sighted musician can learn correct posture by visual example and by looking in a mirror, but for a visually impaired musician, often the only clues to a faulty posture will be in the sound quality or in some difficulty with a particular aspect of technique.

Today there are several organizations devoted to the support of blind musicians such as the National Resource Center for Blind Musicians in Connecticut or the Music Education Network for the Visually Impaired. No-C-Notes provides an alternative to Braille music notation with its audio music reading transcription method for those with visual impairment or reading disability.[17]

Musical Learning with Hearing Impairments

It might seem strange that deaf or hearing impaired individuals enjoy music. However, most deafness is not profound and additionally, as discussed in the introductory paragraph, musical experience is an immensely rich sensory, physical and personal experience which relies on many more components than on hearing. Seeing, moving, and the physical feeling of sound vibration and group synchronisation, for example, also contribute to the joy of music-making.

---

[17] For websites see internet resources at the end of the references.

Students with severely impaired hearing have successfully participated in school bands and orchestras for over 100 years. Many reports remark on the exceptional quality of the music, not only from the standpoint of the musicians not being able to hear, but also from the fact that many of them are of young age (for review see Hash, 2003). Deaf and hearing impaired students in today's schools continue to be involved in instrumental music. Instructional methods were modified to teach playing technique and pitch recognition through visual cues and tactile stimulation. Robbins and Robbins (1980), for example, effectively taught instrumental music to hearing impaired students at the New York State School for the Deaf. They suggested that the requirements for learning an instrument are similar for both hearing and hearing impaired children. In terms of specific musical abilities, deaf students are capable of maintaining a steady beat sometimes better than hearing students. With regard to pitch discrimination, these students are more successful at discriminating lower frequencies.

The primary focus of the education in deaf or hard of hearing individuals, and one of the most prominent benefits of music instruction, is the support for the acquisition of language. Properties of music such as rhythm, accent, tempo, and repetition are also present in spoken language, and therefore may be used to facilitate language acquisition. In addition, music can provide profound motivation for positive behaviour, and can become a means for developing positive self-image. Children can also improve body coordination through rhythmic movement, and develop social skills by interacting with hearing students during music participation. Furthermore, many deaf students enjoy music simply for its own aesthetic value.

Instrument selection is a key factor in determining the success of deaf or hard of hearing musicians. Zinar (1987) recommended the harp and guitar, the harp because the strings are close to the ear and the guitar because it is held close to the body, allowing vibrations to be felt. Another suitable instrument is the electric bass since it produces a significant amount of vibration and can easily be incorporated into both concert and jazz bands. Other stringed instruments such as the violin and cello should not be considered, as the needs of precise intonation may be too difficult for deaf students to master. In contrast, woodwind instruments present possibilities. Hearing disabled individuals have successfully learned the clarinet and saxophone, both utilizing one note per finger and thus allowing to map distinct movements more easily to pitches. Furthermore, these instruments possess good resonance and a large frequency range (Zinar, 1987). Larger versions of these instruments, such as the bass clarinet or tenor saxophone, should also be considered, as the lower frequencies they produce may be easier for some to hear.

As with some string instruments, brass instruments may not be suitable for students with hearing loss because of the pitch discrimination required to discern the partials. According to Atterbury (1990), deaf trumpet players can successfully discriminate pitches throughout the overtone series by holding the fingers onto the bell and feeling for differences in vibrations. This technique can also be applied to the euphonium and tuba, which might be easier for some, as vibration can be felt by wrapping both arms around the instruments. The slide trombone and French horn should be avoided, since these require fine grained pitch discrimination.

Provided they learn to feel musical pulse through body vibrations, percussion instruments can also be played by deaf musicians. The teacher, when introducing this concept, should play steady beats on a low-pitched drum as the student touches the instrument. The child can then attempt to count the beats out loud, or indicate the pulse using the other hand. Once a feeling of pulse is established, the exercise should be repeated as the student attempts to feel vibrations indirectly through the table, stand, or floor where the drum is placed. Finally, the student can attempt to sense vibrations by simply standing near the drum.

Deaf individuals who choose to participate in the band or orchestra will require a great deal of support from their director, peers, and parents if they are to be successful. Finding role models within the community and from the music world as a whole may serve as inspiration and motivation for the deaf individual. Evelyn Glennie, perhaps the most active percussion soloist today, is one musician that could serve in this capacity. Although profoundly deaf, Glennie continues to be recognized for her musicianship rather than her disability.

As Hash states in his conclusion, "including deaf or hard [of] hearing students in instrumental organizations can be a highly rewarding and valuable experience, not only for children with impaired hearing, but also for the teacher and others in the ensemble. Through modification of the environment and teaching practices, many of these students can achieve at the same level as their hearing counterparts. Their presence will not only increase the potential of the ensemble as a whole, but may also lead to greater understanding between the deaf or hard [of] hearing community and the hearing population" (Hash 2003).

In the last two decades, the development and constant improvement of implantable hearing aids, termed cochlear implants (CI) have revolutionized the treatment of deaf and hearing impaired individuals. By means of this device, many children now gain excellent language abilities. In general, the benefits of CI-devices with respect to language abilities are the more pronounced the earlier the implantation is undertaken and sensory input is given to the auditory nerve. This is related to the above-mentioned sensitive periods of the auditory system, since brain networks of the auditory pathway are more easily connected during the first two years of life compared to later childhood. These early wiring effects explain why individuals who are deafened after language acquisition-- that is, who had the opportunity to establish an appropriate auditory network in the early childhood--show much better results after implantation compared to deaf born children. CI-implantation today demonstrates powerful brain plasticity: After implantation, the activation patterns of the brain reorganize dynamically in a couple of weeks. Interestingly, the movement-sensitive areas of the visual cortex are predominantly involved, suggesting that successful learning of auditory language after CI implantation results in a refined visual motion perception of mouth movements supporting auditory learning via a synaesthetic experience (Kang et al., 2004).

Only a few studies have focussed on music perception and the ability to decode affective prosody in CI-users. In summary, it seems that CI users rarely enjoy listening to music, probably due to the fact that many acoustic properties of music are not adequately transmitted by the implants (Gfeller et al., 2000). Recent data suggest that particular structural elements of music are differentially accessible to CI users. Whereas rhythm perception is close to normal, CI users have a poor resolution of pitch constituting musical melodies. In a study of music enjoyment, Gfeller et al. (2003) reported that "real-world" music sounds less enjoyable and more difficult to understand to cochlear implant recipients than it does to normal hearing adults. However, there was considerable variability among implant recipients. In a subsequent study, Gfeller et al. (2005) systematically examined the sorts of complex combinations of pitch, harmony, timbre, and rhythm typically heard in music. Unsurprisingly, cochlear implant recipients were less accurate in their recognition of previously familiar (known before hearing loss) musical excerpts than normal-hearing adults, but the frequency of music listening was positively related to an improvement of music perception, suggesting positive effects of eartraining. However, it is difficult to draw firm conclusions from these data, because those CI-users who were able to enjoy music focussed more on the music and listened to it more frequently. In summary, to date the effects of specific ear training on music enjoyment in CI-users requires more systematic exploration. This is perhaps surprising, since it is known that musical training positively affects cognitive processing of prosody (Schellenberg, 2006) and verbal memory, at least in tonal languages (Chan et al., 1998).

Musical Learning in Individuals with Localized Brain Lesions

Many brain disorders have an impact on music learning. In this paragraph I focus only on those small lesions that exclusively affect the ability to perceive and produce music. These comprise the amusias, and as a special case of defective wiring in the motor areas also a disorder termed focal dystonia. The latter is characterized by a loss of motor control while playing a musical instrument.

Receptive and Expressive Amusia

In contrast to deaf and hearing impaired individuals, who are impaired in perceiving any acoustic signal due to damage to the sensory organs or the auditory nerve, individuals suffering from central hearing disorders are unable to integrate acoustic information into an acoustic "gestalt" and to extract meaning from acoustic material. These disorders usually are not accompanied by abnormal hearing thresholds or any disturbance of the sensory organs and can easily be missed when not specifically assessed by clinical tests. Most frequently, they are caused by damage to certain regions of the cerebral cortex, predominantly the temporal lobes. When the auditory deficit is dominated by difficulties in understanding spoken language, it is called "receptive aphasia," which is usually due to damage to the posterior part of the left temporal lobe, known as Wernicke's region. In the rare cases of deterioration of musical faculties the central hearing disorder is referred to as "amusia."

Amusia is defined as a disorder in the processing of musical material due to brain damage. There are three different forms of amusia: 1. *receptive amusia,* a disorder of the perception of musical material, 2. *expressive amusia,* characterized by a disorder of musical expressive skills, such as singing or clapping hands to a rhythm, and 3. *congenital amusia* in which individuals are thought to be born with deficits in musical processing. This term, coined by Isabelle Peretz, infers that brain damage or brain anomaly is present at birth (Peretz et al., 2002).

To exemplify, I will first describe one of the rare cases of amusia: Jeff is a boy of fifteen years of age who at age 13 underwent cardiac surgery with the implantation of artificial heart valves, due to a congenital heart defect. Jeff was obliged to take anticoagulants in order to prevent embolic strokes originating from the clotting of blood platelets at the surface of the artificial valve. Jeff was an avid violin player. He regularly spent the first two weeks of his summer holidays in a music camp. From there, he was referred to our clinic after he had noticed a sudden difficulty in distinguishing the timbre of different instruments. He described a fundamental change in his music perception in the absence of any language deficits. The sound of the orchestra appeared to be "flat." He was no longer able to distinguish different timbres of musical instruments, a task that he had easily managed prior to the incident. Additionally, music had lost its aesthetic, emotional quality for him and, as a consequence, he was deeply depressed. The MRI of the brain obtained one week after the incident shows a lesion located in the anterior and superior right temporal lobe (*Figure 2)*.

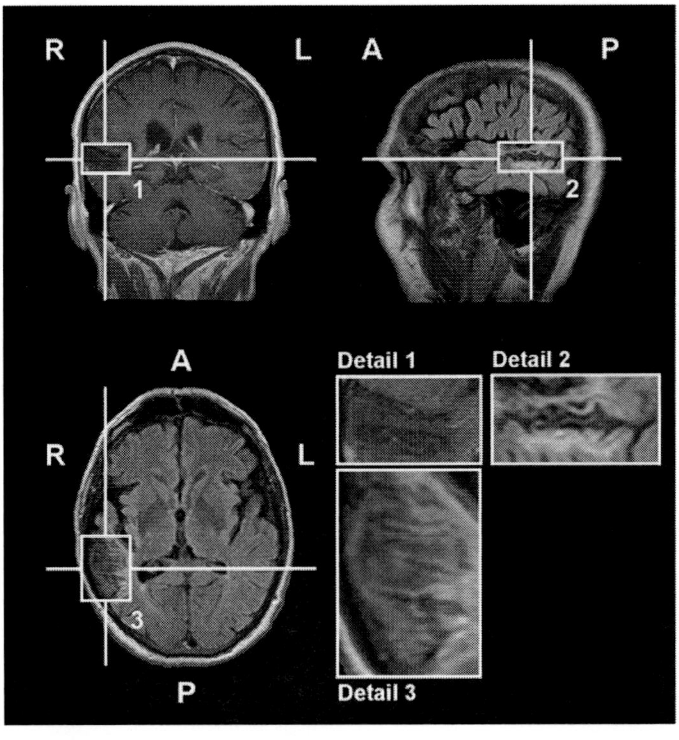

Figure 2. MRI slides of Jeff, obtained one week after his embolic stroke. The small lesion in the right anterior part of temporal lobe is clearly seen in the magnification insets (By courtesy of Prof. Thomas Muente, Hannover).

The lesion had been caused by an embolic stroke which most probably had been caused by the fact that Jeff had forgotten to take his anticoagulant for two days. In neurological and neuropsychological testing, the only deficit Jeff showed was his inability to distinguish the timbre of musical instruments. No other hearing impairment could be detected. When Jeff was re-examined one year later, it was found that his ability to distinguish the timbre of musical instruments had fully recovered and he was able to enjoy music as much as he had prior to the embolic stroke.

This rare case of *receptive amusia* is informative in two respects: First, it demonstrates that different musical abilities can be dissociated and rely on distinct neuronal networks. In the case of Jeff, melody or rhythm perception was not impaired; however, his sense of timbre was deeply affected. Second, it also demonstrates that receptive amusia frequently is only a temporary symptom, suggesting that intact brain regions compensate after a while for the function of the destroyed nerve tissue. Permanent loss of receptive musical abilities is usually only found in bilateral temporal lesions.

In children, receptive amusia after brain lesions is extremely rare and only occasionally reported in the literature. In adults, however, about two thirds of patients with ischemic strokes show signs of receptive amusia when tested with an adequate test battery (Schuppert et al., 2000). Such a widely applied test has been developed by Isabelle Peretz (Peretz et al., 2003). It tests rhythm-, meter-, interval-, melody-, and contour-perception and the memory for simple tunes. Although the miscellaneous patterns of receptive amusia are caused by extremely variable brain lesions, as a rule of thumb, local, analytic strategies, such as listening in an interval-based manner or focussing on simple rhythmic patterns are more frequently processed in the left hemisphere, whereas global structures, such as contour and meter seem to be predominantly processed in the right hemisphere. However, these results can only be obtained when testing is performed shortly after the occurrence of a brain lesion. It has been demonstrated that more than 70% of patients had recovered from the deficits about one year after the stroke (Schuppert et al., 2003).

Expressive amusia is a term used to describe a disturbance of music production, more specifically a reduced ability to reproduce a given rhythm or melody previously heard. Whereas this deficit has been described anecdotally for some outstanding composers such as Maurice Ravel (Amaducci et al., 2002), to date it has not been systematically investigated. In a preliminary study, we developed a test for expressive musical faculties, requiring replay of previously presented melodies, intervals, and rhythms. Patients and controls had to play on a glockenspiel or were asked to tap short rhythmic trails. Performances were recorded and rated by two independent experts. Patients performed significantly worse in the expressive than in the receptive tests, demonstrating that expressive faculties are more sensitive to brain damage than receptive (Steinwede et al., 2000).

Congenital Amusia

Congenital Amusia is a term that was coined by Isabelle Peretz (Peretz et al., 2002) to refer to individuals born with deficits in music processing. The condition was previously labelled "tone deafness." These children and adults have severe difficulties in pitch perception and are, for example, unable to adjust their own pitch while singing in a choir. In the most severe cases they are unable to recognize familiar tunes, such as "twinkle, twinkle little star" and are completely unaware when singing out of tune. When tested with the Montreal Battery for the Evaluation of Amusia, predominantly a severe deficit in the sequential processing of pitch information is found (Ayotte et al, 2002). However, rhythm processing may also be affected, albeit usually to a lesser degree (Peretz et al., 2003). It seems that there are different types of congenital amusia, since variable patterns of deficits are found. Some children are relatively good at interval processing, for example they can distinguish a fifth and a fourth; however, they are completely unable to recognize a melody played previously to them. Other children have better preserved musical memory, but are not able to tap in time with a given meter. Interestingly, congenital amusics have no obvious deficit in processing other complex auditory material, such as language or environmental sounds. In terms of language abilities, it is currently under discussion whether or not a subgroup of congenital amusics may display subtle deficits in processing sentence melody or prosody, a feature that is important for decoding the affective content of a spoken phrase. According to epidemiological data, congenital amusia is not a rare disorder, affecting about 4 % of the population (Kalmus & Fry, 1980).

With respect to the neural correlates of this condition, little information is presently available. With the method of voxel-based morphometry (VBM) it is possible to precisely measure the size of specific brain regions using conventional MRI. In adult congenital amusics, it has been found that a region in the inferior part of the right frontal lobe was smaller relative to musically intact controls of the same age, an effect due to reduced white matter, reflecting a diminished density of nerve fibres (Hyde et al., 2006). It is known from brain activation studies in healthy subjects that this region is crucial for pitch processing, especially when pitch memory is involved (Zatorre et al., 1994). The findings suggest that congenital amusics have fewer connections within this pitch module and with other parts of the brain, explaining, for example, their difficulties in recalling songs that usually are remembered not only by melodies but also by words.

There is still some controversy over whether or not congenital amusia is really "congenital" or due to a lack of adequate musical stimulation in the prenatal and early infancy periods. There are three arguments in favour of a genetic component. First, researchers have identified families with amusia in several family members, even though all grew up in a musical environment and had received music instruction and even formal piano training. Second, amusia occurs more frequently in identical twins than in fraternal twins. Third, it is known from genetic studies on absolute pitch processing that pitch processing abilities that rely on the intact function of the inferior right frontal cortex have a strong a genetic element (Drayna et al., 2001). Thus, the results reported above fit with the idea that reduced neuronal connectivity within this brain region may be related to a genetic component that would contribute to problems in the musical encoding of pitch. However, many questions still remain open. For example, the data do not explain the deficits of amusics in rhythm processing. Furthermore, no amusic relatives can be found for the majority of children with amusia, yet in families with several amusics the gene has not been identified.

With respect to the remediation of congenital amusia, systematic studies are still lacking. According to anecdotal reports, attempts to improve the situation by ear-training don't seem to be effective. Unfortunately, amusic children are not usually encouraged to pursue musical activities and are even excluded from choir singing etc. which contributes to an augmentation of the musical deficits in a vicious circle. It remains a challenge for music educators to develop specific training programs for these children and evaluate their efficacy in long-term follow up studies.

Musicians' Cramp as a Central Nervous Disorder Affecting Musical Performance

Not only music perception, but also specific movement patterns involved in playing a musical instrument can be disturbed as a consequence of diseases of the central nervous system. Evidently, any central nervous system disorder affecting the control of motor function has an impact on playing a musical instrument. Cerebral palsy, for example, is a condition that is usually caused by a lack of oxygen in the infant's motor centres during prolonged delivery. Later in life it is accompanied by an increase in muscle tone and muscular weakness, more frequently in the lower extremities than in the upper limbs. It leads to slow and awkward "spastic" movements, which render the execution of fast passages in music extremely difficult. However, these conditions do not specifically impair music performance and affect fine motor control in daily life in a similar way.

There is one condition that exclusively affects the execution of movements while playing a musical instrument. It is characterised by a loss of voluntary control of these extensively trained, refined and complex sensory-motor skills and is generally referred to as *musicians' cramp* or *focal dystonia.* Usually it occurs in adult professional musicians; however, occasionally it can be diagnosed in highly talented adolescents who spend many hours a day at their instrument. The youngest dystonic patient we have seen was a 16-year-old enthusiastic piano player whose case will be presented here in more detail.

Michael had started piano playing at the age of 5 and was considered to be highly talented. During his first five years he practiced the piano between one and two hours per day, usually supervised by his mother, a professional pianist. At age ten he performed Mozart's piano concerto in d-minor in public and was admitted to a special school for highly gifted musical children. He increased his practice schedule to 5 hours daily and won several first prizes in national and international competitions. He performed in public regularly and gave recitals that included outstandingly difficult works of the romantic period. He also started to compose, and he entered the master class of a renowned German composer. At the first appointment, he reported that during the preparation of a solo recital he had noted subtle difficulties in the regularity of his scale playing with the right hand. Furthermore, he felt as if the right middle finger was sticking on the keys. Trills including the middle finger were slow and irregular, so he had changed fingerings and tried to omit the finger from these tasks. Increased practice had not resulted in an improvement. To the contrary, the condition had grown worse during the last two months. He had noticed that some pieces based on chord-technique, such as Brahms *Intermezzi* and in particular the Schumann *Toccata* op. 7 did not pose any problem. There was no history of pain and he did not feel any difficulty when typing on a computer keyboard or performing other fine motor tasks requiring the control of the right middle finger. When examined at the instrument, his finger position during scale playing showed an involuntary flexion of the middle finger as depicted in Figure 3.

Figure 3. Hand position of Michael, when playing the Schumann Toccata. The involuntary flexion of the right middle finger is clearly visible, a sign of a mild form of hand dystonia.

Michael was diagnosed as suffering from a beginning focal dystonia, in his case exhibiting the same pattern as the first known patient, Robert Schumann (Altenmueller, 2006). Schumann described in his diary the involuntary flexion of the right middle finger and designed a device to stretch the finger during piano playing. He furthermore composed the technically extremely demanding *Toccata*, which in most parts allows the omission of the middle finger of the right hand.

Approximately one in 100 professional musicians suffers from focal dystonia. Subtle loss of control in fast passages, finger curling, lack of precision in forked fingerings in woodwind players, irregularity of trills, sticking fingers on the keys, involuntary flexion of the bowing thumb in strings and impairment of control of the embouchure in woodwind and brass players in certain registers are the various symptoms that can mark the beginning of the disorder. At this early stage, most musicians believe that the reduced precision of their movements is due to a technical problem. As a consequence, they intensify their efforts, but this often only exacerbates the problem, as was the case with Michael. Males, classical musicians of a younger age and instrumentalists such as guitarists, pianists and woodwind players are among the most commonly affected by focal dystonia. The majority of patients have solo positions and often they have a perfectionist, control-type personality. About 20% of such patients report a history of chronic pain syndromes or overuse injury. Preventing these musicians from developing chronic overuse and tendinitis will most probably prevent them from developing focal dystonia (Jabusch & Altenmueller, 2006). However, once focal dystonia is established, the cure of the pain syndrome will generally not eliminate the pathological movement pattern.

Although the neurobiological origins of this disorder are not completely clarified, the link between chronic pain and overuse suggests that focal dystonia, as a cortical sensory-motor mislearning syndrome, may be due to abnormal brain plasticity. A study with trained monkeys (Byl et al., 1996) demonstrated that chronic overuse and repetitive strain injury in highly stereotyped movements can actively degrade the cortical representation of the somatosensory information that guides the fine motor hand movements in primates. A similar degradation of sensory feedback information and concurrent fusion of the digital representations in the somatosensory cortex was confirmed in a magnetoencephalography (MEG) study conducted in musicians with focal dystonia, although these musicians had no history of chronic pain (Elbert et al., 1998). Therefore, additional factors such as a genetic predisposition appear to play an important role in the development of focal dystonia (Schmid et al., 2006).

Unfortunately, there is no simple cure for the condition. Retraining may be successful in a minority of cases, but usually requires several years to succeed. Symptomatic treatment with temporary weakening of the cramping muscles by injecting Botulinum-toxin has proven to be helpful in other cases; however, since the injections need to be applied regularly every three to five months during the professional career, it presents no solution for young patients. Thus, the challenge for music educators is to prevent young musicians from such a disorder. Reasonable practice schedules, economic technique, prevention of overuse and pain, mental practice, avoidance of exaggerated perfectionism and psychological support with respect to self confidence are the components of such a prevention program. Michael has studied medicine and is now a renowned professor in a university hospital, continuing to compose in his free time.

Developmental Disabilities

Attention Deficit Hyperactivity Disorder

Attention deficit hyperactivity disorder (ADHD) is a diagnosis given to children and adults who display certain behaviours over an extended period of time. The most common of these behavioural criteria are inattention, hyperactivity, and marked impulsiveness. There are several other more-or-less synonymously used terms, such as attention deficit disorder (ADD), attention deficit disorder with and without hyperactivity, hyperkinetic impulse disorder, or hyperactive syndrome.

The term attention deficit is inexact, since the disorder is not thought to involve a lack of attention. Rather, there appears to be difficulty in regulating attention, so that attention is simultaneously given to too many stimuli. The result is an unfocused reaction to the world. In addition, individuals with ADHD can have difficulty in disregarding stimuli that are not relevant to the present task. They can also pay so much attention to one stimulus that they cannot absorb another stimulus that is more relevant at that particular time. For many people with ADHD, life is a never-ending shift from one activity to another. Focus cannot be kept on any one topic long enough for a detailed assessment. The constant processing of information can also be distracting, making it difficult for an individual with ADHD to direct his or her attention to someone who is speaking. Personally, this struggle for focus can cause great chaos, which can disrupt or diminish self-esteem.

The neurological manifestations of ADHD are disturbances of what are known as executive functions. Specifically, the six executive functions that are affected include

- the ability to organize thinking,
- the ability to shift thought patterns,
- short-term memory,
- the ability to distinguish between emotional and logical responses,
- the ability to make a reasoned decision,
- the ability to set a goal and plan how to approach that goal.

About half of adults suffering from ADHD additionally meet the criteria for other psychiatric diseases, such as depression, anxiety disorder, substance abuse or antisocial behaviour. ADHD is a common childhood disorder. It is estimated to affect 3–10% of all children, representing in the United States for example up to two million children. Boys are affected about twice as often as girls.

The cause of ADHD is unknown. However, evidence is consistent with a biological cause rather than an environmental cause (for example, home life). The biological roots of ADHD may involve certain areas of the brain, specifically the frontal lobe. One explanation is that the executive functions are controlled by the frontal lobes of the brain. Functional magnetic resonance imaging (fMRI) examination of subjects who are exposed to a sensory cue has identified decreased activity of regions of the brain that are involved in tasks that require attention (Sowell et al., 2003). Another study has documented that the brains of children and adolescents with ADHD are 3–4% smaller than those of their ADHD-free counterparts. Additionally, the decreased brain size is not due to the use of drugs in ADHD treatment, the researchers have concluded (Castellanos et al., 2002).

ADHD symptoms can sometimes be relieved by the use of stimulants that increase the neurotransmitter dopamine, such as ritalin (methylphenidate). Too little dopamine can produce decreased motivation and alertness. These observations have led to the popular "dopamine hypothesis" for ADHD, which proposes that ADHD results from the inadequate supply of dopamine in the central nervous system. The observation that ADHD runs in families (10–35% of children with ADHD have a direct relative with the disorder) points to an underlying genetic origin.

Besides pharmacological intervention, which has proven to be useful in about 40% to 60% of children with ADHD, behavioural treatment is extremely important. Assistance can take the form of special education in the case of those who prove too hyperactive to function in a normal classroom. The child may be seated in a quieter area of the class. A system of rules and rewards for appropriate behaviour may also be helpful. Children and adults can also learn strategies to maximize concentration (such as list making) and strategies to monitor and control their behaviour.

Musical activity in ADHD children has been proven to be useful, although only a few studies addressing the effects of music education exist (Jackson, 2003). There are several mechanisms that might contribute to the positive effects found. The ability to learn and to pay attention is dependent on our ability to integrate and organize our sensory experiences. Thus, musical activity, linking auditory, visual, proprioceptive, motor and emotional experience may constitute a powerful tool to support multi-sensory and behavioural integration. Furthermore, playing a musical instrument can lead to increases in self-esteem and provides a means to excel in a specific area. This is particularly important, since children with ADHD also need emotional and social support. Because they can be "challenging" to both adults and other children, they may lack positive experiences with accomplishment and friendships.

Learning an instrument may be one of the best ways for a child with ADHD to develop a more rhythmic and balanced sense of self. This was confirmed in a recent study testing the impact of both instructional and improvisational music therapy on motor impulsivity in ADHD adolescent boys (Rickson, 2006). In comparison to controls, the groups experiencing music therapy showed slightly reduced impulsive and restless behaviours in the classroom. Furthermore, the music therapy treatment groups significantly improved their accuracy in a rhythmic tapping test, thus documenting the positive effect on sensory-motor integration. These findings are the first hints that music therapy may contribute to a reduction in a range of ADHD symptoms in the classroom.

Dyslexia

Developmental dyslexia is defined as a specific learning disability in the domain of literacy. Children with dyslexia can experience difficulties with the alphabet, reading, writing and spelling in spite of normal or superior intelligence, motivation and schooling. Dyslexia is not rare, since it affects between 5 and 15% of children. The core deficit in dyslexia seems to be a weakness in phonological processing (Liberman & Shankweiler, 1991). However, a wide range of perceptual, cognitive and motor deficits has been found to be associated with dyslexia. For example, there is evidence that difficulties with phonology are the result of a more general problem with auditory processing. In particular, the temporal aspects of auditory decoding have been found to present a specific problem area, both at the rapid temporal processing level and at the rhythm processing level (Tallal et al., 1993; Overy, 2003). Interestingly, this deficit seems not to be restricted to the auditory modality, since timing difficulties in the visual domain (Stein & Walsh, 1997), cognitive domain (Wolf & Bowers, 2000) and motor domain (Wolff, 2002) have also been found, suggesting that general timing skills may be a key underlying difficulty area in dyslexia.

Some researchers have suggested that dyslexia results from a deficit in cerebellar function (Nicolson & Fawcett, 2006). It has also been shown that dyslexic children and adults show less left-hemispheric activation and more right-hemispheric activation for various reading, phonological and auditory processing tasks. An early study by Galaburda and Kemper (1979) found that the language center in a dyslexic brain showed microscopic flaws known as ectopias and microgyria. Both affect the normal six-layer structure of the cortex. An ectopia is a collection of neurons that have pushed up from the lower layers of the cortex into the outermost one. A microgyrus is an area of cortex that includes only four layers instead of six. These flaws affect connectivity and functionality of the cortex in critical areas related to auditory processing and visual processing, which is consistent with the hypothesis that dyslexia stems from a phonological awareness deficit. This has been supported by studies using event related potentials, for example, recently it has been demonstrated that dyslexic children exhibit a delayed left frontal processing in tasks requiring rapid processing of phonemes and phrase structures (Sabisch et al., 2006).

With respect to musical skills, the most apparent deficits in dyslexic children are their problems with learning musical notation, sight-reading and with rhythmic motor skills. To date, no difficulties have been found with musical pitch skills, further indicating that timing may be a particular problem for dyslexic children. It has been suggested that conducting musical training with dyslexic children might lead not only to the improvement of musical timing skills but also to improvements in impaired language functions. A classroom-based musical training program was recently created for this purpose, involving singing games and with a strong emphasis on rhythm and timing skills (Overy, 2003). In a small-scale intervention study, a 15 week training period revealed significant positive effects in rhythm copying, rapid auditory processing, phonological ability and spelling ability, suggesting that the specific training of timing skills might support language capacities. However, further studies, involving longer periods of musical training will be necessary to identify whether such training can lead to long-term improvement in reading ability.

3. Autism and Autism Spectrum Disorder

Autism is a developmental disability that is generally evident before age three, affecting verbal and non-verbal communication and social interaction. About 4 of every 10,000 children are autistic, and 2 per 10,000 have some form of pervasive developmental disorder (PDD). PDD means that some, but not all, symptoms of autism are present. The terminology has even become more complicated, because health care providers also refer to autism spectrum disorders (ASD) which include Autistic disorder, Asperger syndrome, and Pervasive Developmental Disorder. All of these conditions are characterized by varying degrees of deficiencies in communication skills and social interactions, along with restricted, repetitive, and stereotyped patterns of behavior. Autism spectrum disorder ranges in severity from mild, in which the autistic person can live independently, to severe, in which the patient requires social support and medical supervision throughout his or her life. A small proportion of autistic adults, usually those with high-functioning autism or Asperger's syndrome, are able to attain higher professional positions, frequently integrating their specific skills (e.g. mathematics) into their professional activity.

Autism affects boys four times more often than girls. The symptoms vary greatly but follow a general pattern. Autistic infants may act relatively normal during their first few months of life before becoming less responsive to their parents and other stimuli. They may not smile in recognition of their parents' faces, and may put up resistance to being cuddled. As they enter toddlerhood, it becomes increasingly apparent that these children have a world of their own. They do not play with other children or toys in the normal manner; rather they remain aloof and prefer to play alone. Verbal and nonverbal communication skills, such as speech and facial expressions, develop peculiarly. Symptoms range from mutism to prolonged use of echoing or stilted language. When language is present, it is often concrete, unimaginative, and immature. Another symptom of autism is an extreme resistance to change of any kind. Autistic children tend to want to maintain established behaviour patterns and a set environment. They develop rituals in play, oppose change (such as moving furniture), and may become obsessed with one particular topic.

The causes of autism are areas of debate and controversy; there is currently no consensus, and researchers are studying a wide range of possible genetic and environmental causes. Since autistic individuals are all somewhat different from one another, there are likely multiple causes that interact with each other in subtle and complex ways, thus producing slightly differing outcomes in each individual. A possible explanation for the characteristics of the syndrome is a variation in the way the brain itself reacts to sensory input and how parts of the brain then handle the information. In an electroencephalographic (EEG) study it was found that adults with autism show differences in the manner in which neural activity is coordinated (Rippon et al., 2007). The implication seems to be that there is poor internal communication between different areas of the brain. Specifically, the study indicated that there were abnormal patterns in the way the neurons were connected in the temporal lobe, a brain region which is crucial for language and emotional processing. Another theory features the mirror neurons (Williams et al., 2006) of the brain. It states that autism may involve a dysfunction of specialized neurons in the brain that should activate when observing other people. In typically-developing people, these mirror neurons are thought to perhaps play a major part in social learning and general comprehension of the actions of others. Schultz (2005) has proposed that a network connecting the emotional centers of the brain, mainly the amygdala, and the fusiform gyrus (a small region in the temporal lobe which is involved in emotional processing of faces) and the orbito-frontal cortex, is impaired in individuals with autism.

Genetic influence comprises a significant aspect of research in the causes of autism. Originally hinting toward this was the observation that there is about a 60% concordance rate for autism in identical twins, while fraternal twins and other siblings only exhibit about 4% concordance rates.

With respect to treatment, appropriate early intervention is important. Once the diagnosis has been made, the parents, physicians, and specialists should discuss what is best for the child. In most cases, parents are encouraged to take care of the child at home. Special education classes are available for autistic children. Structured, behaviourally-based programs have shown some promise.

Most behavioural treatment programs include

- clear instructions to the child,
- prompting to perform specific behaviours,
- immediate praise and rewards for performing those behaviours,
- a gradual increase in the complexity of reinforced behaviours, and
- definite distinctions of when and when not to perform the learned behaviours.

Parents should be educated in behavioural techniques so they can participate in all aspects of the child's care and treatment. The more specialized instruction and behaviour therapy the child receives, the more likely it is that the condition will improve.

Musical activity has proven to be particularly beneficial in children with ASD. Many of them apparently have preserved emotional awareness in musical domains and show a particular affinity to music. There are numerous cases of autistic "musical savants" (see below) and more than one report of enhanced pitch sensitivity in individuals with autism

(Heaton, 2003). Furthermore, music therapy has been found to be particularly useful for autistic children. It addresses the typical characteristics of autism listed above in the following ways:

- Music is considered a "universal language" which provides bridges in a non-threatening setting between people and/or between individuals and their environment, facilitating relationships, learning, self-expression, and communication.
- Music captures and helps maintain attention. It is highly motivating and engaging and may be used as a natural "re-inforcer" for desired responses. Music therapy can stimulate ASD children to reduce negative and/or self-stimulatory responses and increase participation in more appropriate and socially acceptable ways.
- Music therapy can enable those without language to communicate, participate in group interactions and express themselves non-verbally. Music therapy also very often assists in the development of verbal communication, speech, and language skills. The interpersonal timing and reciprocity in shared play, turn-taking, listening and responding to another person are augmented in music therapy with autistic children and adults to accommodate and address their styles of communication.
- Music therapy allows individuals with diagnoses on the autism spectrum the opportunity to develop an appropriate expression of their emotions.
- Music provides concrete, multi-sensory stimulation (auditory, visual, and tactile). The rhythmic component of music helps organizing the sensory systems of individuals diagnosed with autism. As a result, auditory processing and other sensory-motor, perceptual/motor, gross and fine motor skills can be enhanced through music therapy.
- Musical elements and structures provide a sense of security and familiarity in the music therapy setting, encouraging clients to attempt new tasks within this predictable, but malleable, framework.

Usually, music therapists work individually or in small groups, using a variety of both, music and techniques to engage children and adults with diagnoses on the autism spectrum. They involve clients in singing, listening, moving, playing instruments, and creative activities in a systematic, prescribed manner to influence change in targeted responses or behaviours and help clients meet individual goals and objectives. They create a musical, familiar environment that encourages positive interpersonal interaction and allows clients freedom to explore and express themselves. They utilize music that is preferred by and reinforcing to clients and is appropriate for ages, cultures, and environments in which the clients interact. It has been suggested that participating in music therapy allows individuals with autism to experience and explore their emotions on a far deeper level than would otherwise be possible.

With respect to the efficacy of music therapy, several studies have demonstrated that improved verbal and gestural communicative skills are outcomes of music therapy with autistic children, regardless of the type of therapy (Whipple, 2004; Gold et al., 2006).

Savants

Savant syndrome is a rare but extraordinary condition in which individuals with serious intellectual and/or emotional disabilities, including autistic disorder, have some "island of genius" that stands in marked, incongruous contrast to the overall intellectual condition. A more extensive paragraph on musical savants is found in Hassler and Miller (chapter 8). In the following section, a case report shall be briefly discussed and the relation of savants to developmental disorders will be focused on.

It is estimated that as many as one in 10 autistic persons have such remarkable abilities in varying degrees. Savant syndrome also occurs in other developmental disabilities, brain injuries, or other brain diseases. The 1989 movie *Rain Man* made the "autistic savant" familiar to a broad public. In the following we briefly describe the case of Peter, a 16 year old boy who was seen in our clinic a couple of years ago. He came with his parents, who wanted to know how his musical talent could be further developed. Peter was diagnosed as having a developmental disorder at age 5 when he was still unable to talk or move in a coordinated manner. He was very social and had good emotional relations with his parents and siblings, keeping eye-contact, smiling in response and playing with other children. Therefore, an autistic spectrum disorder could be excluded. Already at age 5 he showed a strong inclination towards music. When his father played the piano, he used to replay the themes afterwards. He rapidly developed this skill. When he was admitted to a specialized school for intellectually disabled children, he was unable to attain literacy or to learn how to read music. He furthermore could not learn basic mathematical operations; remarkably, he memorized thousands of historical dates relevant to the biography of classical composers. For example, he could recite the year that Chopin left Poland, or when Schumann met Brahms for the first time. During subsequent years he developed his amazing skill to replay piano music from CD-recordings. He took piano and organ lessons, which mainly consisted of the teacher demonstrating the repertoire, which Peter then played immediately without any false notes. Peter possessed perfect pitch and his memory was immense, comprising seven hours of piano repertoire and four hours of organ repertoire. When he visited us, he performed the Beethoven *Waldstein-Sonata*, all four Chopin *Ballades* and an intricate *Fugue* by Shostakovich in a professional manner. Peter is currently a free lance church musician, playing during services and funerals. A couple of months ago, he performed the Bach flute sonatas, having memorized the accompaniment from a CD-recording.

Savant skills typically occur in an intriguingly narrow range of special abilities. They are usually found in 5 general categories:

- Music: usually performance, most often piano, with perfect pitch; composing in the absence of performing has been reported, as has playing multiple instruments (as many as 20).
- Art: usually drawing, painting, or sculpting.
- Calendar calculating.
- Mathematics: including the ability to compute prime numbers, for example, in the absence of other simple arithmetic abilities.

- Mechanical or spatial skills: including the capacity to measure distances precisely without benefit of instruments, or the ability to construct complex models.

Interestingly, all these skills are commonly considered as predominantly right hemispheric abilities and they are all linked to a prodigious memory. This observation supports theories concerning the causes of the savant condition. Geschwind and Galaburda (1987) noted that the savant syndrome is almost exclusively found in male individuals. In explaining this finding, they point out that the left hemisphere normally completes its development later than the right hemisphere and is thus subjected to prenatal influences, some of which can be detrimental, for a longer period of time. In the male foetus in particular, circulating testosterone, which can reach very high levels, can slow growth and impair neuronal function in the more vulnerable exposed left hemisphere, with actual enlargement and shift of dominance favouring skills associated with the right hemisphere. A "pathology of superiority" was postulated, with compensatory growth in the right brain as a result of impaired development or actual injury to the left brain. It should be noted that the savant syndrome can be congenital, or it can be acquired following brain injury or disease later in infancy, childhood, or adult life especially when damage to the left hemisphere occurs. Recent reports of savant-type abilities emerging in previously healthy elderly persons with fronto-temporal dementia are particularly intriguing (see chapter 8; Miller et al., 1998). In recent years, several neuropsychological theories have also directly addressed the abundant reports of savant skills in the autistic population. Weak central coherence theory (WCC) cites a particular cognitive and perception style - focusing on details rather than the whole - as being present in persons with autism, and postulates that such a style of information processing could be an important aspect of those persons with savant abilities. Not being distracted by more global patterns, the savant can focus on a single item or skill and perfect it (Frith & Happe, 1994).

There is emerging evidence that prodigies and savants may share certain underlying mental processes when carrying out their specialized, expert tasks. Event-related potentials (ERPs) can measure very early components of brain activity reflecting initial, "pre-conscious" stages of mental processing. This fast, low-level preconscious mental activity contrasts sharply with that seen when higher level "executive" functions are accessed during typical information processing. Birbaumer (1999) compared ERPs of a "human calculator" - a non-autistic arithmetic whiz - to healthy controls who were the same age and IQ. The expert calculator showed evidence of enhanced automatic low-level processing early on in the calculating processes compared to controls.

Teaching a musical savant child is particularly challenging for any music educator. As in the case of Peter, the teacher has to act in a supportive and adaptive way, relying on her or his sensitivity to the specific needs of the individual. In this specific case, the willingness and ability of the teacher to carefully and repetitively demonstrate musical pieces of increasing complexity was the prerequisite of Peter's accomplishments.

CHROMOSOMAL DISORDERS

1. Down Syndrome

Down Syndrome is a developmental disability due to an abnormal chromosome number or structure. It is characterized by physical and behavioural features and has been considered the most common form of genetic aberration. Incidence among newborns is estimated at 3 in 1000, and in the general population at approximately 1 in 1000. The difference reflects an early mortality rate. The most common type (trisomy 21) is due to a non-disjunction of chromosome 21 during the original cell division, resulting in an extra chromosome 21. These children have a total of 47 chromosomes instead of the usual 46. However, the extra material from chromosome 21 can also be attached to another chromosome through translocation; such children have Down syndrome but only 46 chromosomes. More rarely, the trisomy 21 breaks up, giving some cells with 47 chromosomes and some with 46 (mosaicism). The characteristic physical features include almond-shaped eyes, a rounded skull with flattened occipital region, a broad, flattened bridge of the nose, an enlarged fissured tongue, broad hands with stubby fingers, often a single "simian" palmar crease, hypotonic muscle development, thick lips, dry, rough skin, subnormal height, and infantile genitalia. Not all of these physical signs are present in every case, and some may be observed in individuals without Down Syndrome. However, Down Syndrome is diagnosed when most of the anomalies are present.

The degree of intellectual disabilities is not directly related to the number or gravity of the physical signs, but rather to a combination of these anomalies and the specific chromosomal defect. Few children with Down syndrome are classified today as severely retarded. Most are moderately to mildly retarded and are often educable and highly trainable. They tend to be curious, observant, skilful at mimicry, and usually very affectionate. Aggression and hostility are rare; however, they are often stubborn and compulsive and are not easily frustrated. They are excellent candidates for vocational training.

Research into the neurobiology of the disorder suggests a non-specific, generalized defective brain development. There is a tendency toward thyroid dysfunction and congenital heart defects. There may also be vision problems, but below-average dental caries. Medication has little effect on the physical condition or the intellectual development.

Children with Down Syndrome are frequently very responsive to musical stimuli and love to engage in musical activities. Several studies have explored music perception in these children. When comparing normal children and children with Down Syndrome in their preferences for variables of pitch register (high, low), dynamic level (loud, soft), and rhythmic variety (rhythmic, non-rhythmic) normal children preferred music at the forte level, while the children with Down Syndrome preferred music at the piano level, with a significant difference between groups. All other parameters tested showed no significant difference (Flowers, 1984). In keeping with this study, Down Syndrome children have normal rhythm processing and perform superior to other children with developmental disorders, demonstrating that many musical abilities are spared in this condition (Stratford & Ching, 1983). This is reflected in a study on singing ability in students with Down Syndrome, demonstrating that, with an appropriate choral music curriculum, singing abilities can be significantly improved (Edenfield & Hughes, 1991).

In summary, musical activities with children with Down Syndrome are a rewarding field for music educators. With respect to the choice of instrument, the physical peculiarities have to be considered. Brass and string instruments seem to be less appropriate due to the frequent enlargement of the tongue and to the short and frequently hyper mobile fingers. However, we have encountered percussion and electric guitar playing of adults with Down Syndrome, whose abilities were comparable to a professional level in normal adults.

2. Williams Syndrome

Williams Syndrome is caused by the deletion of genetic material from chromosome 7. Children with Williams Syndrome typically are characterized by a distinctive, "elfish" facial appearance, along with a low nasal bridge, a cheerful, amiable temperament and ease with strangers. Furthermore, intellectual retardation is coupled with an unusual facility with language and a love for music. Other symptoms due to the genetic condition include cardiovascular problems, such as supravalvular aortic stenosis and transient hypercalcaemia. Williams Syndrome individuals share some features with autism (such as difficulty understanding the state of mind of conversational partners), although persons with Williams syndrome generally possess very good social skills, causing this condition to sometimes be called the "cocktail-party syndrome." There also appears to be a higher prevalence of left-handedness and left-eye dominance in those with Williams Syndrome, and cases of absolute pitch appear to be significantly higher amongst those with the condition. Another symptom of Williams Syndrome is lack of depth perception and an inability to visualize how different parts assemble into larger objects (in assembling jigsaw puzzles, for example). This discrepancy in highly developed language and musical skills and astonishingly poor visuo-spatial abilities was recently investigated in a series of studies (Farran & Jarrold, 2005).

When considering the gross anatomy of the brain, whole brain volumes are about 15 percent smaller in individuals with Williams Syndrome than normal. However, the temporal lobe, which is involved in processing sounds and interpreting music and language, is of approximately normal volume. Furthermore, the planum temporale, a region of the temporal lobe which is occupied by the primary and secondary auditory cortices processing acoustic information is augmented in the right hemisphere. According to recent research, neurons in these regions are larger and loosely packed, denoting increased connectivity, which in turn may explain some of the superior auditory skills (Eckert et al., 2006).

With respect to musicality, it appears that musical strength observed in individuals with Williams Syndrome involves less formal analytic skills in pitch and rhythm discrimination and more of a strong engagement with music as a means of expression, play, and, perhaps, improvisation (Hopyan et al., 2001). Furthermore, as in the visual domain, Williams Syndrome children seem to have particular difficulties in assessing the global characteristics of melodies, such as contour. WS individuals demonstrate identical rhythmic abilities as normal controls, but are less accurate than the latter group in melody completion tasks (Levitin, 2005). A recent study applied strict test criteria with respect to language skills in a sample of 32 Williams Syndrome individuals. Surprisingly, special verbal skills could not be detected, thus challenging the notion of a specific giftedness in language acquisition (Carrasco et al., 2005). In summary, it appears that imitative auditory skills and hyper-social behaviour account for many of the savant-like characteristics in this condition. Further studies assessing the receptive, expressive and emotional components of music processing in Williams Syndrome children are required in order to give clear directives as to where music educators should focus their educational efforts.

Conclusions

In concluding this chapter, I would like to add some general remarks concerning the impact of music on the various conditions described above. As pointed out in the introductory paragraph, musical experience is probably the richest human emotional, sensorimotor and cognitive experience. It has powerful effects on brain plasticity and may support many cognitive, emotional and sensorimotor integrative processes. Furthermore, music is a means to share deep social feelings, to express hidden inner worlds, and to gain self-confidence. In general, I believe that there is almost no condition for which music education is not beneficial – except, perhaps, in the case of strong aversive reactions due to abnormal sensitivity towards auditory stimulation. The enormous inventiveness and stamina shown by children and adults with physical disabilities participating in musical activities speaks for itself. However, some limitations of the reported scientific data with respect to the practical work of music educators have to be mentioned:

1. There is a huge variety in the severity and type of the disorders discussed in this chapter. Not every method applies to everyone with a specific disability. In cases in which scientifically based "controlled" studies concerning the effect of music education or music therapy are reported, one has to bear in mind that these data represent results based on group investigations: they give the average picture. However, each individual may require a different approach corresponding to his or her specific profile of skills and disabilities. It is the sensibility, the openness and the inventiveness of the individual music educator that, in many respects, determines the success of the musical training.
2. The brain exhibits a very high plasticity and learning takes place even under unfavourable conditions. Especially in children with intellectual disabilities and developmental disorders, cognitive and emotional maturation may be considerably delayed, leaving long periods of frustration for both the child and the educator. Though cognitive development and the acquisition of skills in disabled individuals may require many more years than in normal children, the former will finally succeed in learning, thus making the developing of skills and behaviours all the more rewarding. On the other hand, music educators are frequently confronted with unrealistic expectations from parents. Here, it is important to convey the necessary information without being discouraging.

3. I have reported interventions of music therapists in many conditions. It should be emphasized that music education is not the same as music therapy, and that music educators are not trained as music therapists. Working as a music educator in a therapeutic setting requires professional training and steady supervision. Otherwise, expecting too much from oneself may lead to burn-out. Music therapists generally are also trained in self-awareness and psychotherapy, giving them resources to cope even when confronted with demanding work conditions. In contrast to music education, which focuses on the student's progress in a specific musical skill, music therapy is commonly used to support communication.
4. Inclusion is the most important factor in classroom teaching. Music is a social experience, providing the opportunity to bring people with very different strengths and weaknesses into a shared, communicative space. For years, social and educational policies impeded rather than nurtured the development of such shared music teaching and learning settings in children with various disabilities. Presently, as knowledge concerning the impact of socio-cultural factors on children's developments grows and conclusive data on the principles of human learning become available, the situation seems to have improved. As pointed out by Judy A. Jellison (2006): "Irrespective of available knowledge, action is required to change music education practices. Individuals of good will, knowledge, and skill can contribute substantially to improving the quality of lives of future generations of people with disabilities by ensuring that music is a prominent component in the lives of all children" (p. 270).

Acknowledgements

The author is extremely grateful to Katie Overy for her many valuable suggestions made during the early stages of writing. He wishes to thank Wilfried Gruhn for helpful discussions on the topic and Mary Sutherland for her careful language editing.

References

Altenmueller, E. (2006): The end of the song: Robert Schumann's focal dystonia. In: E. Altenmueller, M. Wiesendanger, & J. Kesselring (Eds.) *Music, motor control, and the brain* (p 251-264). Oxford, Oxford University Press.

Amaducci, L., Grassi, I., & Boller, F. (2002). Maurice Ravel and right hemisphere musical creativity: influence of disease on his last musical works? *European Journal of Neurology 9,* 75-82.

Atterbury, B. W. (1990). *Mainstreaming exceptional learners in music.* Englewood Cliffs, NJ. Prentice Hall.

Ayotte, J., Peretz, I., & Hyde, K. (2002). Congenital amusia. A group study of adults afflicted with a music-specific disorder. *Brain 125*, 238-251.

Bangert, M. & Schlaug, G. (2006). Specialization of the spezialized in features of external brain morphology. *European Journal of Neurosciece 24*, 1832-1834.

Birbaumer, N. (1999). Neurobiology. Rain Man's revelations. *Nature*. 399, 211-212.

Brook, J.G. (1933). Rebsomen, the one-armed flautist. *The Musical Times 74*, 636.

Byl, N.N., Merzenich, M.M., & Jenkins, W.M. (1996). A primate genesis model of focal dystonia and repetitive strain injury: Learning-induced dedifferentiation of the representation of the hand in the primary somatosensory cortex in adult monkeys. *Neurology 47*, 508-520.

Carrasco X, Castillo S, Aravena T, Rothhammer P, & Aboitiz F. (2005). Williams syndrome: pediatric, neurologic, and cognitive development. *Pediatric Neurology, 32,* 166-172.

Castellanos, F.X., Lee, P.P., Sharp, W., Jeffries, N.O., Greenstein, D.K., Clasen, L.S., Blumenthal, J.D., James, R.S., Ebens, C.L., Walter, J.M., Zijdenbos, A., Evans, A.C., Giedd, J.N., & Rapoport, J.L. (2002). Developmental trajectories of brain volume abnormalities in children and adolescents with Attention-Deficit/Hyperactivity Disorder. *Journal of the American Medical Association 288*, 1740–1748.

Chan, A.S., Ho, Y., Cheung, M. (1998). Music training improves verbal memory. *Nature 396*, 128.

Drayna, D., Manichaikul, A., de Lange, M., Snieder, H., & Spector, T. (2001). Genetic correlates of musical pitch recognition in humans. *Science 291*, 1969-1972.

Eckert, M.A., Galaburda, A.M., Karchemskiy, A., Liang, A., Thompson, P., Dutton, R. A., Lee, A.D., Bellugi, U., Korenberg, J.R., Mills, D., Rose, F.E., & Reiss, A.L. (2006). Anomalous sylvian fissure morphology in Williams syndrome. *Neuroimage 33,* 39-45.

Edenfield, T.N. & Hughes, J.E. (1991). The relationship of a choral music curriculum to the development of singing ability in secondary students with Down syndrome. *Music Therapy Perspectives 9*, 52-55.

Elbert, T., Candia, V., Altenmueller, E., Rau, H., Rockstroh, B., Pantev, C., & Taub, E. (1998). Alteration of digital representations in somatosensory cortex in focal hand dystonia. *NeuroReport 16,* 3571-3575.

Farran, E.K. & Jarrold, C. (2005). Evidence for unusual spatial location coding in Willimas syndrome: an explanation for the local bias in visuo-spatial construction tasks? *Brain and Cognition, 59,* 159 – 172.

Flowers, E. (1984). Musical sound perception in normal children and children with Down's syndrome. *Journal of Music Therapy, 21,*144-154.

Frith, U., Happe, F. (1994). Autism: beyond 'theory of mind.' *Cognition, 50,* 115-132.

Galaburda, A. & Kemper, T.L. (1979). Cytoarchitectonic abnormalities in developmental dyslexia: a case study. *Annals of Neurology, 6,* 94-100.

Geschwind N, & Galaburda AM. (1987) *Cerebral lateralization: biological mechanisms, associations, and pathology*. Cambridge, Mass: MIT Press.

Gfeller, K., Christ, .A., Knutson, J.F., Witt, S., Murray, K.T., & Tyler, R.S. (2000). Musical backgrounds, listening habits, and aesthetic enjoyment of adult cochlear implant recipients. *Journal of the American Academy of Audiology, 11,* 390-406.

Gfeller, K., Christ, A., Knutson, J.F., Witt, S., & Mehr, M. (2003). The effects of familiarity and complexity on appraisal of complex songs by cochlear implant recipients and normal-hearing adults. *Journal of Music Therapy,* 40, 78–112.

Gfeller, K., Olszewski, C., Rychener, M., Sena, K., Knutson, J.F., Witt, S., & Macpherson, B. (2005). Recognition of "Real-World" musical excerpts by cochlear implant recipients and normal-hearing adults. *Ear & Hearing, 26*, 237–250.

Gold, C., Wigram, T., & Elefant, C. (2006) Music therapy for autistic spectrum disorder. *Cochrane Database Syst. Rev. 19(2),* CD 004381.

Gougoux, F., Lepore F., Lassonde, M., Voss, P., Zatorre, R.J., & Belin, P. (2004). Pitch discrimination in the early blind. *Nature 430,* 309.

Hamilton, R.H., Pascual-Leone, A., & Schlaug G. (2004) Absolute pitch in blind musicians. *Neuroreport 15*, 803-806.

Hash, P.M. (2003) Teaching instrumental music to deaf and hard of hearing students. *Research and Issues in Music Education 1, http://www.stthomas.edu/rimeonline/vol1/ hash1.htm*

Heaton, P. (2003). Pitch memory, labelling and disembedding in autism. *Journal of Child Psychology and Psychiatry, 44,* 543-551.

Hodges, D. (2006). The musical brain. In G. E. McPherson (Ed.), *The child as musician. A handbook of musical development* (pp. 51-68). Oxford, Oxford University Press.

Hopyan, T., Dennis, M., Weksberg, R., & Cytrynbaum, C. (2001). Music skills and the expressive interpretation of music in children with Williams-Beuren syndrome: pitch, rhythm, melodic imagery, phrasing, and musical affect. *Child Neuropsychology*, 7, 42 – 53.

Hyde, K.L., Zatorre, R.J., Griffiths, T.D., Lerch, J.P., & Peretz, I. (2006). Morphometry in the amusic brain: a two-site study. *Brain 129,* 2562-2570.

Jabusch, H.C., Altenmueller, E. (2006). Focal Dystonia in musicians: from phenomenology to therapy. *Advances in Cognitive Psychology, 2 (2-3),* 207-220.

Jackson, N.A. (2003). A survey of music therapy methods and their role in the treatment of early elementary school children with ADHD. *Journal of Music Therapy 40(4),* 302-323.

Jellison, J. A (2006). Including everyone. In G. E. McPherson (Ed.), *The child as musician. A handbook of musical development* (pp. 257-272). Oxford, Oxford University Press.

Kalmus, H., & Fry, D.B. (1980). On tune deafness (dysmelodia): frequency, development, genetics, and musical background. *Annals of Human Genetics, 43,* 369-382.

Kang, J., Lee, D.S., Kang, H., Lee, J.S., Oh, S.H., Lee, M.C., & Kim, C.S. (2004). Neural changes associated with speech learning in deaf children following cochlear implantation. *NeuroImage 22*, 1173-1181.

Klenerman, L, & Wood B. (2006). *The human foot. A companion to clinical studies.* Springer, London.

Levitin, D. (2005). Musical behavior in a neurogenetic developmental disorder: evidence from Williams syndrome. *Annals of the New York Academy of Sciences, vol. 1060*, 325-334.

Levitin, D., & Bellugi, U. (1998). Musical abilities in individuals with Williams syndrome. *Music Perception, 15(4),* 357-389.

Liberman, I.Y. & Shankweiler, D.P. (1991). *Learning to read. Basic research and its implications*. Hillsdale, NJ, Lawrence Erlbaum.

Miller, L.K. (1989). *Musical savants. Exceptional skills in the mentally retarded.* Hillsdale New York, Lawrence Erlbaum

Miller, B.L., Cummings, J., Mishkin, F., Boone, K., Prince, F., Ponton, M., & Cotman, C. (1998). Emergence of artistic talent in fronto-temporal dementia. *Neurology, 51,* 978-982.

Nicolson, R. I. & Fawcett, A. J. (2006). Do cerebellar deficits underlie phonological problems in dyslexia? *Developmental Science, 9(3),* 259-262.

Overy, K. (2003). Dyslexia and Music. From timing deficits to musical intervention. *Annals of the New York Academy of Sciences, vol. 999,* 497-505.

Peretz, I., Ayotte, J., Zatorre, R., Mehler, J., Ahad, P., Penhune, V.B., & Jutras, B. (2002). Congenital amusia: a disorder of fine-grained pitch discrimination. *Neuron 33,* 185-191.

Peretz, I., Champod, S., & Hyde, K. (2003). Varieties of musical disorders: the Montreal Battery of Evaluation of Amusia. *Annals of the New York Academy of Sciences, vol. 999,* 58-75.

Révész, G. (1925). *The psychology of a musical prodigy.* New York, Hartcourt, Brace & Company.

Rickson, D.J. (2006). Instructional and improvisational models of music therapy with adolescents who have attention deficit hyperactivity disorder (ADHD): a comparison of the effects on motor impulsivity. *Journal of Music Therapy 43(1),* 39-62.

Rippon, G., Brock, J., Brown, C., & Boucher, J. (2007). Disordered connectivity in the autistic brain: challenges for the new psychophysiology. *International Journal of Psychophysiology, 63,* 164-172.

Robbins, C., & Robbins, C. (1980) *Music for the hearing impaired and other special groups: a resource manual and curriculum guide.* St. Louis, MagnaMusic-Baton

Roeder, B., Stock, O., Bien, S., Neville, H., Roesler, F. (2002). Speech processing activates visual cortex in congenitally blind humans. *European Journal of Neuroscience,* 16, 930 – 936.

Ruthsatz, J., & Detterman, D.K. (2003). An extraordinary memory: the case of a musical prodigy. *Intelligence 31,* 509-518.

Sabisch, B., Hahne, A., Glass, E., v. Suchodoletz, W., & Friederici, A.D. (2006). Auditory language comprehension in children with developmental dyslexia: evidence from event-related brain potentials. *Journal of Cognitive Neuroscience,* 18, 1676 – 1695.

Sadato, N., Pascual-Leone, A., Grafman, J., Ibanez, V., Deiber, M.P., Dold, G., & Hallett, M. (1996). Activation of the primary visual cortex by Braille reading in blind subjects. *Nature 380,* 526-528.

Schellenberg, E.G. (2006) Exposure to music: the truth about the consequences. In G.E. McPherson (Ed.) *The child as musician: A handbook of musical development.* (pp 111-134). Oxford, Oxford University Press.

Schmid, A., Jabusch, H.C., Altenmueller, E., Hagenah, J., Brüggemann, N., Hedrich, K., Saunders-Pullman, R., Bressman, S., Kramer, P.L., & Klein, C. (2006). Dominantly transmitted focal dystonia in families of patients with musician's cramp. *Neurology 67(4),* 691-693.

Schultz, R.T. (2005). Developmental deficits in social perception in autism: the role of the amygdala and the fusiform face area. *International Journal of Developmental Neuroscience 23,* 125-141.

Schuppert, M., Muente, T.F., Wieringa, B.M., & Altenmueller, E. (2000). Receptive amusia: Evidence for cross-hemispheric neural networks underlying music processing strategies. *Brain 123*, 546-559.

Schuppert, M., Muente, T. F., & Altenmueller, E. (2003). Recovery from receptive amusia reveals functional reorganisation of music-processing networks. *Zeitschrift für Neuropsychologie 14*, 113-122.

Sowell, E. R., Thompson, P.M., Welcome, S.E., Henkenius, A.L., Toga, A.W., & Peterson, B.S. (2003). Cortical abnormalities in children and adolescents with Attention-Deficit Hyperactivity Disorder. *Lancet 362*, 1699–1702.

Stein, J. & Walsh, V. (1997). To see but not to read; the magnocellular theory of dyslexia. *Trends in Neurosciences*, 20, 147 – 152.

Steinwede, D., Rottbeck, R., Busse, O., Kohlmetz, C., Muente, T.F., Schuppert M., & Altenmueller, E. (2000). Expressive and receptive amusia in patients with hemispheric stroke. Abstract presented at the German Society of Psychology of Music, Karlsruhe, September 22, 2000.

Stratford, B., & Ching, E.Y. (1983) Rhythm and time in the perception of Down's syndrome children. *Journal of Mental Deficiency Research*, 27, 23 -38.

Tallal, P., Miller, S., & Fitch, R.H. (1993). Neurological basis of speech: a case for the pre-eminence of temporal processing. *Annals of the New York Academy of Sciences, vol. 682*, 27-47.

Treffert, D. (1989) *Extraordinary people: understanding savant syndrome.* New York: Harper & Row.

Whipple, J. (2004) Music in intervention for children and adolescents with autism: a meta-analysis. *Journal of Music Therapy, 41(2),* 90 – 106.

Williams, D.L., Goldstein, G., & Minshew, N.J. (2006). Neuropsychologic functioning in children with autism: further evidence for disordered complex information processing. *Child Neuropsychology 12*, 279-298.

Wolf, M. & Bowers, P. (2000). The question of naming-speed deficits in developmental reading disability: An introduction to the Double-Deficit Hypothesis. *Journal of Learning Disabilities, 33*, 322 – 324.

Wolff, P. H. (2002). Timing precision and rhythm in developmental dyslexia. *Reading and Writing, 15*, 179 – 206.

Zatorre, R.J., Evans, A.C. & Meyer, E. (1994). Neural mechanisms underlying melodic perception and memory for pitch. *Journal of Neuroscience 14*, 1908–1919.

Zinar, R. (1987). *Music activities for special children.* West Nyack, NY: Parker.

Internet Resources:

www.blindmusicstudents.org
www.drakemusicproject.org
www.menvi.org
www.mybreathmymusic.com
www.no-c-notes.com
www.remap.org.uk

In: Neurosciences in Music Pedagogy
Editors: W. Gruhn and F. Rauscher, pp. 267-282
ISBN: 978-1-60021-834-7

Chapter X

The Neurobiology of Learning: New Approaches to Music Pedagogy Conclusions and Implications

Wilfried Gruhn and Frances H. Rauscher

Abstract

Modern societies are strongly interested in scientifically approved strategies of teaching and learning. Therefore, educators and scholars as well as politicians have referred to some spectacular results from brain research in order to infer new methods of teaching and learning and to ground educational strategies on neuroscientific fundamentals. In our concluding chapter, we discuss the difficulties of an immediate causal link between the state of neural processes and the state of learning and knowledge acquisition which are at the core of the so-called brain-based learning or *neurodidactics*. In a first section, we review results from brain research as related to the educational practice in general, namely the estimated function of the recently discovered mirror neurons for learning process. Then, in a second section, we focus on the specific implications of brain research in music learning. As an important aspect we introduce the common mechanisms of audiovocal learning that are crucial for both, language and song acquisition. Finally, we conclude from the previous chapters manifest consequences and implications for music pedagogy.

Brain Research and Educational Practice

Modern societies are strongly interested in basing educational decisions on empirical evidence. Brain imaging technologies have opened a window into the active brain, attracting researchers and fascinating the public. The possibility of observing the working brain has inspired our imagination and driven us to try to understand the incomprehensible, such as cognitive processes in the conscious self, of which learning is a part (Metzinger, 2000). The increasing interest in the learning brain is reflected by recent publications that focus on the connection between brain development and learning (Blakemore & Frith, 2005; Hüther, 2002; Ratey, 2001; Spitzer, 2002, 2006; Stern, 2005; Stern et al., 2005).

This could also explain the enormous interest in the Mozart effect, which seems to offer empirical evidence that even listening to music has an impact on human cognition. It is quite obvious that music has an effect on human mood and emotion. Under certain circumstances, music can manipulate human behaviour, which is reflected through respective brain function. Although the special arousing effects of music have been confirmed, and studies have been conducted to investigate this effect as a possible means to enhance the learning process even in domains other than music (Winner & Hetland, 2000), little is known about the neural processes that constitute *music learning*: how is the brain affected by the steps and phases of music learning and, vice versa, how can specific neuronal mechanisms influence the learning of music so that, consequently, this knowledge could be systematically applied to the teaching of music in terms of developing new approaches to music education?

In so far, we aim to look at the neural correlates of learning. A neural correlate is a neural representational brain activation that directly corresponds with a particular learning activity (for definition see Chalmers, 2000). However, the possible discovery of these neural correlates does not necessarily indicate a causal connection. Moreover, much of the research in this area relies on lesion studies in which an observed disability is traced back to a particular brain dysfunction. However, we know of no "lesion studies" on learning other than those on congenital amusia (Ayotte et al., 2002; Peretz et al., 2005; Peretz et al., 2003), on the hearing impaired (Chasin & Russo, 2004) or deaf patients (Kaiser & Johnson, 2000), and especially on children with cochlear implants (Gfeller et al., 2005; McDermott, 2004; Nakata et al., 2005). In our context it would be interesting to find out whether a musical equivalent to dyslexia or colour blindness exists that affects learning.

During the last few decades, a great deal of research in the neurosciences has been devoted to investigating the perception and cognition of music. Brain imaging technologies are used to locate specific brain areas where particular musical information is processed, and to document changes in the magnitude of activation caused by changes in the acoustic (musical) stimuli. Numerous studies have informed us about inter-hemispheric coherence and interaction of brain areas during the processing of music, and we know about the structural and functional changes in the brain caused by musical activities. For neuroscientists, music advanced to a cogent paradigm of neuroplasticity. However, the neuroscientific demonstration of brain activation tells us little about the functional neurobiological and neurophysiological constraints that are necessarily involved in *music learning*, that could clarify the *learning process*, and result in *musical achievement*. Moreover, for music pedagogy it is also essential to keep in mind the limits of neuroscientific data, because they cannot provide educators with concrete rules and prescriptions for learning. Learning is always domain-specific, and music learning is its own domain, although it is linked with many other domains.

If one understands music learning in neurobiological terms as the development and gradual differentiation of mental *musical* representations and their interaction, which is at the core of neuroplasticity and enables the brain to structurally and functionally adapt to actual environmental challenges, then those functional changes can be investigated in relation to different learning strategies. However, we must consider that in no means can we assume the presence of a causal connection between some neural activation and a particular phenomenal quality. Rather, we must acknowledge that the qualitative dimension of learning cannot be read from brain activation. The only opportunity for research is to observe the possible impact of learning modes on the neural processing under strictly controlled research conditions.

The human brain always learns and is optimized for learning through evolution. However, knowledge acquisition is governed by internal conditions that cannot simply be manipulated by external factors. The intrinsic construction of meaning by means of implicit procedural learning is regulated by internal structures of the limbic system that are unconscious and cannot be manipulated externally (Roth, 2006), whereas the hippocampus organizes the structure of declarative memory. However, the neuro-modulatory function of the mesolimbic system (ventral striatum, nucleus accumbens) regulates the reward circuit and influences learning by emotional evaluation and transfer into long-term storage.

Beyond this, there are some general concerns with regard to scientific results and their immediate adaptability to practical applications. This is particularly true for the connection between neurosciences and music pedagogy in terms of teaching methods and curriculum development. The authors of a recently published study on educational research and neurosciences (Stern et al., 2005) argue that none of the commonly accepted principles of meaningful learning based on cognitive science, developmental psychology, or educational research has been derived from findings in brain research. Stern warns that efforts to incorporate brain research into education could cause false hopes that neuronal and educational processes are strongly and evidently linked (Stern, 2005). Another critical voice (Bruer, 1999) claims that the impact of synaptic growth and density during the early years of life should not be overestimated because synaptogenesis does not necessarily depend solely upon the quantity of incoming sensorial information, but rather on an intact sensorial perception. Cortical deficiencies in early childhood often arise from sensorial dysfunctions (Stern et al., 2005).

The common wisdom that learning is based on the plasticity of the brain requires a clear differentiation between two functionally discernible types of plasticity: *experience-expectant* and *experience-dependent* plasticity (Greenough & Black, 1992). Experience-expectant plasticity is based on the early exuberant overproduction of synapses followed by synaptic pruning, which ensures that synapses that are actually used are stabilized, and that irrelevant connections are eliminated. Thus, the normal wiring of the brain is in part a result of the kinds of general experience that have been present throughout evolution, experiences that every human who inhabits any reasonably normal environment will have. As a consequence, the brain can "expect" input from these reliable sources to selectively activate and stabilize some synapses, simultaneously causing the elimination of inactive ones. On the contrary, with experience-dependent plasticity new synapses are formed by exposure to environmental stimuli. In other words, neural connections are created and reorganized throughout life as a function of an individual's experience. For example, comparisons of the brains of animals (i.e., rats, cats, monkeys) that were raised in either complex environments full of objects to explore or in dull laboratory cages revealed large differences. The brains of the animals raised in enriched environments had more dendritic spines on their cortical neurons, more synapses per neuron, and more synapses overall, as well as a generally thicker cortex and more of the supportive tissues (such as blood vessels and glial cells) that maximize neuronal and synaptic function. This extra hardware seems to have had behavioural consequences as well: Rats raised in enriched environments performed better in a variety of learning tasks (e.g., Juraska et al., 1984; Rauscher et al., 2002). During children's developmental phases both types of plasticity are needed, and focusing only on early stimulation is as misguided as neglecting an enriched environment.

In addition to such general effects of experience on brain structure, highly specific effects can also occur. For example, rats that are trained to use either just one or both forelimbs to get a food reward have increased dendritic material in the particular area of the motor cortex that controls the movement of the trained limb(s) (Greenough et al., 1985; Tomie & Wishaw, 1990). Similar effects seem to occur in humans. For example, a study of violinists and cellists revealed that, compared with control subjects, the musicians had increased cortical representation of the fingers of the left hand (Elbert et al., 1995). In other words, after years of practice, more cortical cells were devoted to receiving input from and controlling the hand used for fingering. Similarly, skilled Braille readers exhibit enlarged cortical representations of the left hand, which is the hand they use to read Braille text (Pascual-Leone et al., 1993).

Another recent finding that evokes new expectations for learning is the discovery of mirror neurons. Rizzolatti and collaborators (Gallese et al., 1996; Rizzolatti, 1996) have discovered a population of neurons in the premotor cortex (F 5) of macaques that discharge both when the monkey performs an action or when it just observes or hears the same action performed by another individual. Researchers have speculated as to whether mirror neurons can be viewed as the missing link between the abilities of primates and language abilities in humans (Arbib, 2005). In fact, a mirror neuron system exists in humans in Broca's area -- the homologue of the monkey's area F 5. Functional magnetic resonance imaging (fMRI) studies have shown that even listening to sentences expressing an action using the hand, mouth or foot can cause activation of different sectors in the premotor cortex where the corresponding action is represented (Binkofski & Buccino, 2006). This has stimulated speculations regarding if and how mirror neurons play a role in imitation learning. In humans, mirror neurons resonate to motor movements of the hand, the mouth, and the foot that are only observed in another individual. When an action that is already present in the mirror neuron system is imitated, the act can be immediately replicated. Data from an fMRI study suggest that the coding of a viewed motor action in the mirror neuron system can be transferred to a recombination of these acts in order to replicate it according to the presented model (Buccino et al., 2004). Imitation, here, is always understood in conjunction with the learning of a novel motor act or motor sequence. The critical and still unsolved question relates to how mirror neurons function in the complex process of imitation learning. Here one has to differentiate between neuronal co-activation as a consequence of repeated practice, which causes a neuronal link between related areas (Bangert & Altenmueller, 2003) on the one hand, and the phenomenon of imitation on the other hand. The activation of the same area both in action and visual perception is different from imitation. With regard to macaques it is known that these monkeys do not imitate. Imitation is an intentional and conscious activity, whereas the activation of mirror neurons is recognized as a spontaneous neuronal reaction. At the moment, there are several conflicting opinions. Some researchers argue that mirror neurons are at the core of imitation learning (Buccino et al., 2004), whereas others concede that the actual function of mirror neurons in humans is not yet well understood (Gaschler, 2006). Nevertheless, it is clear that humans possess a comprehensive system of mirror neurons in Broca's area and the premotor cortex, representing a neuronal substrate of the meaningful processing of words (Binkofski & Buccino, 2006) as well as in the anterior insula and the anterior cingulate cortex (Hutchison et al., 1999). This would confirm and empirically explain the traditionally-established theory regarding how understanding relies on embodiment, but it does not necessarily indicate that mirror neurons enable learning through imitation. However, educators can and should make use of the fact that students activate the same neurons while observing an act as those they activate while actually performing the act themselves. This neural activation consolidates learning.

Although music educators must take into account general concerns regarding an immediate application of neuronal findings to education, they should nevertheless ignore those findings. It might be the first time that pedagogy as a discipline can be founded – at least partly – on science. If educators are informed about the underlying neuronal mechanisms of learning, their minds may open up to new arrangements for teaching so that they can adjust their teaching to the mental state of the children. We cannot rest on the traditional belief and opinion that good teachers know about good teaching. Since the teaching environment has changed through the introduction of new media and the implementation of virtual e-learning formats, we must take advantage of all accessible resources that provide us with a clear insight into the inner structure of learning. Here, neurobiological and neurophysiological research can help shed light on the mental processes and neuronal functioning of the developing brain in order to support and enlighten our own understanding of those structures that underpin learning. A further question, then, is if and to what extent the results of brain research can be introduced into music pedagogy.

BRAIN RESEARCH AND MUSIC LEARNING

There is a large body of observational research on the developing abilities of infants and young children that is relevant to music learning, namely structural (grouping and segmentation), rhythmical and tonal differentiation (for review, see Colwell & Richardson, 2002; Gembris, 1998). It has been demonstrated that infants as young as 6 to 12 months are able to recognize rhythmic distinctions in unfamiliar music after a brief exposure to this music (Hannon & Trehub, 2005). This might be due to the fact that infants are enthusiastic learners, but this learning depends on exposure, and therefore relates to the important impact of learning through acculturation. However, only a few studies focus explicitly on the neuronal substrates of music learning, and most of them are concerned with short-term learning. A neurobiological foundation of the process of learning would be extremely beneficial.

This gap can be filled by the new discipline of *Neuropedagogy,* or in European terms, *Neurodidactics* (Caspary, 2006; Herrmann, 2006; Preiss, 1998). Neurodidactics seeks to establish a brain-based learning strategy. Its goal is to adapt the teaching and learning methods to children's mental state instead of aligning children with the curriculum. The question is: what do teachers need to know about the brain that can help them to enhance their teaching more efficiently? Here, at least a few general aspects of brain-based learning can be highlighted:

- the supportive function of the brain-specific reward system that is turned on when an experience is better or stronger than expected, and the opposite function of stress;
- the importance of experience-expectant plasticity, experience-dependent plasticity, and the effects of specific types of environmental stimulation (e.g., music instruction) on brain plasticity;
- the function of complementary holistic experiences in which the different sensorial modalities interact with one another (e.g., what one experiences aurally can be calibrated by what one sees or feels) and generate an embodied meaning;
- the important impact of repeated actions on the development of mental representations and their developing differentiation.

As to music learning, another basic mechanism is extremely important: auditory-guided vocal learning (Brown et al., 2004; Fitch, 2006; Merker, 2005). Interdisciplinary studies on the evolution of music (song) and language emphasize that music is a cultural phenomenon that incorporates a broad variety of distinct sound patterns that have to be learned through imitation by means of listening and performance (Merker, 2005). Humans, like very few animals (such as birds, dolphins, whales, and seals), possess a neuronal mechanism that enables them to imitate arbitrary sounds to which they are exposed. This is especially well investigated in songbirds' brains. Human vocalizations also build upon the audiovocal ability to produce distinct sounds according to what they hear. Three neural pathways – the posterior vocal pathway, the anterior vocal pathway (with the cortex, basal ganglia, thalamus loop), and the auditory pathway – are believed to build a complex *phonological loop* that enables humans and songbirds to control their vocal production by ear (Jarvis, 2004). It has been shown that the similarities between song learning in birdsongs and language, as well as song acquisition in humans, can be traced back to a molecular level (Scharff & Haesler, 2005).

The phonological loop is also essential for working memory (Baddeley, 2003) and it plays an important role in the acquisition of acoustic signal systems like language and music. The analysis of synaptic function by intracellular recordings presents the possibility of simultaneously probing the activity of a single neuron and the synaptic network. This technique suggests that vocal learning may take place through the activation of a cancellation mechanism initiated by the auditory feedback (Mooney, 2004).

In general, effective and enduring learning in daily life always happens in complex situations with interacting environmental conditions. Results from experiments with seven-months-old infants suggest that their listening preferences to familiar music were affected by the extent to which the musical extracts were removed from the musical contexts within which they were originally presented (Saffran et al., 2000). A multi-sensory representation of rhythm has also been demonstrated, particularly in young children. The perception of meters that are encoded in duple or triple form is strongly influenced by movement and bodily experiences (Phillips-Silver & Trainor, 2005).

During the process of learning, the brain links different modes of perception to executive motor skills. If a novice pianist practices finger-motor patterns every day, EEG recordings show an auditory-sensori-motor integration after just a few days of practice. Even in a silent condition, during which the subject only moves his fingers on a mute keyboard, an activation of the auditory areas occurs (Bangert & Altenmueller, 2003). This has been confirmed by an fMRI study. Pianists who observed finger-hand-movements of a pianist playing the instrument showed stronger activations within a fronto-parieto-temporal network compared to controls. Even the participating pianists who only observed silent piano playing showed activation in auditory areas. However, the extent to which mirror neuron function gets involved in this transmodal interaction must still be discovered. Furthermore, it has been suggested that an observation-execution system links visual and auditory perception to motor performance (Haslinger et al., 2005).

In an early learning experiment using DC-EEG recordings, two groups of high-school students (ages 13 to 14 years) were taught in different ways: one group by verbal instruction (explicit learning) the other group by practical performance (implicit learning). The task was to differentiate aurally between correct and incorrect musical periods (Altenmueller & Gruhn, 1997; Gruhn, 1997). Not surprisingly, both groups did not show any differences at the beginning of the study (pre-test). However, after a six-week learning phase both groups achieved higher scores on the same test items in terms of correct answers, but they differed significantly in their brain activation. In a follow-up measurement one year later, the controls who received a non-musical instruction did not exhibit any change on their test scores. On the contrary, the implicit learners retained their higher scores, but only the explicit learners showed a slight decrease of their scores. However, this result was based on a very small sample, and no statistical analyses could be performed.

This study led to another longitudinal EEG study on music learning that included a larger sample of 23 high-school students (ages 12 – 14 years). Two teaching conditions were compared over more than 6 months: a procedural condition and a declarative condition. The students were taught in these different ways so that they would apply a procedural or declarative learning strategy. The procedural teaching condition was characterized by the suppression of all kinds of visual aids and verbal explanations. Rather, many kinds of vocal and instrumental improvisations were introduced. The declarative teaching condition (calling for declarative learning) was characterized by visual aids and verbal explanations, with a complete suppression of vocal production. This was done to eliminate activation of a phonological loop in this condition. This study found that the two groups exhibited significant differences in the magnitude, localisation, and distribution of brain activation depending on the method of teaching and the strategy of learning. Declarative learning was associated with an increase in brain activation over the left frontal areas whereas procedural learning produced an increase over the right frontal and bilateral parieto-occipital regions. This shows that the mode of teaching significantly affects the neural processing of information and the storage and retrieval of implicit musical knowledge (Altenmueller, 2001; Altenmueller et al., 2000; Gruhn, 1997; Gruhn, 2005a). Another complementary study was concerned with intensive learning within a short time span of only 45 minutes. This study employed a similar research design as the former study. An interesting distinction was found: long-term learning over several months (see above) caused a general decrease of brain activation whereas short-term learning for only 45 minutes resulted in a general increase (Liebert, 2001).

Former research studies have repeatedly demonstrated differences in the brain structure of musicians (Schlaug, 2003). Functional MRI studies exhibited a greater asymmetry in the planum temporale of musicians where the planum temporale was larger in the left hemisphere and smaller in the right than in the brains of the nonmusicians (Schlaug et al., 1995). Structural and functional changes in the brain may be attributed to early and continuous training resulting in a long-term enhancement of spatial-temporal and verbal performance (e.g., Chan et al., 1998; Rauscher et al., 1997). Consistent with findings that verbal memory is mediated mainly by the left temporal lobe are indications that musical training has a significant impact on verbal, but not on visual memory (Chan et al., 1998). A longitudinal learning experiment with young children using fMRI imaging techniques has demonstrated significant functional brain changes in the auditory cortex, the cerebellum, and inferior frontal brain areas for young instrumental students as a consequence of training (Kotynek et al., 2006; Overy et al., 2005; Schlaug et al., 2005).

In general, one can conclude from the neurobiological research on music learning that efficient learning is generally related to a decrease in brain activation, which often goes along with a shift of activation centres from prefrontal regions towards those regions relevant to the processing of particular tasks. This phenomenon is known as the *anterior-posterior shift*. Only very intense – and sometimes forced – short-term learning evokes a momentary increase of activation. In fact, this activation is immediate, but it soon disappears. Only the integration of a new ability into long-term memory generates a stable crystallized knowledge system and, therefore, should be a goal of education.

The highly-specialized forms of neuro-feedback in educational research are often neglected in this context because their application is predominantly designed for many forms of behavioural disorders (e.g. Attention Deficit Hyperactivity Disorder, ADHD) and for patients with pathological disabilities of their body functions, including complete paralysis without any form of communication. In these cases, a brain-computer-interface may help to replace the missing verbal system using an EEG-based tool (Thought-Translation-Device, TTD) (Birbaumer et al., 1999; Stern et al., 2005), but by this we enter the realm of therapy which should be separated from education.

New Approaches to Music Teaching and Learning?

Research studies have demonstrated remarkable differences in the brains of musicians compared with non-musicians in that intensive musical training causes structural and functional changes in the brain (Gaser & Schlaug, 2004; Schlaug, 2003). With respect to music learning as a more general concept of musical understanding, it was found that even the mode of learning can affect functional brain activities. Furthermore, the mode of learning is strongly influenced by the teaching method. Thus, even the teaching method has an impact on how the brain processes music. Consequently, if educators know about appropriate teaching methods that empower the brain to process musical information most efficiently, to facilitate musical representation building, and to keep the gathered knowledge accessible in long-term memory, then they can try to implement this knowledge into their actual practice. Even if not all of the following suggestions are exclusively initiated by the findings of brain research, it is nevertheless important to acknowledge that these principles are at least confirmed by findings of research studies in neuroscience.

If music learning in all its facets is grounded on the development of mental musical representations and their interaction within the neural network, it should be a primary goal of music learning to develop genuine musical representations. We should briefly explain what we mean by this. Because music is based on sound and its intentional structure, musical representations are implicit or genuine representations insofar as they represent musical sounds and sound combinations, but not verbally- or symbolically-encoded knowledge *about* music (Gruhn, 2005b). Therefore, genuine musical representations are representations of music or musical elements (e.g., pitch, volume, duration, meter, timbre etc.). This is achieved when we listen to music and recognize its pulse, meter and tonality, identify themes and variations, follow parts, relate them to each other, build expectations about what could come next based upon the experience of what has been heard before. The best way to achieve this goal is to emphasize procedural and implicit learning, for research has indicated that procedural strategies seem to be more effective for long-term achievement than declarative strategies.

Furthermore, rhythm is strongly connected with movement. Therefore, the elementary teaching of tonal and rhythm patterns should always be accompanied by gentle movements so that children can develop a linked representation of motor activities and metrical weight. Another aspect seems worthwhile to keep in mind. Since vocal learning is specific to humans, and its foundation (i.e., the phonological loop), is essential for the transfer of melodic and rhythmic patterns into working and long-term memory, active music making, singing and moving are the primary modes of teaching and learning prior to any kind of verbal explanation. Furthermore, since all learning takes place in a situated and social context, context-dependent learning plays a crucial role for the meaningful learning *of* music instead of *about* music. All of this calls for a praxial approach (Elliott, 1995) because the underlying principles are strongly supported by brain research studies.

Finally, it might be argued as to whether or not it is appropriate to directly link brain research and music pedagogy and establish a new discipline that deals with the interface of neurosciences and learning. The interest in this relationship is possibly due to an overestimation of a momentarily fashionable trend. Nevertheless, the philosophy behind *Neurodidactics* underpins the necessity for the further advancement of music pedagogy in general, which cannot and should not be isolated from the powerful developments in neurosciences. What is known about brain function and neuronal processing must be shown to contribute to music education – not in terms of an immediate application of singular research findings to "brain-based methods," but in the sense that knowledge from neuroscience may help to establish a better understanding of students' difficulties in music learning and performance. Educators should try out new approaches appropriate to their teaching objectives and their students' mental states. This could enable them to adapt their teaching methods to the needs of their students for greater success and a more reliable profession.

FINAL THOUGHTS

A science that focuses on brain function, by definition, has implications for education. Although neuroscientists cannot tell educators what to teach and how to teach, educators need to know some basic information, for example, that brains have their own reward system (dopaminergic hormones) that supportively facilitates knowledge retention and faculties (Spitzer, 2002; 2006), and that external and internal rewarding complemented by slightly stressful emotions provide an appropriate principle for brain-based learning whereas mere stress and socio-emotional deprivation can affect stress-induced synaptic changes (Braun & Bock, chapter 2).

It is also important to consider that memory can no longer be seen as a passive storage of single elements, but rather as an active process of reconstructing stored templates (Braun & Bock, chapter 2). Memory is an integrative system that is based on hippocampal activities of repetitive presentations, establishing a loop between cortex and hippocampus that causes repetitive presentations so that effective and stable learning is initiated. Learning is not only and not mainly accomplished by the explicit storage of verbal (explicit) knowledge. Particularly in music, the specific musical abilities are non-verbal and don't refer to memory but to representations instead. The hippocampus plays an essential role in the formation of new memories, and functions more as a "trainer" than as a "protector" of consolidated memories.

Since music learning goes far beyond mere memorization, persistent acquisition of skills and knowledge is supported by long-term procedures of acting and practicing. Results from this praxial approach are more effective and economic (Braun & Bock, chapter 2) and cause a decrease of brain activation, whereas intensive short-term memory training calls for much stronger concentration on particular isolated items and, therefore, exhibits a significant increase of brain activation (Altenmueller et al., 2000).

What is important for music teachers to know is how to react to an observed behavior that is called "amusia" (see Altenmueller, chapter 9). In the rare cases where this can be traced back to brain damages it is a clear mental disorder. However, there are many cases that are associated with the inability to sing in tune or to differentiate between pitches. It is not quite clear whether this disability is congenitally determined or whether it is caused by a lack of environmental stimulation. If it is congenital a correction or compensation will be difficult. In this case, neural anomalies should be found that correlate with the observed behavior. Although research has presented first results that point at neuronal correlates, there is in general very little information available to date, and many questions still remain open (Altenmueller, chapter 9). Without neural correlates a teacher can hardly decide whether a deficiency in music perception and performance (singing) has its origins in genetic or social deficits. As the German educator Donata Elschenbroich has put it: "Not to be musical is learned!" (Elschenbroich, 2001, 212). By this statement she wants to make it quite clear that it is a social misbehaviour to restrain children from the best possible opportunities to learn and experience music. This terminates in making them unmusical, i.e. to diminishing or even avoiding their musical potential. Therefore, it is probably more often a social and political issue rather than a genetic deficit that children fail to benefit from an education according to their potential. In music pedagogy we should be very cautious not to use the classification of amusia as an excuse for not teaching these individuals. As animal research has shown, humans are prototypically imitative audiovocal learners, i.e. they are able to match pitches correctly and to integrate them into a metric structure just by ear (Brown, 2007). This is equally basic in language and song acquisition. Therefore, if there is a structural brain anomaly that is responsible for a particular deficiency in the recognition of fine pitch differences, there might be other properly working abilities such as time keeping, rhythm perception, structural differentiation etc. that can also be subsumed under the term of musical aptitude (musicality, *musia*). Because of the complex interaction of neural processes related to music processing, it is clear without doubt that further research on so-called amusia is definitely needed for a better understanding of the actual causes of deficits in music processing in children so that educators can respond to them more appropriately.

This volume has attempted to illustrate how the educational implications of brain science are now being pursued rigorously in attempts to restructure music classrooms. Students do not appear to learn to their full potential, and yet society is demanding more and more from high school graduates and more from the school experience in general. Students in all disciplines much know more than how to repeat historical facts and execute rote mathematical procedures. Students need to acquire learning skills that apply not only to music, but across the curriculum, and beyond school as well. This can be accomplished if educational practice is based on what is known about how people learn and reason. Brain science provides this base for educational practice.

Neurodidactic researchers believe that effective educational practice should incorporate what brain scientists know about learning and instruction. The findings of brain science can, and should, serve as the foundation for instructional innovations and educational reform. The research described in this volume provides compelling examples of how brain science can contribute to educational practice.

During the past several decades, we have learned more about the brain than in all of recorded history, but there is much more to learn. As exciting as the new development in the neurosciences are, the dialogue that has begun between neuroscientists, musicians, cognitive scientists, and educators is even more exciting. For the first time, we're seeing substantive conversations between those who are conducting the research and those who are looking for applications of the research.

A start has been made, but many challenges remain for both educators and neuroscience researchers. First, we believe the research community needs to do a better job of making its methods and results comprehensible and accessible to music teachers, school administrators, and parents. End users of the research need declarative and procedural knowledge about the brain, and they need to know when and why to use that knowledge, if they are to view the enterprise as meaningful. There is a complementary challenge for educators, particularly those who control school systems, buildings, and working conditions. They must provide working and learning environments where teachers can appropriate the cognitive neuroscience perspective. Schools need to cultivate and embrace an environment that cherishes professional development, an environment that provides high-quality staff development, encourages input from teachers, and allows time for changes to occur. Finally, the neuroscience community itself must strive to assure that a coherent research program continues to evolve, a research program that incorporates the insights from cognitive neuroscience, builds on and refines the results and methods in the information-processing tradition, and pursues more recent insights about the importance of context, society, and culture for our understanding of how children learn. Given how little we currently know, research within each of the paradigms should be viewed as complementary rather than competing. We must exploit what we already know about learning and the brain, and we must strive to further our understanding. To the extent that current practices are based on common sense and outmoded theories, neuroscience research provides a needed corrective. The research in this volume illustrates how our attempts to apply research to real-world problems of learning and teaching can also serve to advance neuroscience research. Ideally, there should be continuous feedback from theory to practice to theory, and an institutional structure to support this interaction. With each interaction of this feedback loop, we would then be able to improve educational outcomes for children and deepen our understanding of learning and the brain. Rarely does neuroscience prove that a particular classroom strategy works, but the information coming from the neurosciences certainly can provide a more informed basis for the decisions we make in our schools and classrooms.

REFERENCES

Altenmueller, E. (2001). How many music centres are in the brain? In R. Zatorre & I. Peretz (Eds.), *The biological foundations of music* (Vol. 930, pp. 273 - 280). New York: Annals of the New York Academy of Sciences.

Altenmueller, E., & Gruhn, W. (1997). *Music, the brain and music learning* (Vol. 2). Chicago: G.I.A. Publ. Inc.

Altenmueller, E., Gruhn, W., Parlitz, D., & Liebert, G. (2000). The impact of music education on brain networks: evidence from EEG-studies. *International Journal for Music Education*(35), 47 - 53.

Arbib, M. A. (2005). From monkey-like action recognition to human language: an evolutionary framework for neurolinguistics. *Behavioral Brain Science, 28*(2), 105 - 124; 125 - 167.

Ayotte, J., Peretz, I., & Hyde, K. (2002). Congenital amusia: a group study of adults afflicted with a music-specific disorder. *Brain, 125*(Pt 2), 223 - 224.

Baddeley, A. (2003). Working memory and language: an overview. *Journal of Communication Disorders, 36*, 189 - 208.

Bangert, M., & Altenmueller, E. (2003). Mapping perception to action in piano practice: a longitudinal DC-EEG-study. *BMC Neuroscience, 4*, 26 - 36.

Binkofski, F., & Buccino, G. (2006). The role of ventral premotor cortex in action execution and action understanding. *Journal of Physiology Paris, 99*(4 - 6), 396 - 405.

Birbaumer, N., Ghanayim, N., Hinterberger, T., Iversen, I., Kotchoubey, B., Kubler, A., Perelmouter, J., Taub, E., & Flor, H. (1999). A spelling device for the paralysed. *Nature, 398*, 297 - 298.

Blakemore, S.-J., & Frith, U. (2005). *The learning brain. Lessons for education*. Oxford: Blackwell Publ.

Brown, S. (2007). Contagious heterophony: A new theory about the origins of music. *Musicae Scientiae, 11*(1), 3 - 26.

Brown, S., Martinez, H. J., Hodges, D. A., Fox, P. T., & Parsons, L. M. (2004). The song system of the human brain. *Brain Research. Cognitive Brain Research, 20*(3), 363 - 375.

Bruer, J. T. (1999). *The myth of the first three years. A new understanding of brain development and lifelong learning*. New York: The Free Press.

Buccino, G., Vogt, S., Ritzl, A., Fink, G. R., Zilles, K., Freund, H.-J., & Rizzolatti, G. (2004). Neural circuits underlying imitation learning of hand actions. An event-related fMRI study. *Neuron, 42*(2), 323 - 334.

Caspary, R. (Ed.). (2006). *Lernen und Gehirn. Der Weg zu einer neuen Pädagogik.* Freiburg: Herder.

Chalmers, D. J. (2000). What is a neural correlate of consciousness? In T. Metzinger (Ed.), *Neural correlates of consciousness* (pp. 17 - 39). Cambridge, MA.: MIT Press.

Chan, A. S., Ho, Y.-C., & Cheung, M.-C. (1998). Music training improves verbal memory. *Nature, 396*, 128.

Chasin, M., & Russo, F. A. (2004). Hearing AIDS and music. *Trends in Amplification, 8*(2), 35 - 47.

Colwell, R., & Richardson, C. (Eds.). (2002). *The new handbook of research on music teaching and learning*. Oxford, New York: Oxford University Press.

Elbert, T., Pantev, C., Wienbruch, C., Rockstroh, B., & Traub, E. (1995). Increased cortical representation of fingers of the left hand in string players. *Science, 270*, 305 - 307.

Elliott, D. J. (1995). *Music matters. A new philosophy of music education*. New York, Oxford: OUP.

Elschenbroich, D. (2001). *Weltwissen der Siebenjährigen*. München: Kunstmann.

Fitch, W. T. (2006). The biology and evolution of music: A comparative perspective. *Cognition*, in pr.

Gallese, V., Fadiga, L., Fogassi, L., & Rizzolatti, G. (1996). Action recognition in the premotor cortex. *Brain, 119*(2), 593 -609.

Gaschler, K. (2006). Die Entdeckung des Anderen. *Gehirn & Geist* (10), 26 - 33.

Gaser, C., & Schlaug, G. (2004). *Structural brain differences between musicians and non-musicians.* Paper presented at the ICMPC 8, Evanston, IL.

Gembris, H. (1998). *Grundlagen musikalischer Begabung und Entwicklung* (Vol. 20). Augsburg: Wißner.

Gfeller, K., Olszewski, C., Rychener, M., Sena, K., Knutson, J. F., Witt, S., & Macpherson, B. (2005). Recognition of "real world" musical excerpts by cochlear implant recipients and normal hearing adults. *Ear and Hearing, 26*(3), 237 - 250.

Greenough, W. T., & Black, J. E. (1992). Induction of brain structure by experience. Substrates for cognitive development. In M. Gunnar & C. Nelson (Eds.), *Minnesota Symposia on Child Psychology* (pp. 155 - 200). Hillsdale, NJ: Erlbaum.

Greenough, W. T., Larson, J. R., & Withers, G. S. (1985). Effects of unilateral and bilateral training in a reaching task in dendritic branching of neurons in the rat motor-sensory forelimb cortex. *Behavioral and Neural Biology, 44*, 301 - 314.

Gruhn, W. (1997). Music learning - neurobiological foundations and educational implications. *Research Studies in Music Education* (9), 36 - 47.

Gruhn, W. (2005a). *Der Musikverstand. Neurobiologische Grundlagen des musikalischens Denkens, Hörens und Lernens* (2. völlig neu überarbeitete ed.). Hildesheim: Olms.

Gruhn, W. (2005b). Understanding musical understanding. In D. J. Elliott (Ed.), *Praxial music education. Reflections and dialogues.* (pp. 98 - 111). New York: Oxford University Press.

Hannon, E. E., & Trehub, S. E. (2005). Tuning in to musical rhythms: infants learn more readily than adults. *Proceedings of the National Academy of Sciences of the USA, 102*(35), 12639 - 12643.

Haslinger, B., Erhard, P., Altenmueller, E., Schroeder, U., Boecker, U., & Ceballos-Baumann, A. O. (2005). Transmodal sensorimotor networks during action observation in professional pianists. *Journal of Cognitive Neuroscience, 17*(2), 282 - 293.

Herrmann, U. (Ed.). (2006). *Neurodidaktik. Grundlagen und Vorschläge für gehirngerechtes Lehren und Lernen*. Weinheim: Beltz.

Hutchison, W. D., Davis, D. D., Lozano, A. M., Tasker, R. R., & Dostrovsky, J. O. (1999). Pain-related neurons in the human cingulate cortex. *Nature Neuroscience, 2*(5), 403 - 405.

Hüther, G. (2002). *Bedienungsanleitung für ein menschliches Gehirn.* Göttingen: Vandenhoeck & Ruprecht.

Jarvis, E. (2004). Learned birdsong and the neurobiology of language. In H. P. Zeigler & P. Marler (Eds.), *Behavioral Neurobiology of Birdsong* (Vol. 1016, pp. 749 - 777). New York: The New York Academy of Sciences.

Juraska, J. M., Henderson, C., & Muller, J. (1984). Differential rearing experience, gender and radial maze performance. *Developmental Psychobiology, 17*, 209 - 215.

Kaiser, K. A., & Johnson, K. E. (2000). The effect of an interactive experience on music majors' perception of music for deaf students. *Journal of Music Therapy, 37*(3), 222 - 234.

Kotynek, M., Norton, A., Overy, K., Winner, E., & Schlaug, G. (2006). Functional cerebral correlates of instrumental training in children: A longitudinal study. *Human Brain Mapping.*

Liebert, G. (2001). *Auswirkungen musikalischen Kurzzeitlernens auf kortikale Aktivierungsmuster.* Unpublished dissertation, Med.Hochschule, Hannover.

McDermott, H. J. (2004). Music perception with cochlear implants: a review. *Trends in Amplification, 8*(2), 49 - 82.

Merker, B. (2005). *Between perception and performance: vocal learning as key constraint on the path to music and language.* Paper presented at the The Neurosciences and Music II, Leipzig.

Metzinger, T. (Ed.). (2000). *Neural correlates of consciousness. Empirical and conceptual questions.* Cambridge MA: MIT Press.

Mooney, R. (2004). Synaptic mechanisms for auditory-vocal integration and the correction of vocal errors.*Annals of the New York Academy of Sciences,* vol. 1016, 476 – 494.

Nakata, T., Trehub, S. E., Mitani, C., Kanda, Y., Shibasaki, A., & Schellenberg, G. E. (2005). Music recognition by Japanese children with cochlear implants. *Journal of Physiological Antropology and Applied Human Science, 24*(1), 29 - 32.

Overy, K., Norton, A., Cronin, K., Winner, E., & Schlaug, G. (2005). Examining rhythm and melody processing in young children using fMRI, *Annals of the New York Academy of Sciences*, vol. 1060,210 – 218.

Pascual-Leone, A., Cammarota, A., Wasserman, E. M., Brasil-Neto, J. P., Cohen, L. G., & Hallett, M. (1993). Modulation of motor cortical outputs to the reading hand of Braille readers. *Annals of Neurology, 34*, 33 - 37.

Peretz, I., Brattico, E., & Tervaniemi, M. (2005). Abnormal electrical brain responses to pitch in congenital amusia. *Annals of Neurology, 58*(3), 478 - 482.

Peretz, I., Champod, A. S., & Hyde, K. (2003). Varieties of musical disorders. The Montreal Battery of Evaluation of Amusia. In G. F. Avanzini, C.; Minciacchi, D., Lopez, L.; Majno, M. (Ed.), *The Neurosciences and Music* (Vol. 999, pp. 58 - 75). New York: The New York Academy of Sciences.

Phillips-Silver, J., & Trainor, L. J. (2005). Feeling the beat: movement influences infant rhythm perception. *Science, 308*(5727), 1430.

Preiss, G. (Ed.). (1998). *Neurodidaktik. Theoretische und praktische Beiträge.* Herbolzheim: Centaurus.

Ratey, J. J. (2001). *A user's guide to the brain. Perception, attention and the four theaters of the brain.* New York: Pantheon.

Rauscher, F. H., Bowers, M. K., Dettlaff, D. M., & Scott, S. E. (2002). *Effects of environmental, social, and auditory enrichment on maze learning in rats: implications for arousal.* Paper presented at the Conference on Cognitive Neuroscience, San Francisco.

Rauscher, F. H., Shaw, G. L., Levine, L. J., Wright, E. L., Dennis, W. R., & Newcomb, R. (1997). Music training causes long-term enhancement of preschool children's spatial-temporal reasoning. *Neurological Research, 19*, 1-8.

Rizzolatti, G. (1996). Premotor cortex and the recognition of motor actions. *Cognitive Brain Research*(3), 131 - 141.

Roth, G. (2006). Möglichkeiten und Grenzen von Wissensvermittlung und Wissenserwerb. In R. Caspary (Ed.), *Lernen und Gehirn* (pp. 54 - 69). Freiburg: Herder.

Saffran, J. R., Loman, M. M., & Robertson, R. R. (2000). Infant memory for musical experiences. *Cognition, 77*(1), 15 - 23.

Scharff, C., & Haesler, S. (2005). An evolutionary perspective on FoxP2: Strictly for the birds? *Current Opinion in Neurobiology, 15*, 694 - 703.

Schlaug, G. (2003). The brain of musicians. In I. Peretz & R. Zatorre (Eds.) *The cognitive neuroscience of music* (Vol. 366 - 381). New York: Oxford University Press.

Schlaug, G., Jaencke, L., Huang, Y., & Steinmetz, H. (1995). In vivo evidence of structural brain asymmetry in musicians. *Science, 267*, 699 - 701.

Schlaug, G., Norton, A., Overy, K., & Winner, E. (2005). Effects of music training on the child's brain and cognitive development, *Annals of the New York Academy of Sciences* vol. 1060, 219 – 230.

Spitzer, M. (2002). *Lernen. Gehirnforschung und die Schule des Lebens.* Heidelberg: Spektrum Akademischer Verlag.

Spitzer, M. (2006). *Das Gehirn. Eine Gebrauchsanleitung.* Reinbek: Rowohlt.

Stern, E. (2005). Pedagogy meets neuroscience. *Science, 310*(5749), 745.

Stern, E., Grabner, R., & Schumacher, R. (2005). *Lehr-Lern-Forschung und Neurowissenschaften: Erwartungen, Befunde und Forschungsperspektiven* (Vol. 13). Berlin: Bundesministerium für Bildung und Forschung.

Tomie, J., & Wishaw, I. Q. (1990). New paradigms for tactile discrimination studies with the rat: Methods for simple, conditional, and configural discriminations. *Physiology and Behavior, 48*, 225 - 231.

Winner, E., & Hetland, L. (2000). The arts in education: evaluating the evidence for a causal link. *The Journal of Aesthetic Education, 34*(3-4), 3 – 10.

In: Neurosciences in Music Pedagogy
Editors: W. Gruhn and F. Rauscher, pp. 283-296
ISBN: 978-1-60021-834-7

GLOSSARY

Adrenocorticotropic hormone (ACTH), a hormone secreted from the anterior pituitary in response to corticotropin-releasing hormone from the → hypothalamus. Its role is to stimulate the adrenal cortex, which mediates the stress response through the production of glucocorticoids such as → cortisol.

Amusia, also commonly called tone deafness, describes the inability to recognize or reproduce musical tones or rhythms. Amusia is believed to be congenital (present at birth) or acquired sometime later in life (as from brain damage). The term amusia is composed of the prefix "a" (alpha privative) and "musia", and literally means the lack of music or musical abilities. Because congenital amusical individuals are able to process and recognize speech, congenital amusia is viewed as a sub-class of learning disabilities that specifically affects musical capacity.

Amygdala (pl. amygdalae; from Latin: corpus amygdaloideum; Greek: amygdale) is a group of neurons located deep within the medial temporal lobes of each hemisphere in complex vertebrates, including humans. The amygdalae perform a primary role in the processing and memory of emotional reactions, and are considered part of the → limbic system. Whereas early investigations mapping the function of the amygdala demonstrated its predominant role for fear processing, recent studies have found that the amygdala plays an important role in the formation and storage of memories associated with emotions and is involved in affective information processing in general.

Aphasia, a language disorder caused by a lesion to the language processing area(s) of the brain. If → Broca's area in the language-dominant hemisphere is affected, the person is unable to clearly pronounce and articulate words (motor or Broca's aphasia). If → Wernicke's area is affected, the person has difficulty understanding words and concepts. A massive cerebral disorder of several areas in the language dominant hemisphere produces a general communication disorder that includes difficulties in reading and understanding as well as in articulation. This is referred to as global aphasia.

Arousal, a physiological and psychological state that involves the activation of the reticular activating system in the brain stem, the autonomous nervous system, and the endocrine system. In emotion research, the degree or level of a listener's arousal can be measured by psychophysiological indicators such as increased heart rate, blood pressure, and a condition of alertness and readiness to respond.

Axon, appendix of a → neuron that carries electric impulses from the neuron to other neurons via → synaptic connections.

Basal ganglia, a → bilateral group of nuclei located near the center of the brain and interconnected with the cerebral cortex, → thalamus and brainstem. In mammals, the basal ganglia are associated with a variety of functions: motor control, cognition, emotion and learning. The following nuclei are part of the basal ganglia: caudate nucleus, putamen (which combined together form the striatum), globus pallidus, → nucleus accumbens, subthalamic nucleus and the → substantia nigra. Different parts of the basal ganglia serve completely different functions. The nucleus accumbens has been associated with motivation and reward, both of which are important for playing a musical instrument or singing. The basal ganglia are also responsible for the initiation and precise modulation of motor function and the long-term storage of motor programs.

Bilateral, consisting of two specular halves, such as the two hemispheres of the brain.

Biofeedback, a conscious reaction to autonomous body functions. Biofeedback involves measuring a person's bodily functions (e.g., respiration, heart rate, and muscle tension) and simultaneously conveying this information to the person in order to raise his/her awareness and conscious control of these physiological processes. The technique is sometimes used by psychologists to help tense or anxious clients learn to relax.

BOLD, abbreviation for "blood-oxygenation-level-dependent". The term is typically used as an adjective to modify "effect," "contrast," or "signal" to describe aspects of what is measured by → functional magnetic resonance imaging (fMRI). FMRI detects regional changes in brain blood flow associated with neural activity. Neural activity in the brain requires energy, which is delivered locally to those areas that need it with an increase in regional blood flow. Because oxygenated blood has different magnetic properties than deoxygenated blood, local changes occur in the magnetic field of activated brain areas, and these changes are described as the BOLD signal that is measured by fMRI.

Brain based learning and teaching → Neuropedagogy

Brain imaging, a technique used to visualize the activity of the brain by transforming empirical data (like electric potential or blood oxygenation) into graphs or colored images. → EEG, → MEG, → ERP, → MRI, → PET, → NIRS, → DTI.

Broca's area, a left hemisphere brain area (BA 44) that is responsible for sequential processing in speech production. Although it is located in the left hemisphere, 20% of left-handed individuals have their language dominant areas in the right hemisphere. This area is named after the French neurologist and anthropologist Paul Broca (1824 – 1880).

Brodmann areas (BA), regions in the cortex defined in many different species based on the brain's cytoarchitecture. Different areas are associated with different brain functions. The German neuroanatomist Korbinian Brodmann (1868 – 1918) assigned numbers (BA 1 – BA 52) to various brain regions by analyzing each area's cellular structure starting from the central → sulcus.

Cardiovascular system (circulartory system), includes the heart, blood, and blood vessels. When a person inhales, oxygen is taken into the lungs. Blood is pumped from the heart to the lungs, where it is oxygenated. The oxygen-rich blood then travels back to the heart where it is pumped through arteries and capillaries to the whole body, delivering oxygen to all the cells in the body including bones, skin and other organs. Veins then carry the oxygen-depleted blood back to the heart, where the cycle begins again.

Cell assembly, a spatially distributed set of cells (i.e., neural network) that are activated in a coherent fashion. The idea of cell assemblies was first put forth by Donald Hebb in 1949. Hebb proposed that the connections between large groups of neurons would alter in such a way that if neuron A fired and activated neuron B, which fired and activated neuron C, which in turn fired and activated neuron A again, then the synapses linking these neurons would all strengthen and increase the possibility of future activation. This form of positive feedback leads to the strengthening of neural connections and the formation of cell assemblies consisting of many thousands of cells.

Cerebellum, (from Latin: "little brain") is a region of the brain that plays an important role in the integration of sensory perception and motor output. Many neural pathways link the cerebellum with the motor cortex, which sends information to the muscles causing them to move, and the spinocerebellar tract, which provides feedback on the position of the body in space (proprioception). The cerebellum integrates these pathways, using the constant feedback on body position to fine-tune motor movements. Observations in patients with cerebellar damage demonstrated that these patients predominantly suffered from motor deficits in relation to the precise timing, coordination and adjustment of movement (ataxia). These observations led to the conclusion that the cerebellum is predominantly a motor control structure. However, modern research shows that the cerebellum has a broader role in a number of key cognitive functions, including attention and the processing of language, music, and other temporal stimuli.

Cerebral cortex, → cortex

Cerebral palsy refers to a number of neurological disorders that appear in infancy or early childhood and permanently affect body movement and muscle coordination. Although it affects muscle movement, it does not originate from problems in the muscles or nerves, but rather from abnormalities in parts of the brain that control muscle movements.

Chill, physiological and psychological response to music during intensive emotional arousal when an acoustic signal (a tune, a chord, a sudden outburst of sound etc.) provokes an immediate thrill or shiver typically accompanied by a pilo-erection and other psychophysical reactions. Chills are interpreted as an evolutionary early signal system that accompanies the separation call when female primates lose eye contact with their offspring.

Cingulate cortex, (from Latin: cingulum = belt) the part of the → cortex in the medial area of the brain that surrounds the → corpus callosum.

Cochlear implant (CI), a small electronic device (artificial inner ear) for severely hearing impaired or deaf patients with an intact auditory nerve. The CI consists of a microphone to collect sounds from the environment, a behind-the-ear (BTE) speech processor for processing the sounds picked up by the microphone, and a transmitting system (coil and sender) that receives the signals and converts them into electric impulses. These are then transmitted to 22 electrodes on a thin, self-curling array implanted into the cochlea which stimulate the hearing nerve in the inner ear. Today, several companies offer highly-developed systems (Nucleus; Med-el) which apply different strategies (e.g. SPEAK, TEMPO) to achieve the most natural sound clarity. The strategies make use of the → tonotopic structure of the cochlea and combine locally and temporally staggered patterns of stimulation. After surgery, CI patients must learn how to interpret the sounds created by the implant. Because CI technology is mainly developed for language, the tonal differentiation necessary for music perception is still under development.

Contralateral, the representation of a body part or an activity on or relating to the opposite hemisphere (e.g., left hand, right hemisphere).

Corpus callosum, a thick bundle of millions of nerve fibers that connect the left and right cerebral hemispheres.

Cortex (cerebral cortex), the extensively wrinkled outer layer of the brain in vertebrates containing a high number of unmyelinated fibres (→ gray matter) consisting of cell nuclei and synapses. The cerebral cortex in humans can be differentiated into grooves called gyri and sulci; approximately two thirds of the cortical surface is buried in the sulci. The phylogenetically more ancient part of the cerebral cortex, the hippocampus, consists of five layers of neurons, while the more recent neo-cortex has six basic layers. Relative variations in thickness or cell type (among other parameters) allow researchers to distinguish among different neocortical architectonic fields. Important parts are primary and secondary sensory areas such as the auditory, somatosensory and visual cortex, the motor cortex, tertiary areas processing associative functions (e.g. prefrontal and parietal cortex) and emotional areas located under the surface (e.g. cingular cortex, insular cortex).

Cortisol, a corticosteroid hormone that induces the breakdown of cellular proteins. Cortisol is produced by the adrenal cortex and is involved in the response to stress. It is secreted in higher levels during the body's 'fight or flight' response, and is responsible for several stress-related changes in the body.

DC-EEG, a special type of → EEG, recording the slow negative shifts of cortical direct current (DC) potentials.

Dendrite, tree-like extensions of a neuron with a highly branched structure that receive electric impulses from other neurons. Dendritic spines grow along the branches where → axons can dock..

Diffusion tensor imaging (DTI), a recently developed variation of → magnetic resonance imaging (MRI) that allows scientists to study the connections between different brain areas for the first time. In DTI, radiologists use specific radio-frequency and magnetic field-gradient pulses to track the movement of water molecules in the brain. In most brain tissue, water molecules diffuse in several different directions, typically along the length of axons, whose coating of white, fatty myelin holds them in. Scientists can create images of axons by analyzing the direction of water diffusion.

Dopamine, a → neurotransmitter that affects the brain processes that control movement, emotional response, and ability to experience pleasure and pain.

Dyslexia, a specific learning disability affecting the acquisition of literacy skills in children with normal intelligence. The disability is considered to be genetic in origin and is estimated to affect 5-15% of the population. The specific cause of such unexplained literacy difficulties remains under investigation, but it is widely accepted that difficulties with the phonological aspects of language play a major role. Other deficit areas reported include short-term memory, motor skills and multi-sensory coordination, while creativity is often described as a strength shown by dyslexic children. Research continues into the precise nature of this complex disability.

Electroencephalography (EEG), a method employed to record the electrical activity of the brain using electrodes fixed on the skull. Cortical neurons produce electric field potentials which can be collected from the surface of the skull. EEG mirrors the activation or inhibition of larger groups of neuronal → cell assemblies. In contrast to other imaging methods, EEG directly measures neuronal activity and interactions with a very high time resolution in the range of milliseconds. Although the spatial resolution with 2 to 5 cm^2 is relatively low (depending upon the number of electrodes applied to the skull), the advantage of EEG is that inter-electrode coherence and cross correlation as well as task-related synchronization or de-synchronization of signals provide information of extremely fast neuronal interactions. With these tools, EEG is an appropriate method for understanding the function and dynamics of larger brain networks established in the context of musical learning. It is a safe and relatively inexpensive method that has achieved greater value in recent years due to more sophisticated software that permits the routine calculation of complex neuronal interactions.

Endorphins, endogenous opioid biochemical peptides produced by the pituitary gland and the → hypothalamus in vertebrates. They resemble opiates in their ability to produce analgesia and a sense of well-being. In other words, they might work as "natural pain killers." Endorphins are often found at high levels among individuals afflicted with chronic pain disorders. Prolonged, continuous exercise contributes to an increased production and release of endorphins, resulting in a sense of euphoria that has been popularly labelled "runner's high."

Event related potentials (ERP), a technique derived from EEG that measures the response to a perceived stimulus. When a stimulus is presented several times and the ongoing EEG-signal is analyzed time-locked with respect to the stimulus onset, even minimal changes in neuronal activity caused by the incoming stimulus can be detected and their source localized. This technique allows for the analysis of subsequent rapid cognitive processes during music perception or music production by identifying specific components of the signal.

Focal dystonia, a movement disorder which causes uncontrolled movements or cramps in one part of the body. Musicians mainly suffer from this disorder in hand (cramping of one or more fingers) or embouchure muscles which restrict movement and eventually may prevent them from performing.

Functional magnetic resonance imaging (fMRI), a non-invasive brain imaging technique used to determine which parts of the brain are activated during physical sensation, perceptual processing, motor activity, or cognition. Functional MRI measures regional changes in brain blood flow associated with neural activity specific to a particular behavior. The degree of regional blood flow (see also → BOLD) correlates with the intensity and duration of neural activity in the brain. Functional MRI offers spatial resolution on the order of millimeters and time-resolution on the order of a second. Furthermore, using "sparse sampling" short auditory stimuli may be presented in silence with collection of the imaging data (and the associated scanner noise) occurring just after stimulus presentation. In summary, fMRI is a safe (although expensive) technique which can be used in children and in longitudinal studies.

Glial cells, non-neuronal cells that provide support and nutrition, maintain homeostasis, form myelin, and participate in signal transmission in the nervous system. Glial cells have important developmental roles, guiding migration of neurons in early development and producing molecules that modify the growth of axons and dendrites. Recent findings have indicated that glial cells also play an active role in synaptic transmission.

Gonadotropin-releasing hormone (GnRH), a peptide hormone that is synthesized and released by the → hypothalamus. It is responsible for the release of follicle-stimulating hormone (FSH) and luteinizing hormone (LH) from the anterior pituitary. GNRH secretion is necessary for correct reproductive function. This single hormone controls follicular growth, ovulation, and corpus luteum maintenance in the female, and spermatogenesis in the male. GNRH activity is very low during childhood, and is activated at puberty.

Gray (grey) matter, a major component of the central nervous system, consists predominantly of nerve cell bodies, → glial cells, capillaries, and → dendrites. Gray matter is distributed at the surface of the cerebral hemispheres (→ cerebral cortex) and in the → cerebellum (cerebellar cortex), as well as in the spinal cord. Gray matter is mainly responsible for information processing. It transmits sensory or motor stimuli to interneurons of the central nervous system.

Gyrus (gyri), a ridge on the → cerebral cortex that is generally surrounded by one or more → sulci.

Hebbian learning rule states that the connections between two neurons will be strengthened if the neurons fire simultaneously. In 1949 Donald Hebb wrote: "When the axon of cell A is near enough to excite a cell B and repeatedly or persistently takes part in firing it, some growth process or metabolic change takes place in one or both cells such that A's efficiency, as one of the cells firing B, is increased."

Heschl's gyrus (transverse temporal gyrus), the convolution that runs transversely on the surface of the temporal lobe and is identical with the primary auditory area (BA 41). It is part of a language and music processing network that includes → Broca's area, → Wernicke's area, the → planum temporale, and the anterior superior → insular cortices.

Hippocampus, the seahorse-shaped part of the brain inside the temporal lobes, adjacent to the → amygdala. It forms a part of the → limbic system, playing a role in memory and spatial navigation.

Homunculus (motor, sensory), a human-shaped figure that reflects the motor and sensory representation of the body in the cerebral cortex. The homunculus is characterized by its large hands, lips, and face compared to the rest of the human body, corresponding to the superior sensory and motor innervations of these body parts. The somatosensory strip is located behind the central → sulcus and the motor strip is located in front of the central sulcus.

Hypothalamus (from Greek = under the thalamus), a part of the mammalian brain located below the → thalamus. It occupies a major portion of the ventral region of the diencephalon and regulates metabolic processes. The hypothalamus links the nervous system to the endocrine system and controls body temperature, hunger, thirst, and circadian cycles.

Hypothalamic-pituitary-adrenal axis (HPA axis) consists of the → hypothalamus, the pituitary gland which is located below the hypothalamus, and the adrenal or suprarenal gland, a small organ located at the top of each kidney. The highly complex homeostatic interactions between these three organs constitute the HPA axis, a major part of the neuroendocrine system that controls the levels of → cortisol (the "stress hormone") and other important stress-related hormones and regulates various body processes. There is a general belief that hyperactivity of the HPA-axis is associated with depression and anxiety disorders.

Immunoglobulin E (IgE), an antibody subclass (known as "isotypes") found only in mammals. Although IgE is typically the least abundant isotype, it is capable of triggering the most powerful immune reactions. This class of antibodies is produced in the lungs, skin, and mucous membranes and is responsible for allergic reactions.

Insula (insular cortex), a structure of the → cortex located deep in the brain's lateral surface within the lateral → sulcus which separates the temporal lobe and inferior parietal cortex. These overlying cortical areas are known as → opercula, and parts of the frontal, temporal and parietal lobes form opercula over the insula.

Ipsilateral, the representation of a body part or an activity on or relating to the same hemisphere.

Limbic system (from Latin: limbus = "border" or "edge") includes the structures in the human brain involved in emotion, motivation, and memory of emotional content. The limbic system influences the formation of memory by integrating emotional states with stored memories of physical sensations. It includes many different cortical and subcortical brain structures such as the → amygdala, → cingulate cortex, → hippocampus, → hypothalamus, mammilary body, → nucleus accumbens and the ventral prefrontal cortex.

Magnetic resonance imaging (MRI), a non-invasive brain imaging technique employed to examine brain structure, rather than function (see also → fMRI). MRI uses powerful magnets and radio waves. The MRI scanner contains a magnet that produces a magnetic field approximately 10,000 times greater than the magnetic field of the earth. The magnetic field forces hydrogen atoms in the body to line up in a certain way (similar to how the needle on a compass moves when it is held near a magnet). When radio waves are sent toward the lined-up hydrogen atoms, they bounce back, and a computer records the signal. Different types of tissues send back different signals.

Magnetoencephalography (MEG) is closely related to → EEG. It measures the magnetic fields produced by the electric dipoles of the neurons. MEG has excellent temporal resolution, and its spatial resolution is superior to the EEG due to the fact that the magnetic fields are less distorted by the brain's surrounding liquid and the skull. Furthermore, compared to EEG, MEG is far better at assessing activity of the auditory cortex localized in the upper temporal plain within the Sylvian fissure. However, MEG remains an expensive technique that is often restricted to specialized centers because it requires a powerful antimagnetic shielding to isolate it from the terrestrial magnetic field, as well as very sensitive super-conducting quantum interference devices cooled with fluid helium.

Melatonin, a hormone produced by the pineal gland, a gland about the size of a pea located in the center of the brain. Melatonin has been studied for its effects on circadian rhythm sleep disorders (such as jet lag and delayed sleep phase syndrome), the treatment of cancer, immune disorders, cardiovascular diseases, depression, seasonal affective disorder (SAD), and sexual dysfunction.

Mental imagery, a mental representation of sound that is not physically present.

Mental rehearsal, a practice technique applied by instrumentalists and vocalists which consists of intense imagery of a difficult motor task (e.g. roulades) or a sound sequence (e.g. intervals), or the coordination of independent motor movements without physically playing or singing.

Mental representation, the neuronal correlate of a sensory perception. Specialized neurons fire when they receive a stimulus to which they are determined. In a more general sense, the term is also used in cognitive psychology to refer to a mental state that corresponds to an outside phenomenon.

Mental training → mental rehearsal

Mirror neurons, a class of neurons that resonate to motor acts only as a consequence of mere observation. In the early 1990s Giacomo Rizzolatti and collaborators discovered it in the hand-controlling region of the premotor cortex (F 5) of macaques where the same neurons discharged when the monkeys just observed hand movements of the researcher, but only if the intention of that movement was clear. Therefore, mirror neurons were first associated with action understanding. Since then, research has shown that a mirror neuron system exists in humans as well as in some birds. Although the function of these neurons has not yet been fully determined, it is often associated with empathy and imitation (although macaques cannot imitate). Therefore, some researchers assume that mirror neurons might play a prominent role not only in the visuomanual matching process, but also in the imitative audiovocal process.

Mismatch negativity (MMN), a component of → event related potentials (ERP) during repetitive auditory stimulation. The higher negativity in ERP presents a neuronal signal which refers to an inattentive reaction on a deviation from a standard stimulation.

Mozart effect, a term coined by the popular press referring to the scientific finding that listening to a particular Mozart sonata led to enhanced performance on a → spatial-temporal task. Recent research suggests that the effect may be attributed to arousal or mood. The term "Mozart effect" has been overgeneralized and misapplied to refer to the effects of music instruction on cognitive performance.

Musical savants, individuals who exhibit exceptional musical skills in the context of developmental disability, such as mental retardation or autism. The skills typically appear in childhood (before adolescence) and include pitch recognition, memory for melody, and sensitivity to harmonic structure. Musical savant skills are typically expressed as precocity in vocal and/or instrumental reproduction of musical segments.

Near infra-red spectroscopy (NIRS), a brain imaging technique that measures changes of the cerebral concentration of oxygenated and deoxygenated hemoglobin (red blood corpuscles) induced by a cognitive task. Similar to → EEG, the imaging system uses up to 16 pairs of sensors (optrodes) that are positioned on the skull to measure the reflection of a deep infrared beam. It is a safe and reliable method with rare artifacts that is therefore appropriate for use in young children.

Neurodidactics → Neuropedagogy

Neuron, a nerve cell that consists of a cell body containing a nucleus, an → axon and a → dendritic tree that connects it to other neurons. Dendrites receive and conduct signals whereas axons send signals to other neurons. The contact positions where axons connect with spines of the dendrites are called the → synapses.

Neuropedagogy, a recently-developed branch in education theory that aims to link the procedures of teaching and learning to the mental state of the developing brain. It is also called brain-based learning. The philosophy behind this new approach intends to adapt the teaching method to children's brains instead of adapting children's learning to the curriculum.

Neuroplasticity → plasticity

Neurotransmitters, chemicals that are used in synaptic signal transmission to relay, amplify and modulate electrical signals between → neurons (i.e., transmission of the nerve impulse). As a nerve impulse, or action potential, reaches the end of a presynaptic axon, molecules of neurotransmitter are typically released into the synaptic space. Many types of chemicals act as neurotransmitter substances, including acetylcholine, → dopamine, norepinephrine, → serotonin, histamine, epinephrine, and nitric oxide. The action that follows activation of a receptor site may be either depolarization (an excitatory postsynaptic potential) or hyperpolarization (an inhibitory postsynaptic potential). A depolarization makes it more likely that an action potential will fire; a hyperpolarization makes it less likely that an action potential will fire.

Nucleus accumbens is located in the midbrain at the top of the brainstem. It plays an important role in the reward circuit. Its function is mainly based on two → neurotransmitters: → dopamine, which affects the ability to experience pleasure, and → serotonin, whose effects include satiety and inhibition. Animal studies have shown that all drugs increase the production of dopamine in the nucleus accumbens, while reducing that of serotonin.

Omega sign, an elaborated and reliably identifiable gross-anatomical feature of a subregion of the superior precentral gyrus. The configuration has the shape of an inverted

Greek letter Omega and is also referred to as 'hand knob' or 'genu'. It is associated with increased functional hand/finger movement representation.

Operculum, a part of the brain located in the posterior portion of the inferior frontal gyrus of the frontal lobe. One prominent part of the left operculum is → Broca's area, which plays an important role in speech production.

Peptides, the family of short molecules formed from the linking of various amino acids. Two or more amino acids are chained together by a bond called a "peptide bond," a special linkage in which the nitrogen atom of one amino acid binds to the carboxyl carbon atom of another. Many peptides, such as adrenocorticotropic hormone (ACTH), have physiological or antibacterial activity.

Planum temporale, the posterior superior surface of the superior temporal → gyrus in the cerebrum. It is a highly-lateralized brain structure involved in language and music processing.

Plasticity, the potential of neuronal networks to adapt to environmental conditions and perceived stimuli in order to accomplish particular tasks in the most economic and appropriate way. Neuroplasticity allows the → neurons in the brain to compensate for injury and disease, and to adjust their activities in response to new situations or to changes in their environment. Brain reorganization takes place by mechanisms such as "axonal sprouting" in which undamaged axons grow new nerve endings to reconnect neurons whose links were injured or severed. Undamaged axons can also sprout nerve endings and connect with other undamaged nerve cells, forming new neural pathways to accomplish a needed function. For example, if one hemisphere of the brain is damaged, the intact hemisphere may take over some of its functions. The brain compensates for damage in effect by reorganizing and forming new connections between intact neurons. In order to reconnect, the neurons need to be stimulated through activity. Neuroplasticity is therefore one of the basic tenets of learning.

Positron emission tomography (PET), a brain imaging technique that measures blood flow or brain metabolism indicating brain activity. Using this technique, radioactively labeled glucose or oxygen is injected (in former times also inhaled), after which subjects are asked to perform a given musical task. The radioactive substances accumulate in active brain regions, since firing neurons consume more glucose and oxygen. The decay of radioactivity results in the locally enhanced emission of positrons which can be detected by a PET-camera. An advantage of this method is that it provides excellent spatial resolution. The disadvantage is that it has poor temporal resolution in the range of 30 to 90 seconds, which renders it unable to track fast interactions of multiple cognitive processes. Furthermore, although the dosage of radioactivity is very small, it still is an invasive method, and difficult to apply in longitudinal studies or in studies of children and adolescents. In addition, since PET-technology requires a neighbored cyclotron to produce the isotopes it is costly and only available in a few specialized brain-imaging centers.

Serotonin, a → neurotransmitter believed to play an important role in the regulation of body temperature, mood, sleep, vomiting, sexuality, and appetite.

Somatotopic, the representation of body functions in brain maps according to the surface of the body (→ homunculus).

Spatial-temporal tasks, tasks that require combining separate elements of an object into a single whole by arranging objects in a specific spatial order to match a mental image. Spatial-temporal tasks require both spatial imagery and the temporal ordering of objects, abilities that may be necessary for proportional reasoning used in mathematics and scientific endeavours.

Striatum (ventral, dorsal), a subcortical part of the → basal ganglia system located below the → cerebral cortex. It receives projections from most cortical areas, and is to some extent the input nucleus of the basal ganglia system. The striatum is involved in motor and cognitive planning; it also plays an important role in the reward circuit. Dopamine is released in human ventral striatum in expectation of reward.

Sulcus (sulci), a fissure between two convolutions (→ gyri) of the brain.

Synapse, the contact between axons and dendrites for signal transmission. The synaptic gap is bridged by a chemo-electrical process. Healthy brain functions are based on the balance of synaptic excitation and inhibition which are essential for almost all functions, including the representation of sensory information, cognitive processes, and motor control. The inhibition of neuronal transmission is even more important for conducting selected information according to actual need, and for preventing the brain from getting swamped by too much information at one time. Repeated stimulation strengthens the synaptic connection.

Synaptogenesis, the formation of synapses. Although synaptogenesis occurs throughout a healthy person's lifespan, an explosion of synapse formation occurs during early brain development. As we age, old connections are deleted through a process called synaptic pruning. Synaptic pruning eliminates weaker synaptic contacts while stronger connections are kept and strengthened. Experience determines which connections will be strengthened and which will be pruned; connections that have been activated most frequently are preserved.

Tachistoscopic measurement, a technique used to study perception, memory, and learning in which a stimulus is displayed for a controlled duration using an apparatus called a tachistoscope. A tachistoscope presents the stimulus for such a short period of time (on the order of milliseconds) that the viewer does not have time to consciously process it, and so is therefore not consciously aware of having ever seen the message.

Testosterone, an anabolic steroid hormone from the androgen group. It is the principal male sex hormone. In both males and females, it enhances libido, energy, immune function, and protection against osteoporosis. On average, the adult male body produces about twenty to thirty times the amount of testosterone produced by the adult female.

Thalamus, (from Greek: bedroom, chamber), a large mass of → gray matter deeply situated in the forebrain. It is considered "the gate of consciousness" because almost all incoming sensory stimuli pass through the thalamus before reaching the cerebral cortex. It is the most important mass of neurons in the vertebrate brain. The thalamus is comprised of four parts: the hypothalamus, the epithalamus, the ventral thalamus, and the dorsal thalamus. Lesions within the thalamus usually result in severe somatosensory disturbances.

Thyroid hormone is the major hormone secreted by the thyroid gland. It is involved in controlling the body's metabolism, physical development, and the growth of neural tissue.

Tomography, an imaging technique that depicts the body or parts of the body and brain in precisely defined slices. There are many different types of tomography, including → computer tomography (CT); → positron-emission-tomography (PET); → magnetic resonance tomography (MRT, MRI);

Tonotopic, the representation of pitch frequencies in the cochlea from high to low. High pitches are located at the bottom of the cochlea (basal) whereas low pitches are represented at the end (apical).

Transcranial magnetic stimulation (TMS), a brain imaging technique based on the fact that neurons in the brain can be stimulated by applying strong magnetic pulses either singly or repetitively (rTMS). A specialized electromagnet is placed on the individual's scalp. This electromagnet generates short magnetic pulses, roughly the strength of an MRI scanner's magnetic field but much more focused. The magnetic pulses pass easily through the skull. Because they are short pulses and do not cause a static field (unlike MRI scanner fields), they can stimulate the underlying cerebral cortex (brain). Low frequency (once per second) TMS has been shown to induce reductions in brain activation; stimulation at higher frequencies (> 5 pulses per second) has been shown to increase brain activation. For example, reactions of the muscles of the hand can be analyzed for motor reactivity while stimulating the hand area of the motor cortex. Furthermore, cortical interactions can be assessed by applying two or three pulses in different cortical areas. Using this technique, researchers have demonstrated reduced trans-callosal (inter-hemispheric) inhibition in professional musicians compared to non-musicians. The use of TMS is not restricted to the motor cortex, but can also be applied to other brain areas in order to increase or decrease the excitability of neurons. With new TMS-devices, it is possible to "map" the excitability of the brain regions with a spatial resolution of about 2 to 4 cm^2. The presence of changes in the excitability of neurons in the motor cortex during a motor task in musical practice illustrates the impressive potential of this method.

Transfer effect, the ability to extend what has been learned in one context to new contexts. The amount of transfer that can occur between two domains appears to be dependent upon the similarity of the elements of the domains. The more equivalent the elements of the two domains, the greater the likelihood of positive transfer.

Valence, one of the fundamental psychological dimensions of emotion. It refers to the positive or negative quality of an emotional state. It is unclear as to whether valence is a unipolar construct representing varying degrees of pleasantness, or whether it is a bipolar dimension with positive and negative poles. The latter interpretation of valence suggests that negative emotions are not necessarily unpleasant. According to the valence hypothesis, hemispheric asymmetry for expression and perception of emotions depends on emotional valence; the right hemisphere is dominant for negative emotions and the left hemisphere is dominant for positive emotions.

Ventral striatum → striatum

Voxel, a three-dimensional volume unit used in brain imaging techniques, such as → functional magnetic resonance imaging (fMRI)

Voxel-Based Morphometry (VBM), a technique used to compare the brains of two groups of subjects by performing a voxel-wise comparison of the local concentration of → gray matter between each group. The procedure involves spatially normalizing high-resolution images from all the subjects in the study into the same stereotactic space. The gray matter from the spatially normalized images is then segmented, and the gray-matter segments are then smoothed. Voxel-wise parametric statistical tests that compare the smoothed gray-matter images from the two groups are then performed. Corrections for multiple comparisons are made using the theory of Gaussian random fields. This method has been the subject of substantial criticism and has produced a number of contradicting results, especially in the fields of psychiatry and schizophrenia research.

Wernicke's area, a left hemisphere brain area (BA 22) that is primarily responsible for the understanding of spoken words (i.e., language comprehension). This area is adjacent to the primary and secondary auditory cortex (BA 41, 42) and is located in the posterior part of the superior temporal gyrus. It is named after the German neurologist and psychiatrist Karl Wernicke (1848 – 1905).

White matter, nerve tissue of the brain that contains myelinated nerve fibers, or → axons, that connect various → gray matter areas of the brain to each other and carry nerve impulses between neurons. White matter does not contain → dendrites, which can only be found in gray matter. Generally, white matter can be viewed as the part of the brain and spinal cord responsible for information transmission (axons), whereas, gray matter is mainly responsible for information processing (neuron bodies). White matter is white because it is the color of myelin, the insulation covering the nerve fibers.

About the Contributors

Eckart Altenmueller - Institute of Music Physiology and Musicians Medicine, University of Music and Drama, Hannover, Germany.
phone +49-511-3100.552
altenmueller@hmt-hannover.de

Marc Bangert - Neuroimaging Lab, BIDMC and Harvard Medical School, Boston, MA, USA.
Current address for correspondence: Max Planck Institute for Human Cognitive and Brain Sciences Leipzig, Germany.
phone +49-341-9940.130
bangert@cbs.mpg.de

Joerg Bock - Institute of Biology, Department of Zoology and Developmental Neurobiology, Otto-von-Guericke Universität Magdeburg, Germany.
phone +49-391-6263-521
joerg.bock@nat.uni-magdeburg.de

Anna Katharina Braun - Institute of Biology, Department of Zoology and Developmental Neurobiology, Otto-von-Guericke University Magdeburg, Germany.
phone +49-391-6263617
katharina.braun@nat.uni-magdeburg.de

James S. Catterall - Graduate School of Education & Information Studies, University of California, Los Angeles CA, USA.
phone +1-310-825.5572; 825-0267
jamesc@gseis.ucla.edu

Richard D. Edwards - Ithaca College, Ithaca NY, USA.
phone: (+1) 607-274-3857
mail: redward@ithaca.edu

John W. Flohr - School of the Arts, Texas Woman's University, Denton, Texas & Walden University, Minneapolis, Minnesota, USA.
phone +1-940-898.2500
jflohr@twu.edu, jflohr@waldenu.edu

Wilfried Gruhn - University of Music Freiburg and Gordon Institute for Early Childhood Music Learning, Freiburg, Germany.
phone +49-7661-1624
mail@wgruhn.de

Marianne Hassler - Institute of Psychology, University Tuebingen, Germany.
phone +49-30-826.1683
mariannehassler@snafu.de

Donald A. Hodges - School of Music, University of North Carolina, Greensboro, USA.
phone +1-336-334.5176
dahodges@uncg.edu

Gunter Kreutz - Royal Northern College of Music, Manchester, UK.
phone: +44-161- 907.5263
Gunter.Kreutz@rncm.ac.uk

Martin Lotze - Institute for Diagnostic Radiology and Neuroradiology, Ernst-Moritz-Arndt-University Greifswald, Germany.
phone +49-3834-866899
martin.lotze@uni-greifswald.de

Gary E. McPherson - School of Music, University of Illinois, Urbana-Champaign, USA.
phone +1-217-333.8381
gem@uiuc.edu

Leon K. Miller - Waisman Center, University of Wisconsin, Madison, USA.
phone +1-608-246-0583
lkmiller@facstaff.wisc.edu

Frances H. Rauscher - Department of Psychology, University of Wisconsin, Oshkosh, USA.
phone +1-920-424.7172
rauscher@uwosh.edu

Gottfried Schlaug - Department of Neurology, Music and Neuroimaging Laboratory, Beth Israel Deaconess Medical Center and Harvard Medical School, Boston, USA.
phone +1-617-632.8912
gschlaug@bidmc.harvard.edu

Colwyn Trevarthen - Department of Psychology, University of Edinburgh, UK.
phone +44-131-650.3436
c.trevarthen@ed.ac.uk

INDEX

A

B

C

D

E

F

G

H

I

J

K

L

M

N

O

P

Q

R

S

T

U

V

W

Y